Environmental Science and Engineering

The ultimate goal of this series is to contribute to the protection of our environment, which calls for both profound research and the ongoing development of solutions and measurements by experts in the field. Accordingly, the series promotes not only a deeper understanding of environmental processes and the evaluation of management strategies, but also design and technology aimed at improving environmental quality. Books focusing on the former are published in the subseries Environmental Science, those focusing on the latter in the subseries Environmental Engineering.

Sushma Dave · Jayashankar Das
Editors

Trends and Contemporary Technologies for Photocatalytic Degradation of Dyes

Springer

Editors
Sushma Dave
Department of Applied Science
Jodhpur Institute of Engineering
and Technology
Jodhpur, Rajasthan, India

Jayashankar Das
Valnizen Healthcare
Mumbai, Maharashtra, India

ISSN 1863-5520 ISSN 1863-5539 (electronic)
Environmental Science and Engineering
ISBN 978-3-031-08990-9 ISBN 978-3-031-08991-6 (eBook)
https://doi.org/10.1007/978-3-031-08991-6

This Springer imprint is published by the registered company Springer Nature Switzerland AG
The registered company address is: Gewerbestrasse 11, 6330 Cham, Switzerland

Contents

Chapter 1
Dyes and Pigments: Interventions and How Safe and Sustainable Are Colors of Life!!!

Sushma Dave, Jayashankar Das, Bhoomika Varshney, and V. P. Sharma

Abstract We are fond of varied colors and they are strongly bonded in our life. In most cases, even a child looks eagerly for colorful toys, bottles and with passage of time desire to have attractive textiles to psychological imprints for colorful life with individual perceptions. Thus, dyes and pigments are integral part of our civilization and documented since prehistoric days. They may be classified as natural and synthetic, application or solubility based with continual efforts to have best combinations. There is growing demand as well as concern over the adverse implication of synthetic colorants on both the consumer, environment, ecosystem and thus need for safe, nontoxic, sustainable coloring alternatives have increased multifold for varied applications in food additives, cosmetics, textiles, pharmaceuticals, packaging, automobiles, defense sectors, etc. They may be amine, nitro, Sulfur based azodyes, or others with established reports on usage at levels above than the permissible regulatory limits. The usage of bacterial and fungal biobased pigments is increasing gradually due to advantages over the synthetic pigments. They are preferred because of simple culturing, pigment extraction procedures, ease of scale-up, and economic viability in less time amidst few limitations. Liposome-dependent innovative pathways may be potential to meet the demand of next generations and developing cost-effective, environment-friendly biobased products with economical feasibility in a time bound manner. For effective decision-making in terms of market share, it is necessary to address market intelligence, consider Covid-19 or similar impacts for assessing the market trends, forecast, and human behavior for comprehensive scientific mapping. The strict implementations of updated holistic, industry-driven standards with compliances to international norms are vital from safety and quality

S. Dave
Department of Applied Science, JIET Group of Institutions, Jodhpur, India

J. Das
Valnizen Healthcare, Vile Parle, Mumbai, India

B. Varshney · V. P. Sharma (✉)
CSIR-Indian Institute of Toxicology Research, Lucknow, India
e-mail: vpsitrc1@rediffmail.com; vpsharma@iitr.res.in

S. Dave and J. Das (eds.), *Trends and Contemporary Technologies for Photocatalytic Degradation of Dyes*, Environmental Science and Engineering,
https://doi.org/10.1007/978-3-031-08991-6_1

perspective. We must strategically solve environmental issues faced by the current markets as a challenge and safeguard the environment for future generations.

Keywords Colorants · Dyes · Innovative · Pigments · Regulatory · Sustainability

1.1 Introduction

The pleasure of seeing a rainbow is awesome and with knowledge we understand that it is a meteorological phenomenon which is result of reflection, refraction, and dispersion of light in water droplets leading to a wonderful spectrum of light appearing in the sky. The multicolored circular arc is sufficient to create thrill with colors from nature to artificial colors of life. These are important as they impact our decisions during the purchase of an item and create attraction while purchase in market. Marketing icons and branding experts have taken advantage of our cognitive thinking for specific products (Masters 2005). Several colorants of natural or synthetic origin provide a shade or tint to items ranging from textiles to packaging materials and even camouflaging the material scientists have helped to formulate new colors through nanomaterials, designing, and tools of innovation. They affect on human health, environment, and even personality interpretation. The acceptability of consumable items is generally perceived by our senses for texture or taste. All synthetic dyes and pigments are chemically not stable and pose poor bioavailability.

The acceptability of consumable items is generally perceived by our senses for texture or taste. All synthetic dyes and pigments are chemically not stable and pose poor bioavailability. Color plays an important role and it is dependent on age and individual perceptions for selection of food and clothes. Man has been using varied range of natural colorants originated from both animals and plants in different sectors like textiles, food (annatto curcumin, cochineal), painting, packaging, cosmetics (henna, catechu), pharmaceuticals (saffron, rhubarb), etc. In view of the increased popularity of eco-safety and health concerns, environmentally benign and nontoxic bio-resourced colorants are material of choice (Yusuf et al. 2017).

In 1856, while attempting for synthesis of quinine in respect to treatment of malaria, accidentally the synthetic dyes, mauveine were prepared from aniline. This was discovered by William Henry Perkin and thereafter several new combinations of dyes were developed and used in different sectors like foods, pharmaceuticals, and cosmetics. Synthetic dyes also known as coal-tar dyes because they were made from the byproducts of coal processing (Gürses et al. 2016a).

Colorant is coloring materials and coloration matter and includes pigments and dyes but both are different from each other in chemical composition, function, and properties. Pigments alter the appearance and scattering of light and they are usually dispersed in substrates for applications. They retain particulate structure throughout the coloration process (Index 2017).

Dyes are colored fluorescent substances imparting color to a substrate through absorbance of light. They are commonly soluble and its process of application

Table 1.1 Colorants of natural origin. Some common dyes for staining

Basic dyes	Acidic dyes
Safranin	Eosin
Basic Fuchsin	Acid Fuchsin
Crystal violet	Congo red
Methylene blue	

destroys the crystal structure (Pachade 2020). Market demands of natural dyes are expected to increase by $5 billion by 2024 (Arizton 2019).

Scientifically, synthetic colors are not always safe and regulatory requirements pose the necessity to assess the suitability. This necessitates monitoring using analytical equipments, spectrophotometers, to ensure the safety, efficacy, compatibility, etc. of a colorant in food, beverages, plastics, photography, printing, ceramics, cosmetics, and other product ranges. Few antioxidants found in food, vegetables, fruits are colored with positive health effects and rendering protection from diseases like cancer. Colorants may be classified according to their structure, source, color, and solubility or application method. Two main categories are established according to solubility: dyes and pigments. The major types of dyes and pigments are mentioned in Table 1.1.

1.1.1 Ancient History of Uses of Dyes

Mauviene dye was prepared from sea snail, *Bolinus brandaris* (Cova et al. 2017), and indigo blue *Indigofera tinctoria* (Adeyanju and State 2011). Dyes, e.g., picric acid, eosin, fuchsin, oil red, malachite are synthetic organic compounds that are hydro or oil-soluble and they may be found in cosmetics such as skin care products or toiletries. Pigments may be insoluble and they remain in particulate form, and they are mainly employed in toothpastes or decorative make-up. Since the prehistoric era color clothing was done using extracts of botanical sources-plants, trees, roots, seeds, nuts, fruit skins, berries, lichens, or from animal sources viz crushed insects, molluscs, etc. (Rivers and Revolution 2007). Pigments for paints were obtained from colored minerals (haematite which are mostly based on iron oxides) earth digging converting to fine powder and mixing into a crude binder Clay (Hradil et al. 2003). Charcoal from burned wood provided the early forerunners of carbon black pigments. The Egyptian Prussian blue color has chemically the structure of iron (III) hexacyanoferrate (II). In textiles and ceramics of India, the usage of mordant's or fixing agents started to give metallic look of copper, tin, chromium, and others (Fox 1999).

The pigments may be incorporated through dispersion process into products viz paints, printing inks, and plastics (Pct et al.). Plant pigments viz chlorophyll is the dominant natural green pigment and depict significant role in photosynthesis in plants, and thus vital to our civilization. Generally, synthetic dyes are manufactured from organic moieties (Montero 2015). Prior to discovery of synthetic dyes in 1856, dyestuffs were manufactured from natural products such as flowers, roots, vegetables, insects, minerals, wood, and molluscs.

Dyes or pigments, depending on composition may be toxic to humans and biodiversity including various flora and fauna (Synthetic and Perkin 1856). The environmental concern of dyes is dependent on absorption and reflection of sunlight entering the water (Samchetshabam et al. 2017). The absorption of light may diminish photosynthetic activity of algae and impact on the food chain. The intended usage of solvatochromic dyes are as molecular probes of serially diluted and agitated solutions in therapeutic formulations (Cartwright 2016). The azo dyes may demonstrate toxic implications at higher doses ranging from dermal toxicity, carcinogenicity to mutagenicity (Gičević et al. 2020). Although conjugation is well known as an important contributor to color, there are sporadic data concerning involvement of imine and iminium functions in the physiological effects of dyes and pigments (Khattab et al. 2019). The dyes viz. methylene blue, rhodamine, malachite green, fuchsin, crystal violet, auramine, and cyanins, are quite common to the pigments consisting of pyocyanine, phthalocyanine, and pheophytin with physiological effects in varied concentrations. The concept of electron transfer-reactive oxygen species-oxidative stress is used as the rationale and toxicity may be prevented or alleviated by antioxidants (Sharifi-Rad et al. 2020).

The soluble pigments and polymer dyes may be extensively used for the coloration of the collected resins which may not decompose rapidly or generate any toxic substances during the compounding process.

1.2 Classification of Dyes

1.2.1 Natural Dyes

Color has played significant role in the formation of different cultures of society all over the globe. It affects every phase of our life through strong influences on our clothing, furnishing of homes and offices, tapestry in hotels, etc. Natural dyes are color materials derived from plants, animals (molluscs, insects), microorganisms, and minerals. These are used in textiles, leather, bakery, cosmetics, and packaging to reduce the toxic effect of synthetic dyes. Alizarin and indigo are most usable natural dyes. Alizarin is a red color natural dyes which is extracted from root of Rubia tinctorium (Patel 2011; Scheer and Scheer 2002).

1.2.1.1 Plant-Derived Dyes

Plant dyes are extracted from different parts of plants and trees11. Blue indigo dyes are natural dyes that are extracted from the leaves of Indigo fera tinctoria. Vegetables and fruits are the main sources of vegetative dyes. Plant-natural dyes have applications in food ingredients and traditional medical systems. The vegetative dyes are extracted from plant aqueous, alkaline, acidic, and /or alcoholic methodologies (Verma and Gupta 2017).

1.2.1.2 Dyes Derived from Animal Sources

The main sources of animal-derived dye are insects, and most of the red-colored dyes are extracted from animal sources. Tyrian purple is the most ancient animal-derived dye extract, derived from the mollusc Murex. It is a very expensive dye because thousands of molluscs are used for the extraction of this dye. This dye acts as a symbol of royalty. One of the important dyes is cochineal dye, derived from insect species known as Dactylopius and used in textiles (Prelini 1974).

1.2.2 Synthetic Dye

Synthetic dyes are the best alternatives to natural dyes, and these dyes are derived from organic molecules. These synthetics offer a wide range of new color combinations at a low cost. Synthetic Dyes are used all most every industrial sector such as textile, paper, printing, leather, and food.

Synthetic dyes are classified according to their composition, color, and applications. Synthetic dyes are carbon-based chemical molecules made primarily from petrochemical derivatives, i.e., hydrocarbon benzene (aromatic compound) (Chaudhary 2020).

Synthetic dyes are made by combining various chemical reactants under carefully controlled conditions. This process is performed on glass or stainless steel at the specified temperature (Calogero et al. 2015).

1.2.3 Pigments

Pigments are often dispersed on substrates constituting of paper, plastics, sheets, etc. During the coloring process, the pigment retains its crystal or particle shape20. Pigments come in two types: organic and inorganic. They may be present in several living organisms, rendering attractive colors and helping camouflaging. The plant-based pigments namely carotenoids, anthocyanins which are water-soluble vacuolar pigments (>500), and Betalain have received greater attention. Carotenoids generate in isomers and oxidized. The anthocyanins (>500) may undergo hydrolysis with nucleophilic attack of water, polymerization during thermal processing and sometimes ring fission. The biotechnology general of carotenoids as food colored as be preferred due to non-toxicity. The recent stabilization study is related to microencapsulation and encapsulation. The studies are continuing progress for assessing health implications including cell, animal, epidemiological and human. The sporadic clinical studies are reported and need throw investigation (Alappat and Alappat 2020). The pigments generated microbial interventions have been used in textiles, paper, and food. As sustainable alternative to synthetic dyes. The red pigments have been isolated from Talaronyces aurantiacus. Based on acetyl cholinesterase activity assay, the red pigments generated by T. Aurantiacus is environmentally friendly.

For putative health-promoting effects and positive bioactivity of bioavailability anthocyanins they are pigments of choice with more than 635 Anthocyanins in nature. The dietary consumption of compatibility more than flavonoids due presence in plant material moreover, based upon clinical trials and in-vitro (cell lines) study. It can be inferred that they passives anti-inflammatory, anti-carcinogenic activity cardiovascular disease prevention, OBCD control with their potent antioxidant characteristics.

Bacterial pigments have different pharmacological viz ant allergic, antimicrobial, anticancer, antioxidant, and anti-inflammatory properties with large economic potential from there immense scope in food and Pharma sector for economical production of natural colored in all seasons from bacterial sources, compare to plant sources.

Commercial production of environmentally suitable and cost-effective high-value metabolites in a shorter duration through microbial biosynthesis highly advanced omics techniques and metabolite engineering tools for polyphenols. They may serve natural preservatives with well-established health-promoting benefits. The global market share is rapidly growing due to the increased awareness of health and the discovery of novel pharmacological benefits of several natural pigments including curcuminoids, flavonoids (>800), and carotenoids.

Chemical interactions of few pigments are stable than others in reference to heat stability or migration behavior. The regulatory expertise are needed to select a safer range of pigments for application in targeted sector with intensive studies as per OECD guidelines prior to approval by regulations of the land. The toxicological profile may serve as a point of reference or base for new products designing and safety assurance. Quinacridone pigments which are based on bio-succinic may be effective as sustainable and cost-effective pigments beside the commonly used petrochemical-based pigments. The bio-succinic acid is sourced from natural renewable sugars extracted from carbohydrates, for example, corn, sorghum, and any other readily available, high-volume agricultural crop. Liposome technology is being implemented as cost-effective and environmentally sensitive manner. The synthetic colors have been widely used in various industries including food, textile, cosmetics, and pharmaceuticals and in due course of time toxicity problems generated by synthetic pigments have triggered intense R&D related to natural colors and dyes. Among the natural sources, pigment-producing microorganisms hold a promising potential to meet present-day challenges. Moreover, natural colors not only improve the acceptability and market share of the product. The prominent dyes may be from cluster of azo, acid direct, dispersible, oil-soluble, reactive, or sulfur-based with use as food additive, paper, and leather.

Based on product types, the pigments market may be classified into:

- Optical whitening Agents
- Organic Pigments
- Pigment Emulsion
- Inorganic Pigments.

1.3 Application of Dyes

1.3.1 In Textile Sector

The admirability of a fabric is also dependent on synthetic dyestuffs, colorfastness, and bright hues. Environmental friendly remediation technologies, innovations (bicolor, natural mordents, supercritical carbon dioxide assisted waterless dyeing) are becoming essential to be exploring for textile dyeing industries. Existing methods of treatment fail to degrade dyes and have few limitations due to long-term impacts and requirement of reclamation of soil.

In the textile industry, many synthetic dyes, especially reactive dyes, are used in many types of textile dyeing, especially cotton fiber dyeing. It requires large amounts of water, resulting in large amounts of wastewater that severely pollute the environment. Synthetic dyes have been found to have dangerous properties in the event of a natural disaster. To eliminate these environmental and health risks, it is important to consider alternatives to synthetic dyes that are safe for animals, plants, and human life (Khattab et al. 2019).

Liposome-based technology for dyes are advanced technology. According to the carrier role of liposomes, they can be used in several textile processes such as textile finishing and dyeing with several types of dyes and fibers. They have a remarkable level-dyeing promoting effect on cotton fibers, although the dyeing promoting effect was not as good as that of sodium chloride. The optimum dyeing and level-dyeing effects were achieved at a dye-fixing temperature of 85 °C, sodium carbonate concentration of 10 g/L, and dye dosage of 2% (on the basis of oven-dry cotton fibers) when liposomes were used as the dyeing and level-dyeing promoters (Radhakrishnan 2014; Barani and Montazer 2008).

1.3.2 In Food Industry or Food Supply Chain

In the textile industry, many synthetic dyes, especially reactive dyes, are used in many types of textile dyeing, especially cotton fiber dyeing. It requires large amounts of water, resulting in large amounts of wastewater that severely pollute the environment. Synthetic dyes have been found to have dangerous properties in the event of a natural disaster. To eliminate these environmental and health risks, it is important to consider alternatives to synthetic dyes which may be safe for the target organisms and health of the stakeholders. Environmentally safe ingredients/additives may be potential option for the food industry or processors. FSSAI guidelines need to adhere for compliance to ISO 22000 norms or HACCP requirements fulfillment. Natural colorants are found in many plants and vegetables around the world and can be used to supplement the colorants produced. Different types of dyes used in food coloring and their applications are described in Table 1.2. The high demand for various processed food products and carbonated drinks in the market is propelling the worldwide market forward. Because substantial amounts of food colors are used in the manufacture of soft drinks, and there is a growing demand for them, the market is likely to experience significant growth throughout the forecast period (Document G, Safety F, Systems M).

Food dyes are used in many foods that we consume such as fruit, meat, fish, cheese, and more. Colorants are used to preserve the freshness of food and make it more visually appealing to us. When we go to the grocery store to buy food, the first thing we see is the food's color. Based on the color of the food we buy, we have expectations of that food's freshness, taste, and flavor. Foods such as fruits are dyed brighter to seem more attractive to us and our palate. Over time, brighter colored foods were deemed more "natural" than duller colored foods. Unfortunately, this is not the case at all. For example, we're used to oranges being the color orange. However, in many equestrian communities, oranges grow naturally green. We've grown to associate colors with the many expectations we have of our food.

The issue is that food dyes have negative effects on our environment and health. During production, approximately 10–15% of food dyes are discharged as effluents into the environment (OEHHA 2021). Effluent is liquid waste that is sent to a river or the sea. An example of this can be seen in food dye factories in China. The river water becomes contaminated with the food colorants discharged from these factories. When farmers try to use the river water to farm, the soil becomes toxic with the chemical waste. Many plants and fish that live in the area surrounding the factory also die because they can't survive the chemical contaminants that stain the water. From an environmental perspective, the discharge of dye effluents from dye manufacturing or consuming units into the water bodies poses potential threats to the quality of water and induces serious health problems to humans, plants, and animal life in particular and aquatic biota in general (Cova et al. 2017).

Not only do aquatic ecosystems suffer due to food dye effluents, but we do as well. Food dye effluents contaminate the fresh water we drink. Beyond that, common food dyes such as Red 40 and Yellow 5 are contaminated with carcinogens. The carcinogens in these dyes degrade the cells in our bodies and may put us at a higher risk for health issues such as cancer. Food dyes have also been shown to increase hyperactivity in kids and cause allergic reactions.

Food dyes in products such as breakfast cereals, juice and soft drinks, frozen dairy desserts, candies, and icings were linked to adverse neurobehavioral outcomes in children including inattentiveness, hyperactivity, and restlessness. Animal studies also revealed effects on activity, memory, and learning. The Food and Drug Administration (FDA) will continue to engage in the scientific and regulatory review of

Table 1.2 A list of common exempt colors and their FDA*a* and INS*b* numbers

Natural colorant	FDA	E number
Curcumin	73.600	100
Tumeric Oleoresin	73.650	None
Riboflavin	73.450	101(i)
Carmine/Cochineal	73.100	120
Chlorophylls	None	140i to ii
K-Cu-chlorophyll	None	141ii
Na-Cu-chlorophyll	73.125	141ii
Cu-chlorophylls	None	141i
Caramel I	73.85	150a
Caramel II	73.85	150b
Caramel III	73.85	150c
Caramel IV	73.85	150d
Vegetable carbon	None	153
Carotenes	None	160(a) iv
Carrot Oils	73.300	None
b-carotene	73.950	160(a)i to iii
b-apo-8′-carotenoic acid (caroteneal)	73.900	160e
Annatto-bixin	73.300	160(b)i
Annatto-norbixin	73.300	160(b)ii
Paprika	73.340	None
Paprika oleoresins	73.345	160(c)
Lycopene	73.585	160(d)I, ii, iii
Lutein	None	161(b)i
Canthaxanthin	73.750	161 g
Beet red	73.40	162
Anthocyanins	None	163
Grape color extract	73.169	None
Fruit juice	73.250	None
Vegetable juice	73.260	None
Enocianin (Grape skin extract)	73.170	163ii
Saffron, crocetin, and crocin	73.500	None
Titanium dioxide	73.575	171
Iron oxides	73.200	172
Spirulina blue	73.530	None
Spirulina green	None	None
Gardenia blue	None	None
Gardenia yellow	None	None

(continued)

Table 1.2 (continued)

Natural colorant	FDA	E number
Gardenia red	None	None
Monascus red	None	None
Monascus yellow	None	None
Carthamus yellow	None	None
Carthamus red	None	None
Flavoanthin	None	None
Huito (*Genipa americana*)/Jagua	None	None

[a]Food and Drug Administration from the Federal Food, Drug, and Cosmetic Act, 21 C.F.R §73
[b]Intl. Numbering System (European Union) by Codex Alimentarius

color additives to evaluate their potential impact on various populations, including children, and act when necessary to ensure that the products marketed to consumers are safe and properly labeled.

Food dyes have made more and more of our foods colorful-from breakfast cereals to ice creams. While natural colorants made from foods like beets are available, many manufacturers opt for synthetic dyes which may have dangerous health consequences, particularly for children (Munasinghe et al. 2002).

1.3.2.1 Permitted Synthetic Color Standards

The permitted colors are Carmoisine/Azorubine (E 122), Ponceau 4R (E 124), Erythrosine (E 127), Allura Red (E 129), Tartrazine (E 102), Sunset yellow FCF (E110), Indigotine/Indigo Carmine (E 132), Brilliant Blue FCF (E 133), and Fast Green FCF (E 143) (Dilrukshi et al. 2019).

1.3.2.2 Nonpermitted Color Standards

The nonpermitted colors are Fast red, Rhodamine B, Metanil yellow, Bromocresol purple, Green S, Sudan 1, Sudan 2, Sudan 3, and Sudan 4.

Color additives are applied to many food, drug, and cosmetic products. With up to 85% of consumer buying decisions potentially influenced by color, appropriate application of color additives and their safety is critical. Color additives are defined by the U.S. Federal Food, Drug, and Cosmetic Act (FD&C Act) as any dye, pigment, or substance that can impart color to a food, drug, or cosmetic or to the human body. Under current U.S. Food and Drug Administration (FDA) regulations, colors fall into two categories as those subject to an FDA certification process and those that are exempt (Table 1.2) from certification often referred to as “natural” colors by consumers because they are sourced from plants, minerals, and animals. Certified colors have been used for decades in food and beverage products, but consumer interest in natural colors is leading market applications (Dyes 2021).

However, the popularity of natural colors has also opened a door for both unintentional and intentional economic adulteration. Whereas FDA certifications for synthetic dyes and lakes involve strict quality control, natural colors are not evaluated by the FDA and often lack clear definitions and industry-accepted quality and safety specifications. A significant risk of adulteration of natural colors exists, ranging from simple misbranding or misuse of the term "natural" on a product label to potentially serious cases of physical, chemical, and/or microbial contamination from raw material sources, improper processing methods, or intentional postproduction adulteration.

Consistent industry-wide safety standards are needed to address the manufacturing, processing, application, and international trade of colors from natural sources to ensure quality and safety throughout the supply chain (Simon et al. 2017).

1.3.3 In Packaging and allied Sectors

Due to the optical property of dyes and colorants, they are widely used in the plastic and packaging sectors to enhance the outlook and demand for materials. There are lots of methods that are used to incorporate colors and pigments into plastic and its packaging. For example, the direct dry method is used to incorporate colorants and fluorescent materials into thermoplastics. Colorant heat stability properties are used in smart and intelligent plastic packaging. Used in electrical PVC wire because of their electrical property (Christie 1994).

In the field of beverages, perfumery or cosmetics packaging the metallic effect of packaging attracts a greater attention and seduces several stakeholders. Based in prating process, ink technology the use of metallic paste (gold or silver pigments) or substrates may pose variable possibilities for polyethylene Terephthalates (PET) or metalized paper/sheets. In luxury packaging, we may fine highly reflective surfaces with metallic sparkles.

The manufacturing of multicolor displays, logos, and polymer-based electronic parts are using inkjet printing or screen printing which is highly promising polymer deposition technique. The 3D printing with light-emitting diode displaying system is the demand of the modern customers. The UV curable, inkjet printability of polymer solutions is based on viscoelastic fluid jets and droplets under gravity with strain hardening.

Colorants used in polymers in food packaging are required to comply with strict regulations for food packaging coloration. The wide range of high-performance organic pigments, polymer soluble dyes, and optical brighteners fulfill strict regulations and safety laws for food packaging coloration while ensuring the product remains eye-catching and attractive to the end-consumer product specification and performance in terms of quality, color strength, and fastness properties. Bluish red pigment, Pink E with excellent fastness properties and outstanding dispersibility.

Some are suitable for high-end applications and for most plastics and recommended for fiber, film, and thin-wall applications (Goods Consumer Goods).

1.3.4 In Cosmetic Sector

Cosmetics may contain colorants to color certain parts of the product or body (skin, hair, nails, and eyelashes). They may be of two categories viz. leave-on and rinse-off cosmetics. Most of the colors used in cosmetics are micronized (5–30 μm) and homogenized pigments of organic nature or oxidized of different metals, e.g., iron (III) oxidized for production of red lipstick, black iron oxide used in mascara, eye pencil, etc.

Few of the contaminants (Parabens, mercury, o-phenylene diamine, methylene glycol, formaldehyde, quaternion 15, per fluoroalkyl substances) in cosmetics and personal care products may pose risks that are linked to serious health problems, birth defects or reproductive disorders. Some of the chemicals are endocrine-disrupting chemicals and thus categorized as development toxicants Phthalic acid esters (PAEs) may affect the developing fetus. Regulatory agencies have already banned few ingredients.

Leave-on Product: These are cosmetics that are developed for prolonged contact with the skin, mucus membrane, and nails for defined period of time.

Rinse-off Product: cosmetics are those that are clean after usage, such shower gel, Conditioner, or soap (Guerra et al. 2018).

1.4 Toxic Implications Production and Discharge to Effluents

Natural dyes have long been used in a variety of applications, especially in the fields of textile dyeing, paper, leather, painting, and food. The invention of synthetic dyes from petroleum chemical compounds was driven by increasing demand and high prices for natural dye extraction. Synthetic dyes dominate the textile market, producing approximately 8×105 tons annually. The textile industry uses high amounts of water for dyeing textiles. They have harmful effects when released into the environment in an untreated or partially treated form. Textile industrial effluents have the worst impact on the environment because they are discharged directly into the water bodies without any type of pretreatment procedure and cause lots of problems for not only the water bodies but also the surrounding environment like humans, animals, plants, and so on. Textile synthetic dyes, along with industrial pollutants, are toxic and have been shown to cause cancer, mutation, and other diseases. Textile-influenced synthetic dyes are toxic not only to humans but to fishes and algae also. Every time this fish comes into contact with textile influents, its ionic regulation will be upset because it will increase the concentration of potassium and magnesium and decrease the concentration of sodium. Dyes and pigments remain in the atmosphere for many years due to its high thermal and photosensitive quality, absorbing the sunlight and reflecting the sun's light. If textiles are present in large amounts in water bodies, it will stop their reoxygenation capacity.

Synthetic dyes are a necessary component of food because food loses its natural color during processing and storage. On the one hand, synthetic dyes are very important to making food effective according to human demand, but on the other hand, they are very toxic to human health and will cause many food-borne diseases. Like hypersensitive diseases in children, Tetrazine, a synthetic dye used as a food colorant, will cause depression, restlessness, and insomnia in children. Red, yellow, and yellow contains toxic compounds like benzedrine, 4-aminobiphenyl, and amino azo benzene, which have wide application in the food sector. These dyes have the potential to cause cancer, and Yellow 5 contains tetrazine and many other synthetic dyes that cause allergies. Rhinitis Asthma Angioedema, atopic dermatitis, liver disorders, disability in oral tolerance, disruption in the functioning of digestive enzymes, imbalance in intestinal permeability, etc. in human beings (Lellis et al. 2019; Ben et al. 2021).

To overcome these problems, lots of methods are used, like chemical ones (like Fenton's Reagents), ozonation, sodium hypo chloride, and electrochemical destruction). Chemosorption, activated charcoal, and coal mixture silica gel, filtration by different membranes, ion exchange and coagulation methods, biological methods (Bio-remediation), leaner production technologies, and air dying technology are examples of physical methods (Lellis et al. 2019).

It is estimated that over 10,000 different dyes and pigments are used industrially and over 7×10^5 tons of synthetic dyes are annually produced worldwide. Textile materials can be dyed using batch, continuous or semi-continuous processes. The kind of process used depends on many characteristics including type of material as such fiber, yarn, fabric, fabric construction, and garment, as also the generic type of fiber, size of dye lots, and quality requirements in the dyed fabric. In the textile industry, up to 200,000 tons of these dyes are lost to effluents every year during the dyeing and finishing operations, due to the inefficiency of the dyeing process (Drumond Chequer et al. 2013).

The textile dyes significantly compromise the aesthetic quality of water bodies, increase biochemical and chemical oxygen demand (BOD and COD), impair photosynthesis, inhibit plant growth, enter the food chain, provide recalcitrance, and bioaccumulation, and may promote toxicity, mutagenicity, and carcinogenicity. The textile industry is one of the important industries that generate a large number of industrial effluents. Color is the main attraction of any fabric. Manufacture and use of synthetic dyes for fabric dyeing have therefore become a massive industry. Synthetic dyes have provided a wide range of colorfast, bright hues. However, their toxic nature has become a cause of grave concern to environmentalists. Use of synthetic dyes has an adverse effect on all forms of life[4] A. Presence of Sulfur, naphthol, vat dyes, nitrates, acetic acid, soaps, enzymes chromium compounds, and heavy metals like copper, arsenic, lead, cadmium, mercury, nickel, and cobalt and certain auxiliary chemicals all collectively make the textile effluent highly toxic. These organic materials react with many disinfectants, especially chlorine, and form byproducts (DBPs) that are often carcinogenic and therefore undesirable (Gürses et al. 2016a).

1.5 Guidelines

The safety of colorants is vital and experts have identified critical issues related to compounders and processers to match type color requirements. We need their usage in a consistent manner and approved permissible limits under specific field situations. Generally, organic pigments provide stronger colors but may be comparatively less dispersible than inorganic. With plastic resin or base material (polyethylene film, metalized sheet, polypropylene bottles), the judgment of the coloration also depends on metamerism, i.e., a phenomenon wherein the color of an object shifts depending in the angle and kind if incident light (Downham and Collins 2000).

1.5.1 Market Scenario and Growth of Dyes and Pigments

The global dyes and pigments market size was estimated at USD 32.9 billion in 2020 and is expected to exponentially to compound annual growth rate (CAGR) of 5.1% from 2021 to 2028. In our country, dyes and pigments market is being driven by several usage sectors and may reach nearly 133.52 million tons by volume, i.e., CAGR of 11% approximately between 2021 and 2026 (Dyes 2021). The dyes industry in India is expected to witness a steady growth in the coming years due to environmental crackdowns in China, resulting in a shutdown of several domestic dye companies. India is better placed due to the availability of the ecosystem, feedstock, technology, and compliance required for the industry. Thus, the consumer base of China is likely to shift to India due to these reasons in the coming years.

In view of rising demand for organic pigments, they are the dominant type of pigments being produced in India, accounting for 58% of the total pigment production in India. The growth of the pigment market is aided by the cosmetic industry, improved living standards, evolving lifestyles. The demand for cosmetic products viz personal care products such as shampoos, conditioners, skin care, hair care, and perfumes are realizing a high boost with different applications in beauty clinics or centers (Pct et al.).

1.5.2 New Designing of Dyes Using Computed Programming for Functional Applications

For these new applications, but also for traditional uses, the concepts of computer-aided molecular design of dyes for specific properties have become increasingly important. The modern functional dyes are being developed using molecular, quantum, and computer-aided methods for synthesis and recent applications. Reasonable degree of confidence is also ascertained for assessing the desired properties (Gastaldi et al. 2021). The morphological characteristics are also ascertained through crystallographic computing data.

The modern manufacturers such as tinctorium are using bacteria as they are powerful multipliers, when you put them in the right conditions, we can grow these organisms to create dye products in a much more scalable and sustainable manner that isn't reliant on petroleum (Images G 2020). We may be able to create beautiful colors without toxicity which was the critical issue of concern for dyes and pigment industry. No doubt there are challenges of techno feasibility and circular economy.

In few cases of food items, we are also getting reports of illegal contamination and adulteration in sweets, savories, etc., especially during festive seasons. The studies indicate that six banned dyes have been found in the sweets tested for rhodamine B, orange II, metanil yellow, malachite green, quinoline yellow, and auramine (Martins 2020; Dixit et al. 2013).

1.5.3 Bacterial Pigments with Reduced or Minimal Toxicity Implications

Non-toxicity, biodegradability, and non carcinogenicity of the natural pigments, dyes, and colorants make them an attractive source for human use. Bacterial pigments are colored metabolites secreted by bacteria under stress. The industrial uses of bacterial pigments have increased many folds because of several advantages over the synthetic pigments. Among natural resources, bacterial pigments are mostly preferred because of simple culturing and pigment extraction techniques, scaling up, and being time economical. Generally, the bacterial pigments are safe for human use and therefore have a wide range of applications in pharmaceutical, textile, cosmetics, and food industries (Numan et al. 2018).

The use of marine-derived bacterial pigments displays a snowballing effect in recent times, being natural, environmentally safe, and health-beneficial compounds. Although isolating marine bacteria is a strenuous task, these are still a compelling subject for researchers, due to their promising avenues for numerous applications. Marine-derived bacterial pigments serve as valuable products in the food, pharmaceutical, textile, and cosmetic industries due to their beneficial attributes, including anticancer, antimicrobial, antioxidant, and cytotoxic activities. Biodegradability and higher environmental compatibility further strengthen the use of marine bio-pigments over artificially acquired colored molecules (Nawaz et al. 2021).

In view of the health-related safety issues, there are increasing attention to assess the toxicity of additives used in food. The European Parliament and the council have published the regulation EC No. 1333/2008 on food additives establishing that the toxicity of food additives must be re-evaluated by European Food Safety Authority (EFSA) (Parliament Thee, Council The, The of, Union P 2008). Despite the widespread usage of food colorants adverse reactions related to their consumption, including reactions triggered by immune (immediate and delayed-type hypersensitivity) and non-immune (intolerance) mechanisms, are considered rare. In few cases, allergenicity/hyperreactivity are reported through sporadic data. Holistic region-specific detailed studies are needed in recent contexts with quantitative surveys and experimentation. The uniform maximum permissible limit of synthetic colors (in most cases at 100 mg/kg) under the regulations needs to be reviewed, revisited and may be governed by the technological necessity and the consumption profiles of food commodities so that the vulnerable population may not unnecessarily be exposed to excessive amounts of synthetic colors to pose health risks (Feketea and Tsabouri 2017).

The growing concern over the harmful effects of synthetic colorants on both the consumer and the environment has raised a strong interest in natural coloring alternatives. As a result, the worldwide demand for colorants of natural origin is rapidly increasing in the food, cosmetic and textile sectors. Natural colorants have the capacity to be used for a variety of industrial applications, for instance, as dyes for textile and non-textile substrates such as leather, paper, within paints and coatings, in cosmetics, and in food additives. Currently, pigments and colorants produced through plants and microbes are the primary sources exploited by modern industries. Among the other non-conventional sources, filamentous fungi particularly ascomycetous and basidiomycetous fungi (mushrooms), and lichens are known to produce an extraordinary range of colors including several chemical classes of pigments such as melanins, azaphilones, flavins, phenazines, and quinines.

Policymakers and regulators must be stricter to enforce the regulations at all levels to safeguard human health and punish the adulterators. We need the implementation of industry-driven standards that will translate to both safety and quality solutions for issues facing the natural color markets. By utilizing the historical and applied knowledge of the impacted industries with respect to the identification of hazards and associated risk assessment strategies, the resultant standards should mitigate risk, increase consumer and industry confidence, and provide a framework for self-regulations.

1.6 Future Prospects

Internet of Things (IoT) based management systems for remote monitoring and control of textile industries effluents are complex and need effective management planes. The effluents contain multiple chemical entities in dyes and chemicals released from critical industries and are generally the target for non-compliances to regulatory norms. The sensor-based low-cost adsorbents may serve effective way of removal of dyes from the effluents (Hynes et al. 2020).

1.7 Conclusion

The environment-friendly solutions for usage of varied azo or amine-based dyes are vital for safer processes and circularity through use of natural colorants. The biogenic dyes are on the higher demand due the structural diversity and application of green chemistry approaches.

We need a detailed systematic comparison between specifically between biologically derived dyes and synthetic counterparts. Most of the industrial effluents and landfill migrants may contain dyes, heavy metals, and aromatic amines characterized through toxicological parameters viz. Cytotoxicity, mutagenic and carcinogenic (Chavan 2011).

The bioremediation, Phytoremediation, or depolarization using appropriate technologies like Micro capsulation and nano capsulation for such locations is essential to prevent adverse implications on flora and fauna. Encapsulation technology is used not only in the food industry, but also in textiles, cosmetics, packaging, pharmaceuticals, and many other industries. Liposome-based encapsulation technologies are widely used in the textile industry and have lots of advantages viz. eco-friendly, nontoxic, and safe for human health, and also enhancing the absorbing capacity of natural pigments. Innovators may explore cleaner production strategies for economical, value-added newer textile dyes through the advancement of nano textile materials of quality (Vankar 2000). Colorants derived from cleaner bio-researches having minimal ecological negative impacts are being investigated globally for serving the next generations (Yusuf et al. 2017).

References

Adeyanju O, State E (2011) Int J Phys Sci 6(1):137–143
Alappat B, Alappat J (2020) Anthocyanin pigments : beyond aesthetics
All BTO effects of food dye on the environment, pp 1–9
Al-Tohamy R, Ali SS, Li F et al (2022) A critical review on the treatment of dye-containing wastewater: ecotoxicological and health concerns of textile dyes and possible remediation approaches

for environmental safety. Ecotoxicol Environ Saf 231:35026583. https://doi.org/10.1016/j.ecoenv.2021.113160
Arizton (2019) Natural dyes market—global outlook and forecast 2019–2024, pp 1–147
Barani H, Montazer M (2008) A review on applications of liposomes in textile processing, 18770074. https://doi.org/10.1080/08982100802354665
Calogero G, Bartolotta A, Di Marco G et al (2015) Vegetable-based dye-sensitized solar cells. Chem Soc Rev 44:3244–3294. https://doi.org/10.1039/c4cs00309h
Cartwright SJ (2016) Solvatochromic dyes detect the presence of homeopathic potencies. Homeopathy 105:55–65. https://doi.org/10.1016/j.homp.2015.08.002
Chaudhary B (2020) Chemistry of synthetic dyes: a review. J Interdiscip Cycle Res 12:390–396
Chavan RB (2011) Environmentally friendly dyes. Handb Text Ind Dye Princ Process Types Dye 1:515–561. https://doi.org/10.1533/9780857093974.2.515
Christie RM (1994) Pigments, dyes and fluorescent brightening agents for plastics: an overview. Polym Int 34:351–361. https://doi.org/10.1002/pi.1994.210340401
Class F pigments what are pigments ? pp 1–7
Cova TFGG, Pais AACC, Seixas De Melo JS (2017) Reconstructing the historical synthesis of mauveine from Perkin and Caro: Procedure and details. Sci Rep 7:1–9. https://doi.org/10.1038/s41598-017-07239-z
Dilrukshi PGT, Munasinghe H, Silva ABG, De Silva PGSM (2019) Identification of synthetic food colours in selected confectioneries and beverages in Jaffna District, Sri Lanka. J Food Qual. https://doi.org/10.1155/2019/7453169
Dixit S, Khanna SK, Das M (2013) All India survey for analyses of colors in sweets and savories: exposure risk in Indian population. J Food Sci 78:1–9. https://doi.org/10.1111/1750-3841.12068
Document G, Safety F, Systems M guidance document food safety management systems food industry guide to implement GMP/GHP requirements and risk assessment health supplements and food industry guide to implement GMP/GHP requirements health supplements
Downham A, Collins P (2000) Colouring our foods in the last and next millennium. Int J Food Sci Technol 35:5–22. https://doi.org/10.1046/j.1365-2621.2000.00373.x
Drumond Chequer FM, de Oliveira GAR, Anastacio Ferraz ER et al (2013) Textile dyes: dyeing process and environmental impact. Eco-Friendly Text Dye Finish 1–25. https://doi.org/10.5772/53659
Dyes G (2021) Global dyes, pigments market valued at $32.9 Billion in 2020, pp 4–9
Engel (2014) 済無No title no title no title. Pap Knowl Towar a Media Hist Doc, 1–17
Feketea G, Tsabouri S (2017) Common food colorants and allergic reactions in children: myth or reality? Food Chem 230:578–588. https://doi.org/10.1016/j.foodchem.2017.03.043
Fox MA (1999) Dyes and pigments
Gastaldi M, Cardano F, Zanetti M et al (2021) Functional dyes in polymeric 3D printing: applications and perspectives. ACS Mater Lett 3:1–17. https://doi.org/10.1021/acsmaterialslett.0c00455
Gičević A, Hindija L, Karačić A (2020) Toxicity of azo dyes in pharmaceutical industry. IFMBE Proc 73:581–587. https://doi.org/10.1007/978-3-030-17971-7_88
Give I (2018) William Henry Perkin : how an 18-year-old accidentally discovered the first synthetic dye, pp 1–5
Goods C Consumer goods sensitive and non-sensitive search here for pigments and more colorants for sensitive consumer goods, pp 1–6
Guerra E, Llompart M, Garcia-Jares C (2018) Analysis of dyes in cosmetics: challenges and recent developments. Cosmetics 5. https://doi.org/10.3390/COSMETICS5030047
Gürses A, Açıkyıldız M, Güneş K, Gürses MS (2016a) Dyes and pigments: their structure and properties. In: SpringerBriefs in green chemistry for sustainability, pp 13–29. https://doi.org/10.1007/978-3-319-33892-7
Gürses A, Açıkyıldız M, Güneş K, Gürses MS (2016b) Colorants in health and environmental aspects, pp 69–83. https://doi.org/10.1007/978-3-319-33892-7_5
Gürses A, Açıkyıldız M, Güneş K, Gürses MS (2016c) Historical development of colorants, pp 1–12. https://doi.org/10.1007/978-3-319-33892-7_1

Hradil D, Grygar T, Hradilová J, Bezdička P (2003) Clay and iron oxide pigments in the history of painting. Appl Clay Sci 22:223–236. https://doi.org/10.1016/S0169-1317(03)00076-0
Hynes NRJ, Kumar JS, Kamyab H et al (2020) Modern enabling techniques and adsorbents based dye removal with sustainability concerns in textile industrial sector—a comprehensive review. J Clean Prod 272:122636. https://doi.org/10.1016/j.jclepro.2020.122636
Images G (2020) Making beautiful colours without toxic chemicals, pp 1–12
Index C (2017) Definitions of a dye and a pigment ETAD defines dyes as CPMA defines pigments as, pp 3–4
Khattab TA, Abdelrahman MS, Rehan M (2019) Textile dyeing industry : environmental impacts and remediation, 1
Kovacic P, Somanathan R (2014) Toxicity of imine-iminium dyes and pigments: electron transfer, radicals, oxidative stress and other physiological effects. J Appl Toxicol 34:825–834. https://doi.org/10.1002/jat.3005
Lellis B, Fávaro-Polonio CZ, Pamphile JA, Polonio JC (2019) Effects of textile dyes on health and the environment and bioremediation potential of living organisms. Biotechnol Res Innov 3:275–290. https://doi.org/10.1016/j.biori.2019.09.001
Martins JMJ (2020) Advances in information retrieval. Springer
Masters J (2005) The 360-degree rainbow. Weather underground. the weather company. Archived from the original on 29 January 2015
Montero F (2015) Photosynthetic pigments. Encycl Astrobiol 1883–1884. https://doi.org/10.1007/978-3-662-44185-5_1205
Munasinghe H, Silva ABG, De SPGSM (2002) Journal of food quality: editorial. J Food Qual 25:1–14
Nawaz A, Chaudhary R, Shah Z et al (2021) An overview on industrial and medical applications of bio-pigments synthesized by marine bacteria. Microorganisms 9:1–24. https://doi.org/10.3390/microorganisms9010011
Numan M, Bashir S, Mumtaz R et al (2018) Therapeutic applications of bacterial pigments: a review of current status and future opportunities. 3 Biotech 8:1–28. https://doi.org/10.1007/s13205-018-1227-x
Office C, Health E, Assessment H (OEHHA) (2021) Food dyes linked to attention and activity problems in children "Most consumers have no idea that something that is allowed in the food supply by the FDA could trigger adverse behaviors", pp 3–6
Organic P, Dyes P, Resources U Organic and inorganic colorants. 1–11
Pachade V (2020) Differences between dyes and pigments in flooring, pp 30–32
Packaging F packaging food packaging and non-food search here for pigments and more colorants for food packaging, pp 1–6
Parliament THEE, Council THE, The OF, Union P (2008) L 354/16, pp 16–33
Patel BH (2011) Natural dyes. Woodhead Publishing Limited
Pct W, Darren I, Ortalano M et al Patents pigment dispersions and printing inks with improved coloristic properties WO2014062227A1, pp 1–18
Prelini G (1974) Dyeing and printing, p 113831. https://doi.org/10.5040/9781501317484.ch-023
Radhakrishnan S (2014) Roadmap to sustainable textiles and clothing
Rivers T, Revolution I (2007) Pigment 2007:1–5
Samchetshabam G, Hussan A, Choudhury TG, Gita S (2017) Impact of textile dyes waste on aquatic environments and its treatment wastewater management view project tribal sub plan view project. Environ Ecol 35:2349–2353
Scheer H, Scheer H (2002) Chapter 2 the pigments, pp 29–81
Sharifi-Rad M, Anil Kumar NV, Zucca P et al (2020) Lifestyle, oxidative stress, and antioxidants: back and forth in the pathophysiology of chronic diseases. Front Physiol 11:1–41. https://doi.org/10.3389/fphys.2020.00694
Simon JE, Decker EA, Ferruzzi MG et al (2017) Establishing standards on colors from natural sources. J Food Sci 82:2539–2553. https://doi.org/10.1111/1750-3841.13927

Ben SH, Bouket AC, Pourhassan Z et al (2021) Diversity of synthetic dyes from textile industries, discharge impacts and treatment methods. Appl Sci 11:1–21. https://doi.org/10.3390/app11146255

Synthetic C, Perkin WH (1856) Dyes, chemical and synthetic classes of dyes, pp 1–7

Timilsena YP, Haque MA, Adhikari B (2020) Encapsulation in the food industry: a brief historical overview to recent developments

Vankar PS (2000) Chemistry of natural dyes. Resonance 5:73–80. https://doi.org/10.1007/bf02836844

Verma S, Gupta G (2017) Natural dyes and its applications: a brief review. IJRAR Int J Res Anal Rev 4:57–60

Yusuf M, Shabbir M, Mohammad F (2017) Natural colorants: historical, processing and sustainable prospects. Nat Products Bioprospect 7:123–145. https://doi.org/10.1007/s13659-017-0119-9

Chapter 2
Recent Advances in Photocatalytic Degradation of Dyes Using Heterogeneous Catalysts

Bubul Das, Hirendra Nath Dhara, Anjali Dahiya, and Bhisma K. Patel

Abstract Synthetic dyes are playing a vital role in day-to-day life as different products ranging from textiles to leather to furniture contain dye for colouring purposes. It is reported that 12% of these dyes are wasted during processing. However, these industrial non-biodegradable dyes often enter water bodies such as groundwater, river, and lake and pollute them. Removal of these hazardous pollutants or dyes from wastewater has gained attention due to environmental concerns. Various techniques have been developed for the removal of these carcinogenic dyes from the natural environment. Degradation of dyes and eventually their removal from the mainstream and aquatic media using UV or visible light in the presence of photocatalyst (PC) are some reasonable technologies. Photocatalytic degradation (PD) could convert bio non-degradable dye complex molecules into smaller, non-carcinogenic, low molecular species. The process of PD is based on the generation of highly reactive, hydroxyl and superoxide anion radicals, which target the dye molecules and convert them into H_2O and CO_2. The chapter focuses on the principle and mechanism of dye degradation using heterogeneous photocatalyst. A brief discussion of the various important heterogeneous photocatalyst in the degradation of dye will be discussed in the latter part of the chapter.

Keywords Dyes · Degradation · Heterogeneous photocatalyst · Radicals · Advanced oxidation processes (AOPs)

2.1 Introduction

The development of science and technology leads us towards the new chemicals which are raw materials in various industrial processes. Organic dyes turned out to be one such chemical that is useful in different industrial activities for colouring various products. Though the textile dyeing industry has progressed a lot, initially

B. Das · H. N. Dhara · A. Dahiya · B. K. Patel (✉)
Department of Chemistry, Indian Institute of Technology Guwahati, Guwahati, Assam 781039, India
e-mail: patel@iitg.ac.in

S. Dave and J. Das (eds.), *Trends and Contemporary Technologies for Photocatalytic Degradation of Dyes*, Environmental Science and Engineering,
https://doi.org/10.1007/978-3-031-08991-6_2

nature was the sole source of dyes (Ferreira et al. 2004). Natural dyes can be obtained from animals, plants, minerals. Natural dyes are less harmful compared to synthetic dyes and they generate biodegradable wastewater (Silva et al. 2020). Synthetic dyes are organic compounds that are hydro or oil-soluble (Vázquez et al. 2020). The chemical structure of dyes is diverse, which include azo, nitro, phthalocyanine, and diarylmethane dyes having different chemical and physical properties (Belpaire et al. 2015). They are unsaturated complex organic compounds and can absorb light and emit colour in the visible region (Rehman et al. 2020). At present more than 100,000 dyes are commercially available and are being used increasingly in textile, rubber, plastic, cosmetics, food industries, and pharmaceuticals. Industrial effluents pollution is mainly caused by the increasing use of these dyes in the dyeing process. Biological treatment processes can be used for the treatment of wastewater containing high concentrations of biodegradable pollutants. However, low biodegradable toxic pollutants released from the textile (Touati et al. 2016), pharmaceutical (Olama et al. 2018), and agricultural industries (Kushniarou et al. 2019) are found in groundwater, surface water, and in municipal sewage treatment plant. Due to the incomplete removal of these dyes from wastewater, persistent organic pollutants with complex compositions remain in waterbodies. The wastewater from the dye industry is distinguished from others by high chemical oxygen demand (COD), total solids (TS), pH, biological oxygen demand (BOD), water usage, and colour. A ratio of 0.25 between biological oxygen demand to chemical oxygen demand indicates that industrial wastewater contains vast non-biodegradable organic materials. Water pollution related to dye contamination is life-threatening and harmful to the surrounding environments. Dye accumulation significantly affects aquatic life by increasing BOD and COD. This also inhibits plant growth as it impairs photosynthesis and human being by entering the food chain (Rehman et al. 2019; Tayade et al. 2009). The bulky size and complex structure of these dye molecules make them non-biodegradable, mutagenic as well as carcinogenic (Saeed et al. 2015). The annual worldwide production of synthetic dyes is around 7×10^5 tons for the textile industry. Out of these, 10–15% of these dyes do not bind to the fabrics during the dyeing process and enter directly into wastewater (Chandanshive et al. 2018). These unbound portions of dyes are found in very high concentrations in the textile effluents and directly enter into the wastewater (Drumond et al. 2013). The release of these dye-contaminated effluents into waterbodies is undesirable due to the non-biodegradable nature of dyes and their breakdown into hazardous intermediate products. The high tinctorial value of less than a ppm of dye in water could produce highly coloured elements and lead to aesthetic issues. In addition, the uncontrolled contaminations also retard the penetration of light through water, thus significantly affects the photosynthesis function of aquatic plants (Sahoo et al. 2012; Akpan and Hameed 2009).

Dye pollutants threaten the equilibrium of the food chain as it hampers the growth of photoautotrophic organisms by increasing the BOD of the waterbodies. These recalcitrant dyes are claimed to promote toxicity, carcinogenicity, mutagenicity, genotoxicity, teratogenicity, and their long-term exposure might affect the structure of ecosystems (Holkar et al. 2016). For example, crystal violet, acid violet, and malachite green which belong totriphenylmethane class of dyes are found to be phototoxic to plants, cytotoxic to mammalian cells and even induce tumour growth in some aquatic species. Oral ingestion, inhalation, or workers in the field of dye manufacturing and processing may have skin and eyes abnormality, insomnia, immune suppression, asthma, allergic conjunctivitis, contact dermatitis or nasal problems on prolonged exposure to these dyes (Jadhav et al. 2012). The potential toxicity of these dyes and their high concentrations in the industrial effluents need to be carefully addressed.

Elimination of these dyes from contaminated water is a key environmental concern. Access to clean and toxic-free, carcinogenic agents free and harmful microbes free water is the need of the hour and draw special research attention to ensure smooth survival of human being in the near future. The conventional wastewater treatment is unsuccessful in the degradation and subsequent removal of these dyes from water bodies. The complex chemical structure and strong photocatalytic stability of dyes make them resistant to temperature, water, soap, and microbial attack. As a large variety of dyes are produced worldwide, various physical, chemical, and biological methods are available for their degradation and removal. These include adsorption (Wang et al. 2020), biosorption (Sharma et al. 2018), ozonation (Sripiboon et al. 2018), coagulation/flocculation (Zahrim and Hilal 2013), reverse osmosis (Abid et al. 2012), ultrafiltration (Ben et al. 2019), nanofiltration (Kong et al. 2019), electrochemical degradation (Moura et al. 2016), biodegradation (Ajaz et al. 2019), hybrid biological treatment (Zhang et al. 2017), advanced oxidation processes (AOPs) (Rauf and Ashraf 2009), etc. All these processes have their advantages and disadvantages over other methods. The high chemical stability of these dyes makes the conventional treatment method ineffective. Practical employment of these methods has limitations considering their efficiency, practicability, sludge production, reliability, feasibility, pre-treatment requirements, operation difficulty, and by-products generation (Crini and Lichtfouse 2019). A balanced approach is therefore necessary for choosing an appropriate method for dyes degradation. Out of these techniques mentioned above, the AOPs found to be promising green methods and have been reported to be effective for removing soluble organic pollutants. The AOPs can provide an almost total degradation of contaminant dyes and it is recognized as one of the most accepted techniques for controlling environmental pollution.

Heterogeneous photocatalysis is considered as one of the most recognized AOPs. Photocatalysis is a green approach over other traditional dye removal techniques for its simple, cost-effectiveness, high performance at even ambient temperature and complete mineralization of pollutants to ppm or ppb level (Zheng et al. 2019). Since the discovery of photoelectrochemical water splitting reaction by Fujishima and Honda (1972), photocatalysis has developed rapidly (Fujishima and Honda 1972). Photocatalysis is defined as the process in which the rate of the chemical reaction increases in the presence of some photocatalyst via the generation of electron–hole pairs when irradiated in the presence of UV/visible light. Photocatalysis is advantageous for low biodegradable, highly complex, and high concentrations of pollutants in wastewater. Photodegradation (PD) harnesses solar or UV light for the breakdown of large complex molecules into smaller, non-toxic, and lower molecular weight species. Oxidation/reduction of organic pollutants is induced by light via some simple redox reactions. The process is triggered by the initial absorption of a photon by photocatalysts in the UV, visible or IR spectral range. PD has been considered as an energy-saving technology as it directly converts light energy into chemical energy. UV photolysis, oxidants such as hydroxyl radicals, H_2O_2, ozone-assisted photooxidation, and UV photocatalytic oxidation are some of the adopted technologies for PD (Ray 2000). Being green in nature PD supersedes conventional methods in the field of environmental purification. Some unique advantages of PD are listed in Fig. 2.1.

PD is carried out in the presence of heterogeneous catalysts. These catalysts should be stable, non-toxic, recyclable, highly photoactive, and readily available. Appropriate bandgaps in semiconductors make them good photocatalysts as they can utilize the energy of the photon (Serpone and Emeline 2002). Different metal oxides are used as heterogeneous photocatalysts for the decolouration of dyes. Catalysts TiO_2 and ZnO are frequently used for photo degradation. The physical state of the metal oxides is different from the reactant phase, for which it is termed as heterogeneous photocatalysis. Limited thermal stability, separation problems after

Fig. 2.1 Key advantages and efficiency of PD technologies for dye removal

the reaction are some obvious challenges associated with homogenous photocatalysis (Haibach et al. 2012). Therefore, most of the researchers focus on heterogeneous PC, which is going to be addressed in this chapter.

2.2 Classification of Dyes

Dye is a substance with the property of absorbing part of visible light and shows complementary colour. They are complex unsaturated aromatic compounds and characterized by intense colour, high solubility, and stability. As the annual global production of dyes is increasing, their classification is important to have a comprehensive study. Classification of dyes is a very difficult process because of their diverse chemical structure. At present, around 10,000 structurally different dyes are used annually (Chequer et al. 2013). Large structural variations amongst dyes challenge their classification from one parameter's view. They can be classified into different groups depending on their source, chemical structure, solubility, and fibre type for which they are most compatible. The more or less colouration of certain dyes is linked with the chemical composition. The textile industry is highly dominated by azo dyes. Azo dyes are characterized by a double bond between two nitrogen atoms [–N=N–]. Furthermore, the presence of auxiliary functional groups such as amino, hydroxyl, sulfoxyl, and carboxyl make them amphoteric (Benkhaya et al. 2020). Dyes can be further sub-divided as acidic, basic (cationic), direct (substantive), reactive, disperse (non-ionic), sulfur, and vat dyes. Various types of dyes are discussed in Table 2.1. Dyes mainly consist of three important components viz. chromophores, auxochromes, and matrixes. The chromophores are some functional groups that help the dye in the absorption of light in the form of electromagnetic waves. The functional groups may be azo (–N=N–), nitro ($-NO_2$), nitroso (–N=O), carbonyl (–C=O), thiocarbonyl (–C=S) as well as alkenes (–C=C–). On the other hand, auxochromes are the colour-enhancing groups and help in the modification of the colour of the dyes. When present in the dye they increase the fibre affinity for colour and decrease the water solubility either by electron-withdrawing or donating nature. They may be acidic (–COOH, $-SO_3H$, –OH) or basic ($-NH_2$, –NHR, $-NR_2$). The rest of the atoms of the molecule is called matrix. In general, dyes are a complex combination of chromophoric, auxochromic, and conjugated aromatic structures (benzenes, anthracenes, perylenes, etc.). Conjugation of the dyes increases with an increase in the number of aromatic rings. This conjugation leads to a redshift in absorption. Similarly, an electron donor auxochromic group when situated along with a conjugated system joins in the conjugation and the molecule absorbs in the longer wavelength and gives darkintense colour (Benkhaya et al. 2020).

Dyes are also classified based on solubility in aqueous media. Soluble dye (acidic or anionic, basic or cationic, direct, reactive) and insoluble dye (sulfur, vat, disperse, pigments) are two major categories in this class. About 8000 chemically distinct synthetic dyes are listed in the Colour Index (C.I.) under 40,000 trade names (Yuan et al. 2020). They are identified under a code name representing their class, colour, and

Table 2.1 Classification and example of various dyes (Javaid et al. 2021)

Types of dye	Description	Fibres applied to	Typical fixation (%)	Example	References
Acid or anionic dye	Water-soluble anionic dye, usually have –OH, –COOH, $–SO_3H$ functional group	Wool, polyamide, silk, leather	80–93	Picric acid, mauritius yellow	Lam et al. (2012)
Basic or cationic dye	Water-soluble anionic dye, brilliant colours,	Cotton, wool, acrylic	97–98	Methylene blue, malachite green	Benkhaya et al. (2020)
Reactive dye	Water-soluble, excellent fastness property, largest dye class	Cotton, cellulose, wool	60–90	Reactive red 3, reactive blue 19	Lewis (2011)
Direct Days	Water soluble anionic dyes, second largest class of dye, no need of affixing agent, cheapest price	Cotton, rayon, other cellulosic	70–95	Congo red, directive blue 1	Benkhaya et al. (2020)
Vat	Water insoluble, better colourfastness and excellent brightness, expensive dyes	Cotton, linen, other cellulosic	80–95	Indigo	Benkhaya et al. (2020)
Sulfur	Water insoluble, mostly have S-containing heterocycles, cheaper and easy to apply	Cotton, cellulosic fibres	60–70	Leuco sulfur black 1	Benkhaya et al. (2020)
Disperse	Water insoluble, non-ionic dye	Polyester, cellulose acetate	80–92	Disperse yellow 42, disperse blue 6	Benkhaya et al. (2020)
Pigment	Water insoluble, do not contain any functional group, most are benzoic derivatives and metal derivatives Used in printing			Prussian blue	Es-sahbany et al. (2018)

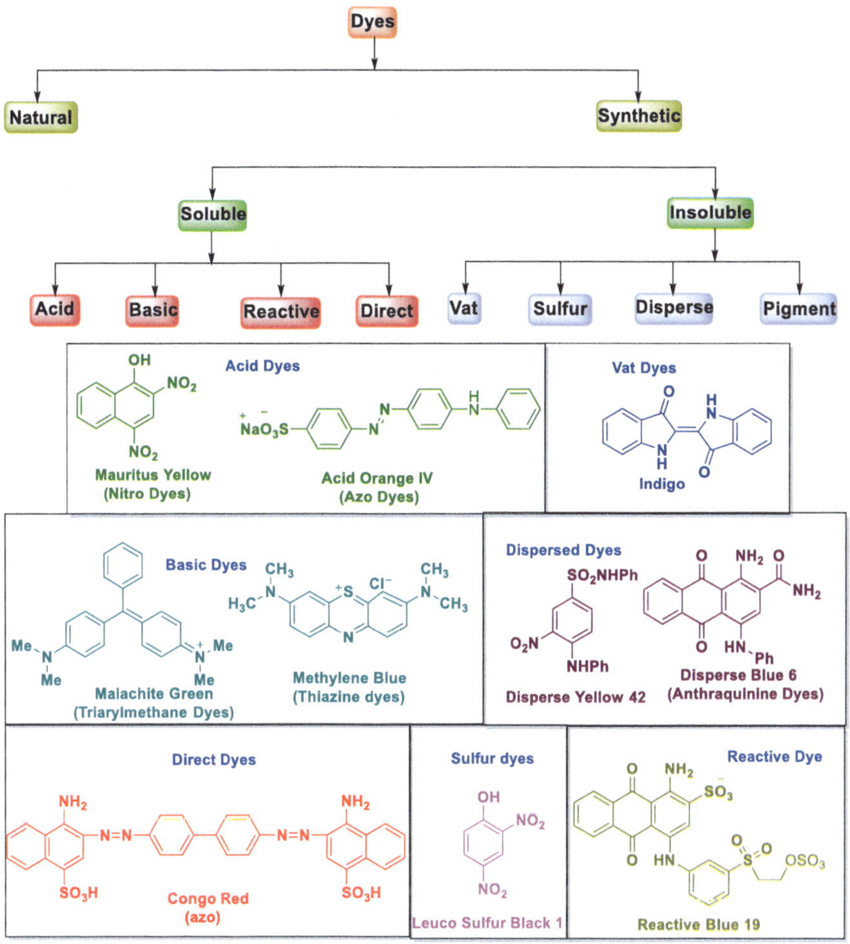

Fig. 2.2 Classification and chemical structure of dyes

order number. A general classification of dyes and chemical structure is represented in Fig. 2.2 to gain some idea of classification (Dahiya and Patel 2021).

2.3 Impact of Textile Dyes on the Environment

Dyes are referred to as recalcitrant pollutants as they are complex chemical substances and less biodegradable, persistent in the environment for a longer duration and eventually transmitted widely within diverse geographical regions. Higher stability, toxicity, high molecular mass and the presence of unusual bonds are some characteristic features of these recalcitrant. The toxicity of a dye is due to its structure rather than the dyeing process.

Organic wastes from textile industries contain a huge amount of coloured material and are released into nearby water bodies without any appropriate treatment causing major environmental pollution. Over usage of dyes increases the toxicity of the entire ecosystem. Dye material increases pH, and BOD and COD in the released water bodies (Solís et al. 2012). Many coloured substances reduce the photosynthetic ability in aquatic plants, affect protozoa colonization and the aquatic ecosystem. The by-products of the bio treatment of azo dyes are some amines that may be harmful, carcinogenic, and more toxic than the original dye (Franca 2020). Acute toxicity involves oral ingestion and inhalation which creates problems like skin irritation and skin sensitization. Reactive dyes can cause respiratory trouble and allergic reactions in workers on occupational exposure (Tkaczyk et al. 2020). Heavy metal-containing dyes harm aquatic organisms as well as human health. These dyes produce heavy metal cations in the aquatic ecosystem which is deposited in fish gills and ultimately enter the human food chain (Ito et al. 2016). This affects human organs with a series of pathogens. Chromium-containing dyes are highly carcinogenic and impair the growth and development of plants. Apart from being carcinogenic for humans, synthetic dyes cause skin disease and affect the central nervous system. For example, 2-naphthylamine, Azure-B are known to cause bladder cancer, mutagenic disorders by interacting with DNA and RNA respectively (Khan and Malik 2018; Haq et al. 2018).

The risk associated with synthetic dye can be minimized by avoiding exposure to dry dye. Increasing the application of liquid dyes, low-dusting formulations, and using appropriate personal protective equipment are some preventive measures, and the best of them is the use of proper degradation technology of synthetic dye. The rest of the chapter is devoted to photocatalytic degradation and removal of these synthetic dyes from wastewater and soil.

2.4 Principle of Photocatalysis and Mechanistic Pathways

As described earlier, various techniques are employed for the degradation and removal of organic dyes. The AOPs are the most popular and frequently used technique amongst them. AOPs have been reported to be operative for the degradation of soluble organic pollutants from water and soils and can provide almost total degradation. The AOPs include the following techniques.

i. photolysis (UV or visible)
ii. hydrogen peroxide (include Fenton-like-reagent: $H_2O_2 + Fe^{2+}/Fe^{3+}$)
iii. ozonation/photoozonation
iv. photocatalysis (include heterogenous photocatalysis).

The photocatalytic degradation of dye using heterogeneous catalysts is the main focus of this chapter.

2.4.1 Photocatalysts

Photocatalysts harness solar energy for the PD of organic contaminants which makes the photocatalysis treatment process to be economically feasible. Photocatalysts mostly consisted of chemically stable semiconductors with a specific characteristic of sensitizers for a photocatalytic redox reaction. In a semiconductor, the bandgap (E_g) separates the lowest occupied (valence band, VB) from the highest unoccupied (conduction band, CB) energy bands. Semiconductors should be photoactive, chemo-stable, photo-stable, biologically and chemically inert, recyclable, high activity in the UV–vis region and inexpensive. The bandgap energy, irradiation wavelengths, and adsorption of a species under light determine the photocatalytic activities of a semiconductor. Thus, a semiconductor with a large bandgap absorbs in the UV region whilst a smaller bandgap semiconductor can perform a better photocatalytic activity in the visible light region. Though catalysts possess active sites for catalytic conversion, the use of the active site may not be appropriate for photocatalyst since the reaction activity of these sites is dependent on the light source. It is the preferential defects in the crystal of catalyst which act as active sites (Zhang and Jaroniec 2017).

The large numbers of reported photocatalysts can be categorized in the family of metal-oxides, metal-nitrides, metal-sulfides, and also from metal-free components like graphenes or polymers. Inorganic semiconductors have achieved a lot of success in the field of photocatalysts. Herein, the principle of photocatalytic activity of metal oxides shall be discussed mainly. Different metal oxides like TiO_2, ZnO, ZrO_2, Fe_2O_3, SnO_2, MgO, Sb_2O_3, GeO_2, V_2O_5, Cu_2O, In_2O_3, Nb_2O_5, and perovskites are the most studied materials for photocatalytic dye degradation. Photocatalytic activity of TiO_2 and ZnO has been known since 1970. Photocatalytic activity of other oxides like Fe_2O_3, ZrO_2, MgO, SnO_2, GeO_2, Sb_2O_3, V_2O_5 though known from the 70s yet their photocatalytic activity is weak. Since then, TiO_2 and ZnO have attracted the research community because of their high photocatalytic activity, non- or low toxicity, good chemical stability, inexpensive, and more or less long-term photo durability. However, the lack of activity of these catalysts in the visible spectrum is one of the main concerns.

2.4.2 Basic Principle of Photocatalysis

Different experimental techniques of dye degradation are analysed by various instrumental methods such as GC/MS, UV–vis spectrometry, ion chromatography, HPLC, capillary electrophoresis, radiometry, etc. (Rauf and Ashraf 2009). During PD, the ultimate fate of the dye molecules is to break down into smaller molecules like H_2O,

CO_2, etc. So the evaluation of total organic carbon (TOC) of the system under investigation is very important (Tanaka et al. 2000). Vigorously stirred batch photochemical cells, glass dishes, cylindrical flasks or vessels are some typical laboratory-scale experimental setups (Wu et al. 2006).

Inexpensive semiconductors as a catalyst and easy mineralization of various organic compounds have proven photocatalysis as one of the promising AOP technologies. The basic mechanism of PD consists of the following steps. When light having photon energy equivalent to or greater than the bandgap of semiconductor strikes on the surface, an electron from the valence band of the catalyst is promoted to the conduction band. This results in the generation of electron–hole pairs.

Photoexcitation:

$$\text{Catalyst} + h\nu \rightarrow e^{-}(CB) + h^{+}(VB) \tag{2.1}$$

Here e^{-}(CB) and h^{+}(VB) represent electrons in the conduction band and hole in the valence band respectively. These species can migrate to the surface of the catalyst and may undergo redox reactions with other species on the surface.

Trapping of electrons on the surface:

$$e^{-}(CB) \rightarrow e^{-}(TR) \tag{2.2}$$

Trapping of holes:

$$h^{+}(VB) \rightarrow h^{+}(TR) \tag{2.3}$$

$e^{-}(TR)$ and $h^{+}(TR)$ indicate surface trapped CB electron and VB hole respectively. In most cases, $h^{+}(TR)$ reacts with the surface-bound H_2O and produces hydroxyl radical. On the other hand, electron in the conduction band $e^{-}(TR)$ can react with the dissolved O_2 to form superoxide radical anion of oxygen, $O_2^{\bullet-}$.

Oxidation of water:

$$H_2O + h^{+}(\text{VB}) \rightarrow OH^{\bullet} \tag{2.4}$$

Photoexcited electron scavenging:

$$O_2 + e^{-}(\text{CB}) \rightarrow O_2^{-} \tag{2.5}$$

However, the activity of these two steps {steps (2.4) and (2.5)} or the activity of the photocatalytic process is reduced by the recombination of trapped carriers. This recombination of electrons and holes should be delayed so that they take part in the redox reaction. Recombination of electrons and holes takes place in the absence of oxygen, as the latter captures electrons to form a superoxide radical anion.

Charge carrier recombination:

$$e^{-}(TR) + h^{+}(\text{VB}) \rightarrow e^{-}(\text{CB}) + \Delta \tag{2.6}$$

The $OH^{\bullet}$ produced in step (2.4) is the most powerful oxidizing agent and reduces dye into intermediate compounds. The same reaction is repeated till all dye molecules are degraded into CO_2 and water and eventually colour fades away. This reduction reaction is more feasible in the organic medium than in water. A higher concentration of organic matter increases the number of positive holes. This enhances photocatalytic activity by reducing the electron–hole recombination (Chen et al. 2017).

Photodegradation by OH radicals:

$$OH^{\bullet} + \text{dye} \rightarrow \text{dye}_{ox} \quad (k = 10^9 - 10^{10} M^{-1} s^{-1}) \tag{2.7}$$

On the other hand, the superoxide radicals decompose pollutants by undergoing a series of reactions to produce hydrogen peroxide, hydroxy radicals, and hydroperoxyl radical anions as shown in the following reactions.

Protonation of superoxide anion radical:

$$O_2^{\bullet -} + H^+ \rightarrow HOO^{\bullet} \tag{2.8}$$

Recombination of electron–hole pairs:

$$\begin{aligned} HOO^{\bullet} + e^-(CB) &\rightarrow HO_2^{\bullet -} \\ HOO^{\bullet} + h^+(VB) &\rightarrow H_2O_2 \end{aligned} \tag{2.9}$$

A schematic presentation of the above mechanism is shown in Fig. 2.3. The overall steps of the mechanism of mineralization of dye can be summarized in the following steps (Ahmed and Haider 2018).

(i) Mass of the organic pollutants is loaded on the surface of the photocatalyst from the bulk liquid phase
(ii) Active sites of the photocatalyst adsorb the pollutant

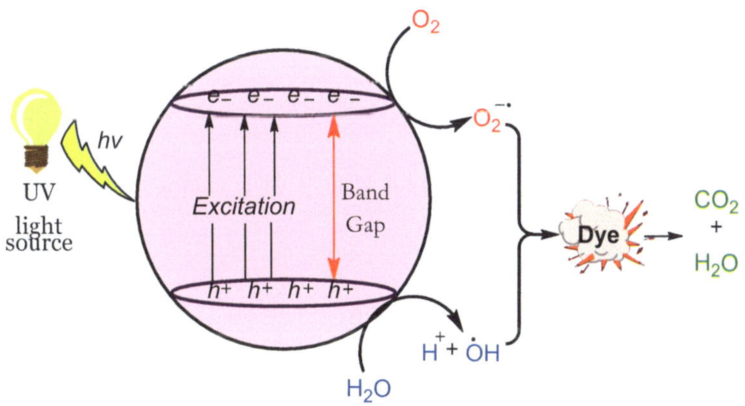

Fig. 2.3 Schematic representation of PD of dye by a photocatalyst

(iii) Hydroxyl radical ($OH^{\bullet}$) and superoxide anion radical ($O_2^{\bullet -}$) generated degrade the pollutants
(iv) The final product is detached from the catalyst surface
(v) Mass of the final product is transferred into the bulk liquid phase.

Direct (or Type-I) and indirect (or Type-II) are two suggested mechanisms for photocatalytic degradation of dyes which are discussed below.

2.4.2.1 Direct (or Type-I) Mechanisms

According to this mechanism, photoexcitation of the catalyst produces electrons and holes. The light-generated electron and hole are then captured by the dye molecules which are already chemisorbed on the catalyst surface. This results in the excitation of dye molecules (Dye*) which finally decay by the combination with the electrons. During the process, the excited electron is captured by some oxidant which enables completion of the photocatalytic cycle. Dye sensitization increases the photocatalytic activity (as it offers an expansion of the light absorption region into visible light), enhances excitation efficiency, and helps in charge separation leading to the modification of the photocatalysts (Zhao et al. 2005). The Langmuir–Hinshelwood process and the Eley–Rideal are two processes that can explain this mechanism. The former process states that electrons and holes are chemisorbed by the adsorbed dye whilst according to the later process it is the surface defects (S), that act as active centres. The degradation process as suggested by the Eley–Rideal process is shown below (Prihod'ko and Soboleva 2013).

$$\text{Catalyst} + h\nu \rightarrow e^- + h^+ \text{(generation of charge carriers)} \tag{2.10}$$

$$\text{S} + h^+ \rightarrow \text{S}^+ \text{(trapping of hole by surface defects)} \tag{2.11}$$

$$\text{S}^+ + e^- \rightarrow \text{S(decay of active centres)} \tag{2.12}$$

$$\text{S}^+ + \text{Dye}^+ \rightarrow (\text{S} - \text{Dye})^+ \text{(chemisorption of S on dye)} \tag{2.13}$$

$$(\text{S} - \text{Dye})^+ \rightarrow \text{S} + \text{CO}_2 + \text{H}_2\text{O} + \text{(degradation of dye)} \tag{2.14}$$

2.4.2.2 Indirect (or Type-II) Mechanism

In this mechanism, upon absorption of UV light by the photocatalyst, electrons from the VB are promoted to CB, creating holes in the VB. In the next phase, the photogenerated electrons and holes participate in the redox reaction on the surface of the catalyst. Holes are trapped by water molecules giving $OH^{\bullet}$ and H^+ whilst electrons form $O_2^{\bullet-}$ by reacting with oxygen in the medium. The reactive hydroxyl radical ($OH^{\bullet}$) and oxygen radical anion ($O_2^{\bullet-}$) finally decompose the dyes into CO_2 and H_2O.

There is an alternative mechanism for degradation in which dye molecules get self-photosensitized leading to photolysis of the dyes. The photolysis process is dominant in the natural environment but this is inefficient in removing organic pollutants from the wastewater (Zhang and Zhang 2020; Pingmuang et al. 2017). The dye molecules get excited to the triplet excited state by absorption of light. The energy released in this process excites the ground state oxygen into the first excited state which eventually oxidizes dye molecules. The $OH^{\bullet}$ and $O_2^{\bullet-}$ generated in this process are very reactive which is essential for dye degradation. Methyl Orange (MB) degrades in the natural environment via this process. However, N groups containing dye like Rhodamine B (RhB) are resistant to photolysis and undergo anaerobic photodegradation leading to carcinogenic products (Lakshmi Prasanna and Rajagopalan 2016).

2.5 Effect of Key Operational Parameters

Lately, scientists have put lots of effort to study the key operational parameters that affect photodegradation. The primary motive of all these studies is to obtain some optimum and effective conditions that are best for photo-degradation of organic pollutants. Factors like catalyst loading, exposure duration, pH, oxidizing agent, and temperature can influence the rate of PD. The effect of these factors is explained by taking TiO_2, ZnO, and some other commonly used catalysts as a reference which is discussed below.

2.5.1 Effect of pH

Evaluation of the effect of pH on photocatalytic dye degradation is a challenging task. The complication arises from different degradation reaction mechanisms for each dye. The pH of the solution is considered as one of the essential parameters owing to its impact on the performance of photocatalysts. The pH of the discharged effluent controls the surface charge of photocatalysts and influences its oxidation potential. The pH also determines electrostatic interaction between catalysts surface, substrates, and radicals during the degradation process. The separation of the photo-generated electrons and holes, surface charge, and aggregation of the catalysts are pH-dependent (Zangeneh et al. 2015). When the pH is raised beyond the isoelectic point (pH_{pzc}) of the photocatalyst, its surface acquires a negative charge. At a lower pH, the functional groups get protonated and increase the surface charge of the catalyst. Above pH_{pzc}, cationic molecules get adsorbed on the surface of the photocatalyst which is negatively charged at this pH. The reverse occurs at lower pH as anionic

Table 2.2 Optimum pH for different photocatalysts and organic dyes (Gnanaprakasam et al. 2015)

Light source	Initial concentration	Catalyst used	Organic pollutant	Optimum pH	Irradiation time (min)	Degradation efficiency (%)
UV light	200 mg/l (COD)	TiO_2	Coking water	7	60	30
100 W mercury lamp	50 ppm	Codoped TiO_2	2-Chlorophenol	12	180	95
UV light	1.2×10^{-4} M	TiO_2 and cement	Reactive yellow dye 17	5–7	480	85
40 W fluorescent lamp	6×10^{-6} M	$MnWO_4$	Methyl orange	9	480	50

molecules are easily attracted by positively charged photocatalysts (Li et al. 2020). The photodegradation mechanism has the following three important steps which can be influenced by the pH of the solution: (i) attack by hydroxyl radical, (ii) direct oxidation by the positive hole, and (iii) reduction by the valence band electron. The contribution of each step is determinant to the functional group of the substrates and pH. Though higher pH increases the hydroxyl ion concentration yet higher pH is not suitable for PD of dye because it competes with the organic molecules for getting adsorbed on the photocatalyst surface (Rajabi et al. 2013). In contrast, lower pH reduces adsorption of cationic organic dyes, therefore, reduces the efficiency of the PD in acidic media (Vinu et al. 2010). Hence, for degradation of different organic pollutants, there should be some optimum pH using photocatalyst and the same is focussed in Table 2.2. The discharge effluents of various industries are released at different pH so their effect on the PD is needed to be necessarily scrutinized (Saeed et al. 2015).

Being an amphoteric photocatalyst TiO_2 can develop both positive and negative charges on its surface. Thus, variation in the pH can affect the adsorption of dye molecules on the TiO_2 surface and thereby causing an impact on the PD reaction rate (Wang et al. 2008). Protonated and deprotonated form of titania under acidic and basic condition respectively proceeds as outlined below:

$$TiOH + H^+ \rightarrow TiOH_2^+ \tag{2.15}$$

$$TiOH + OH^- \rightarrow TiO^- + H_2O \tag{2.16}$$

Thus, the surface of the TiO_2 catalyst is positively charged in acidic conditions (or in lower pH) and negatively charged in the basic medium (or in higher pH). Reports suggest that titanium dioxide has higher oxidizing activity in acidic medium or lower pH. However, a very higher H^+ concentration slows down the reaction rate. The optimum pH for Acid Yellow 17, an acidic dye is 3 whilst Orange II and Amido

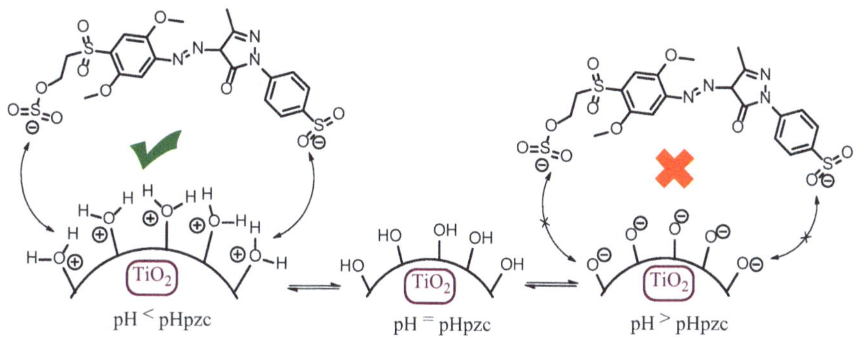

Fig. 2.4 Electrostatic interaction between RY17 and TiO_2 as a function of pH

Black 10B has shown efficient degradation at pH 9 (Behnajady et al. 2004; Qamar et al. 2004). It is reported that due to the cationic nature of Methylene Blue (MB), adsorption enhances at higher pH resulted in the high efficiency of photocatalytic degradation (Saeed et al. 2015). Anionic dye, Methyl Orange (MO) is more efficiently photodegraded in the acidic medium than in the basic medium. Research suggests that the optimal pH for MO photodegradation is 2. It is found that at this optimal pH, MO is photodegraded with complete efficiency in 20 min whilst in the basic medium no photodegradation was observed even if the exposure time is raised to 1 h. The negatively charged sulfonate groups of MO dye strongly interact with the positively charged surface of the catalyst in an acidic medium which enhances the PD of MO dye in an acidic medium compared to the basic medium. In 2017, Alahiane et al. studied the degradation of anionic Reactive Yellow-17 (RY17) dye on TiO_2 catalyst surface. The pH of zero-point charge (pH_{pzc}) for TiO_2 is 6.3. Thus, below pH_{pzc}, the TiO_2 surface acquires a positive charge which is shown by the attraction of the skeleton of negatively charged RY17 dye by the catalyst surface. Thus, dye molecules get concentrated on the catalyst surface which leads to efficient PD at pH 3. On the contrary, electrostatic repulsion between the negatively charged dye molecules and the negative charge of the catalyst surface above pH_{pzc} delays the accumulation of the dye molecules resulting in lower PD activity as proposed in Fig. 2.4 (Alahiane et al. 2017).

2.5.2 *Effect of the Dose of Semiconductor*

The efficiency of the PD process is highly dependent on the photocatalyst dosages. An increase in the photocatalyst dosages increases the dye degradation. Catalyst surface area increases with the increase in the amount of catalyst dosages/loading. A large surface area means the number of available active sites is higher leading to the formation of more $OH^•$ *radicals, which are involved in the dye degradation. However, this positive correlation between semiconductor dosage and efficiency of dye degradation is limited up to certain semiconductor concentrations. Beyond the optimum amount of the catalyst, the rate constant for the mineralization of the dye diminishes. Again, this optimum amount of the catalyst is different for different catalysts and determined by the nature of the catalyst, types of contaminants and the operating conditions.*

In 2019, López-Ramón et al. *evaluated the concentration effect of a nickel organic xerogel (X-Ni) as photocatalyst in the degradation of the herbicide diuron (DRN) in the aqueous phase. They varied the concentration of X-Ni photocatalyst from 100 to 5000 mg/L, the results explained that the degradation rate for constant initial DRN concentration increases with an increase in the X-Ni concentration and reaches the maximum in the 4167–5000 mg/L concentration range of the photocatalyst* (López-Ramón et al. 2019). *Further, an increase in the concentration to 5000 mg/Ls hows no improvement in the degradation rate. This effect of the catalyst concentration can be understood as the surface area and the active catalyst sites increase with the increase in the catalyst concentration up to a certain limiting concentration of the catalyst. However, above 5000 mg/L of xerogel concentration, catalyst agglomeration starts which leads to the decrease in the concentration of active sites of the catalyst for light absorption. Excess concentration of the catalyst turned the solution turbid and increase light dispersion resulting in the lower light penetration for effective degradation (*Fenoll et al. 2013*). The narrower bandgap of CdS semiconductors compared to* TiO_2 *makes them better photocatalysts under visible light irradiation.* The nanocomposites of CdS on graphene support make the former even better photocatalyst due to retardation in the charge recombination rate. Lu et al. in (2014) studied the effect of loading amount of this catalyst system for the photodegradation of Rhodamine B (RhB) under visible light irradiation recorded at 270 min. The report suggests that the photodegradation of RhB keeps increasing from 4 to 20 mg of the catalyst and after which it becomes constant (Lü et al. 2014). Similar trends were obtained when methylene blue (MB) is photodegraded on the TiO_2catalyst surface (Zhang et al. 2002). The effect of catalyst loading on the degradation of the dyes Reactive Red 2(RR2), Reactive Blue 4 (RB4), and Reactive Yellow 17 (RY17) was further examined by Neppolian et al. by taking TiO_2 as a catalyst. They varied the catalyst loading from 100 to 600 mg/100 mL of the dye solution. Results showed that the PD rate increases rapidly with an increase in the amount of TiO_2 up to 300 mg/100 mL for all three types of dyes. Beyond 300 mg of TiO_2, RR2 and RB4 dyes show only a slight increase in the degradation rate but for RY17, percentage degradation was found to be almost constant and rather decreases for catalyst loading 500–600 mg/100 mL. This is attributed to a decrease in the photoactivated volume of the suspension with an increase in the catalyst loading (Neppolian et al. 2002).

2.5.3 Effect of the Initial Concentration of Dye

The optimal dye concentration for better photodegradation of various dyes is needed to be considered carefully. Depending upon the initial concentration the irradiation time needs to be adjusted accordingly to arrive at appropriate degradation under standard operating conditions. The adsorption of the dye molecules on the surface of the catalyst is dependent on the nature of the photocatalyst and functional groups present in the dyes. In general, the rate of photodegradation is higher at a lower initial dye concentration. The rate rises to a certain dye concentration and then starts falling. Increasing dye concentration would lead to more adsorption of dye molecules on the surface of the photocatalyst, thus decreasing the light penetration to the catalyst surface. The dye molecules would compete for the absorption of UV radiation with the photocatalyst. The collision frequency between the dye molecules and the active sites is higher at low initial dye concentration. The path length of the photons reaching the photocatalyst surface also increases when dye concentration is increased. Therefore, following Beer-Lambert law, this will lower the photon adsorption on the catalyst surface and eventually lowers the kinetics of PD. Haile Hassena in 2016 studied the effect of initial concentration of Methylene Blue (MB) using Al_2O_3/Fe_3O_4 semiconductor. The concentration of MB was varied from 0.5×10^{-5} to 1.4×10^{-5} M and it is observed that PD efficiency starts decreasing beyond 1.1×10^{-5} M of MB after 90 min. As more active sites are available for the cationic MB dye at low concentration, PD increases. But at higher MB concentrations, more and more molecules of MB dye get adsorbed on the Al_2O_3/Fe_3O_4 catalyst surface thus, reduces the efficiency of photogeneration of $OH^{\bullet}$ (Hassena 2016). Research conducted by Kiran et al. in (2016) explained that the rate of PD of methyl orange (MO) using Mg-doped titania under visible light raises upto 10 mg/L and then starts decreasing (Kiran et al. 2016). In (2017), Zada et al. studied the degradation of Malachite Green using bimetallic zinc and manganese oxides nanoparticles (Zn-Mn NPs). The report suggests that the percentage degradation is inversely affected by the dye concentration and about 88.41%, 80.95%, and 55.43% dye degraded in 12.5, 25, and 50 ppm dosage respectively (Zada et al. 2017). In (2019), Babu et al. examined the effect of initial dye concentration of MO in an ultrasound-assisted photo catalytic degradation. They tested the MO dye degradation by varying initial dye degradation from 0.01 to 0.04 mM in presence of $CuO\text{-}TiO_2/rGO$ photocatalyst amount ($1.0\ gL^{-1}$) under the mentioned technique. Results obtained from this study suggested that 98% dye degradation is obtained irrespective of initial dye concentration in 90 min through the present technique (Babu et al. 2019).

The photodegradation of dyes is dependent on the concentration of $OH^{\bullet}$ radicals on the surface of the photocatalyst and their interaction with dye molecules. At initially low concentration, the chances of collision between the dyes and the free radical are less. However, beyond the optimized concentration, the collision rate decreases due to the fewer availability of active sites resulting in the decrease of the degradation rate. As the concentration of dye increases, more dye molecules get adsorbed on the catalyst surface making the photogeneration of hydroxy radicals difficult. Degradation efficiency is not enhanced further as active sites and hydroxyl radical concentration remain constant even at higher substrate concentrations.

2.5.4 Effect of Additives

In dye manufacturing, some inorganic ions are usually added to the dye solution as ionic compounds make the industrial process feasible. However, inorganic ions like Fe^{2+}, Ag^{+}, Zn^{2+}, Na^{+}, Cl^{-}, $SO_4{}^{2-}$, BrO^{3-}, $PO_4{}^{3-}$, $CO_3{}^{2-}$, HCO^{3-} and persulphate ions; which are an integral part of the effluent have a diminishing effect on the dye degradation in the solution. *As a result, different additives are added during PD to increase the efficiency of the process. These additives are some oxidants like* H_2O_2 *which enhance the degradation efficiency and reduce the reaction time.*

The addition of H_2O_2 can increase the concentration of $OH^{\bullet}$ *and its ability to trap electrons make PD more efficient. Rahmat* et al. *described the role of* H_2O_2 as an initiator in the degradation of crystal violet in the presence of MnO_2-based nanofibrous mesh (MnO_2/NF) (Rahmat et al. 2019). They took 0.04 mL of H_2O_2 and measured the absorption spectra to check its impact within 1.5 h. Lowering in the absorption peak indicated the degradation of the dye. It is found that efficient degradation was obtained in 280 min in the absence of the initiator whilst the presence of the initiator reduced the timing to 60 min. As an electron capturer, H_2O_2 generates $OH^{\bullet}$ under irradiation of visible light and thus initiates the photocatalytic chain reaction (Lin et al. 2014). At optimal H_2O_2 concentration, it inhibits the complex effect of photo-generated electron hole-pairs. How H_2O_2 helps in the degradation of the dye is shown in the reaction *below:*

$$OH^{\bullet} + H_2O_2 \rightarrow H_2O + HO_2^{\bullet} \tag{2.17}$$

$$OH^{\bullet} + HO_2^{\bullet} \rightarrow O_2 + H_2O \tag{2.18}$$

$$OH^{\bullet} + OH^{\bullet} \rightarrow H_2O_2 \tag{2.19}$$

For the detection of the main species in the PD of dyes, various researchers add scavengers to the photocatalytic reaction condition. A thorough examination of these species and their way of generation and utilization in dye degradation is a matter of prime importance in the photocatalysis study. Alcohols, *p*-benzoquinone, EDTA-2Na, and Ar gas, $AgNO_3$, KI are some frequently used scavengers. Daneshvar et al. examined the effect of radicals in the PD of Acid Red 14 dye using TiO_2 suspension under irradiation of UV-C lamp (30 W) (Daneshvar et al. 2003). They used ethanol to quench the hydroxy radical. Upon addition of 0.02% (v/v) of quencher, the PD efficiency reduced to 12.4% after exposing for 2 h but the degradation did not stop completely. This indicates that hydroxy radicals are not the only ones that assist in the degradation. Positive holes h^{+}(VB) formed on the irradiated photocatalyst which reacts with the adsorbed dye, may assist in the degradation.

As explained above, ions have a diminishing effect on photocatalytic dye degradation. The effect of these ions on the reduction of degradation efficiency is explained below based on their chemical reactions in solutions. For example, the presence of

Fe^{2+} ion easily converted $OH^{\bullet}$ to OH^- and thus decrease the concentration of $OH^{\bullet}$ in the reaction medium. This eventually leads to a decrease in the dye degradation (Yoon et al. 2001).

$$OH^{\bullet} + Fe^{2+} \rightarrow OH^- + Fe^{3+} \quad (k = 3.5 \times 10^8\,M^{-1}s^{-1}) \tag{2.20}$$

Likewise, the other ions such as $CO_3{}^{2-}$, $HCO_3{}^-$ *which are added to the dye bath for maintaining pH eventually have a negative impact on the dye degradation. The presence of these ions scavenges the* $OH^{\bullet}$ *radicals and causes a percentage decrease in the PD. The presence of oxygen or air also influences the degradation efficiency. Oxygen molecules scavenge CB electrons and this leads to an increase in the production of active oxygen species such as* H_2O_2, $O_2{}^{\bullet-}$, $HOO^{\bullet}$, and $OH^{\bullet}$. This helps the catalysts to restore their original form and thus completes the catalytic cycle (Zhao et al. 2005).

2.5.5 *Effect of Temperature*

The influence of experimental temperature conditions on the efficiency of degradation of organic dyes by photocatalysts has been less investigated. Zazo et al. in (2011) *reported that the decomposition efficiency of phenolic dyes using Fenton oxidation is almost 80% at 120 °C which is reduced to 28% on decreasing the experimental temperature to 25 °C* (Zhao et al. 2005). *Salem et al. examined the influence of temperature on the photodegradation efficiency of Acid Blue 29 (AB29) using* MMTK10-Cu(cn)$_2$ as the catalyst. They varied temperatures in the range of 20–40 °C. *The decolouration efficiency was raised from 51.8 to 88.2% on rising the experimental temperature from 20 to 40 °C when irradiated for 18 min* (Salem et al. 2014). *Javaid et al. studied the total organic carbon (TOC) removal of* Remazol Brilliant Blue R (RBBR). Keeping the initial dye concentration at 20 ppm, pressure 10 MPa, and using a continuous-flow tubular reactor coated with a thin layer of PdO as a catalyst they increased the temperature from 200 to 250 °C and finally to 300 °C. The catalytic efficiency likewise increased from 89 to 92% and 99.9% respectively (Javaid et al. 2016). Thus, a rise in temperature increases the degradation efficiency. Higher temperature pushes the degradation reaction forward by overcoming electron–hole recombination and generating more free radicals by bubble formation in the solution. Additionally, an increase in the experimental temperature helps in the rapid oxidation of organic dyes at the interface (Vinu et al. 2010).

2.5.6 *Effect of Light Intensity and Wavelength*

Being a photocatalytic process, light has a critical role in the PD of organic dyes. The rate and the efficiency of the PD process are proportional to the intensity of the used light. As stated by Ollis et al. the rate of PD is first order if the light intensity is between 0 and 20 mW/cm^2 and varies linearly with increasing light intensity. At intermediate light intensities, the PD rate is proportional to the square root of the light intensities. Thus, follows half order kinetics at intermediate intensities. However, PD is independent of light intensities for the highly intense light sources. At high light intensity photons per unit time and unit area increases resulting in stronger photocatalytic activity. Though light intensity increases, the number of active sites remains the same on the catalyst surface and hence the reaction rate remains constant after reaching an optimum light intensity (Ollis et al. 1991).

The undesired electron–hole recombination is high when irradiated at high intensity, leading to a decrease in the PD rate of the dye. Muruganandham and Swaminathan investigated the effect of light intensity on the decolouration and degradation of azo dye Reactive Yellow 14 (RY14) using a TiO_2-UV system (Muruganandham and Swaminathan 2006). Their results explained that the decolouration of dyes increases to 87.9% from 35.9 when light intensity is enhanced from 16 to 62 W in 20 min time. Similarly, the degradation of dyes also increases from 26.5 to 59.4% in the same duration. The photooxidation increases linearly between 16 and 32 W and from 32 to 48 W no insignificant improvement in photooxidation was observed (Ollis et al. 1991). However, beyond 48 W, the photooxidation is independent of the light intensity. The optimum photooxidation was achieved at 32 W UV light. Though solar and UV lights are used for PD, yet artificial (UV) light is dominating in this field. Because of the constant stable source of light intensities irrespective of the weather and other environmental factors, artificial light sources are preferred. Moreover, the higher efficiency of UV light in PD over solar irradiation makes them more useful. However, the higher abundance, easy accessibility, and non-harmful nature of solar light make them alternative and economical light sources. Solar light consists of 5% UV light (200–400 nm), 43% visible light (400–800 nm), and more than 52% IR. The artificial UV light for its reproducible nature and high energy compared to solar irradiation found to be highly efficient in the degradation of the dyes. At higher light intensity electron–hole pairs recombination is slow thus makes the PD process faster. TiO_2 has a wide bandgap energy (3.2 eV for anatase, 3.00 eV for rutile, and 3.13 eV for brookite). The wider bandgap restricts its absorption in the UV region of the solar spectrum (Kuang et al. 2009). The photodegradation rate of dyes in the aqueous phase for TiO_2 catalysts is influenced by the intensity of UV light. Neppolian et al. investigated the effect of intensity and source of the light whilst using TiO_2 as photocatalyst and Reactive Yellow 17 (RY17), Reactive Red 2 (RR2), and Reactive Blue 4 (RB4) as dyes (Neppolian et al. 2002). Higher degradation was achieved for all three types of dyes when the intensity of solar radiation was raised.

To study the effect of different light sources Yan-fen et al. took sulforhodamine-B dye for photodegradation employing TiO_2 catalyst under three different light sources. The visible light source was 500 W halogen light with intensity 70.6×10^3 I_x, source of solar light was irradiation of sunlight between 12.00 am and 2.30 pm with corresponding intensity $(80 \pm 5) \times 10^3$ I_x and a 100 W mercury lamp was the UV light source with intensity 105×10^3 I_x. Results obtained showed that the PD rate constant with UV light is the highest and least with visible light (Yan et al. 2006).

Stengl et al. examined the wavelength dependency of PD rate in the degradation of Orange II dye in an aqueous medium using rare earth metal-doped TiO_2 as the catalyst. And explained that the degradation rate constant is affected by wavelength for the TiNd_3 catalytic system. TiNd_3 i.e., titania doped with rare earth metal Nd is found to be the best photocatalytic system for the degradation of Orange II dye under irradiation of visible light. The energy of the photon is interrelated to its wavelength and a shorter wavelength is associated with greater photon energy which is reflected in the PD rate constant. Results showed that rare-earth ion doping in titania is helpful in the visible light photocatalytic activity of TiO_2 as reflected in the redshift of the optical absorption spectrum (Štengl et al. 2009).

2.5.7 *Effect of Irradiation Time*

PD capacity of heterogeneous catalysts towards organic dyes has been analysed considering irradiation time. Reports suggest that PD increases with an increase in the irradiation time in photocatalytic reactions. This attributes to an increase in the formation of more $OH^{\bullet}$ and $O_2^{\bullet -}$ with irradiation time. However, the rate slows down after some optimum time. This optimum time depends on the catalysts as well as the types of dye. Sakthivel et al. used azo dye Acid Brown 14 as a model dye for the investigation of the effect of irradiation time by two different catalysts viz., ZnO and TiO_2 (Sakthivel et al 2003). As per the report, the rate of decolouration and degradation increases with an increase in the irradiation time and slows down after irradiation of light for 1 h. The reduced rate of degradation after a certain time limit is attributed to the difficulty in the oxidation of this nitrogen-containing dye. They found that when all other operating conditions were kept constant, the complete decolouration of acid Brown 14 occurred in 2 h for ZnO and 5 h for TiO_2. However, for the complete degradation, ZnO took 6 h and 7 h for TiO_2. Ramli et al. in (2014) also evaluated photocatalytic activity as a function of irradiation time in the degradation of MB under different catalysts (Ramli et al. 2014). Results explained that the rate of PD is rapid in 0–30 min for TiO_2/AC (TiO_2 with activated carbon) and TiO_2/PANi (TiO_2/Polyaniline). The rate became slower after 30 min for both the samples. However, TiO_2/AC was found to be a better catalytic system with complete degradation in 90 min. Availability of more active sites and a high rate of formation of $OH^{\bullet}$ at the starting of the degradation process is the reason behind this. The repulsive forces between the molecules on the surface with bulk phase hindered the filling of remaining active sites after a certain limiting time. Thus, the rate of degradation tends to be constant after some time.

2.6 Degradation Studies of Dyes

For the removal of toxic dyes, different types of physical, chemical, and biological techniques have been established in the last few years. Amongst all these procedures, the adsorption process is one of the utmost choices for the decolourization of dyes and various types of dissolved colouring materials. Different technologies such as UV–vis spectrophotometry (Al-Mamun et al. 2019; Alnuaimi et al. 2007), HPLC (Abdullah et al. 2007), GC/MS (Benatti et al. 2006), ion chromatography (Blanco-Galvez et al. 2007), radiometry (Dionysiou et al. 2000), X-ray diffraction (XRD) (Andronic and Duta 2007) etc. have been employed to quantify the efficiency of dye removal and characterize the fate PC. After total mineralization, most of the dyes were converted to carbon dioxide (CO_2) and water (H_2O). During degradation, nitrogen is mineralized to N_2, NH_4^+ and NO_3^- ions which depends on the initial oxidation state of nitrogen, the structural conformation of the substrate, and irradiation time. Similarly, chlorine can be transformed to chloride (Cl^-), sulfur to sulfate (SO_4^{2-}), phosphorus to phosphate (PO_4^{3-}), and so on. The extent of dye mineralization can be monitored by measuring total organic carbon (TOC) or the measurement of chemical oxygen demand (COD) and biological oxygen demand (BOD). Usually, with increasing the amount of NH_4^+, Cl^-, SO_4^{2-}, and NO_3^- the values of COD or TOC decrease with irradiation time. In general, the compounds which do not form stable intermediates at low dye concentration can complete their mineralization with half-lives for parent dye.

2.6.1 General Considerations

With the evolution of experimental techniques, several semiconductors have been used according to their effectiveness towards dye degradation, including TiO_2, V_2O_5, ZnO, WO_3, CdS, ZrO_2, and their saturated forms (Andronic and Duta 2007; Hasnat et al. 2007). Titanium dioxide (TiO_2), one of the most popular photocatalysts, has been applied to investigate dye degradation due to its non-toxicity, ready availability, and high chemical stability. Because of the large bandgap (~3.1 eV), TiO_2 is utilized under UV irradiation (250–350 nm). Later, it has been extended to hydrogen production, solar cells, pollutant photo-oxidation, and removal of dye in liquid and gaseous phases (Atarod et al. 2016).

In comparison to TiO_2, the photocatalytic efficiencies of other semiconductors such as ZrO_2 are quite dissimilar. Although the bandgap energy for both TiO_2 and ZrO_2 are the same (3.1 eV) the efficiencies of TiO_2 in the separation of photo generated charges (less e^-/h^+ recombination rate) are higher due to the structural conformation of the material. On the other hand, due to intra-bandgap surface states, ZrO_2 shows a low absorbance in UV ranges, thereby causing its low activity (Bandara et al. 2017; Reddy et al. 2018).

Due to chemical stability, inexpensive, and large absorption spectrum, ZnO has been broadly employed as a photocatalytic material for decolourization of wastewater than that of TiO_2. Methyl Orange (MO), Acid Orange (AO), and Acid Red (AR) were degraded using ZnO under UV illumination (365 nm) and the rate of degradation order of these dyes is: AR > AO > MO (Anwer et al. 2019). The significant advantage of using ZnO is its ability to absorb a large fraction of the solar spectrum compared to the other semiconducting metal oxides and the major drawbacks are its wide bandgap energy and photo corrosion. Due to wide bandgap energy, the absorption of ZnO is limited in the visible light region, resulting in fast charge carrier recombination and ultimately low photocatalytic efficiency. The optical, structural, and magnetic properties of ZnO can be improved and altered by surface modification of ZnO catalysts that had several higher defect states exhibited better dye adsorption but low photocatalytic activity (Liu et al. 2013).

2.6.2 *Photocatalytic Degradation Scheme for an Azo Dye*

Azo dyes constitute a significant class of synthetic and coloured organic compounds which are characterized by one or more nitrogen to nitrogen double bonds (–N=N–). The determination of the colour of azo-dyes fully depends upon the azo bonds and their associated chromophores and auxochromes. The azo bonds are much more active and can readily oxidized by positive hole or hydroxyl radical ($OH^•$) or reduced by the electron in the conduction band (CB), thereby resulting in the cleavage of the –N=N– bonds which leads to decolourization of dyes (Rauf and Ashraf 2009). However, synthetic dyes commonly used on an industrial scale are azo (monoazo, diazo, triazo) anthraquinone, phthalocyanine, triarylmethane, and xanthene derivatives. These types of dyes are released in the aqueous streams from the effluents of various industries such as textiles, paper, leather, plastics, etc. Because of the toxicity and persistency of azo dyes, their removal from textile wastewater has become a matter of interest. Recently, conventional technologies, including primary (adsorption, flocculation), secondary (biological methods), and chemical processes (chlorination, ozonization) are used for degradation of the colour of the water with dye-contamination.

The photocatalytic degradation of aqueous solutions of Acid Orange 7 (AO7) has been analysed with the use of a solar light-irradiated TiO_2 catalyst. AO7 mainly exists in two tautomeric forms, i.e., one is the azo form and the other one is the hydrazone form. However, the hydrazone form is predominant in the aqueous solution and the photo catalytic decolouration leads to complete mineralization of the dye solution (Rauf and Ashraf 2009). AO7 absorbs on the surface of the photocatalyst via the oxygens of its hydrazone and sulfonate group. Initially, the cleavage of the dyes in the vicinity of the azo bond occurs in the presence of solar light, followed by the cleavage of attached aryl rings. This step takes place via a photosensitized mechanism, resulting in decolouration of the solution without significantly affecting the COD. The mechanism for the formation of oxidative species, following excitation

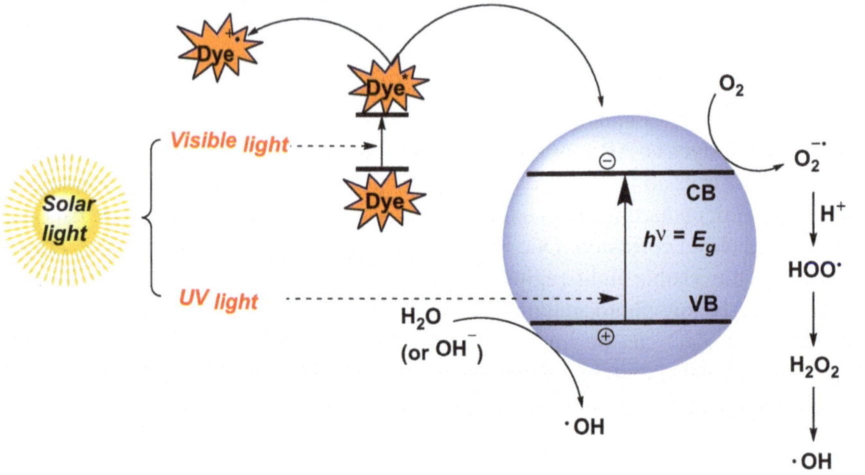

Fig. 2.5 Mechanisms of generation of oxidative species of AO7/TiO_2 system

of AO7–TiO_2 system with solar light is shown in Fig. 2.5. In this process, TiO_2 adsorbs the dye on its surface and undergoes a series of oxidation steps, which lead to decolourization and formation of several intermediates (mainly aromatic and aliphatic acids), resulting in the decrease of pH and an increase in the conductivity of the solution. Further, these molecules undergo oxidation forming low molecular weight species and, eventually, to CO_2 and inorganic ions, such SO_4^{2-}, NO_3^-, and NH_4^+ (Rauf and Ashraf 2009; Stylidi et al. 2003).

In addition to this, the degradation of several azo dyes such as Acid red 18, Methyl Orange, Remazol Black 5 (RB5), and Procion Red MX-5B has also been investigated utilizing TiO_2 photocatalyst (Tanaka et al. 2000; Sahel et al. 2007; Mozia et al. 2005). The examined dyes include both monoazo and diazo dyes and their rate of degradation can be associated with the disappearance of colour and the rate of elimination of total organic carbon (TOC). Based on this, it can be revealed that monoazo dyes are easy to degrade in comparison to diazo dyes and the majority of intermediates observed are aromatic amine, phenolic compounds, and several organic acids. Adsorption of dye on the surface of TiO_2, is one of the significant factors that measures the rate of degradation. Photocatalytic degradation of azo dye mainly proceeds through oxidation by positive hole (or $OH^•$) and reduction by CB electron. The adsorption isotherm of RB5 dye represents an L3 type which indicates the formation of at least two layers of dye on the photocatalyst surface. With increasing the dye concentration the adsorption of MX-5B decreases. This behaviour can be described by the competition of adsorption between impurities and dye or the modification of TiO_2 charge (Sahel et al. 2007).

2.6.3 Effect of Substituents

Dyes with different substituents have a considerable effect on PD which has been investigated using an aqueous solution of TiO_2. The rate of photocatalytic degradation of azo dyes is higher compared to the dyes with anthraquinone structure. The presence of methyl ($-CH_3$) and chloro ($-Cl$) groups in dyes slightly decreases the degradation efficiency whilst the reverse is shown by the nitrite group. The side-chain alkyl groups decrease the water solubility and consequently, disfavour the process of photocatalytic degradation. The dyes having more sulfonic substituents are less reactive compared to hydroxyl-containing ones. The hydroxyl groups enhance the electron resonance in the molecule and the degradation rate of the dye. Also, photocatalytic degradation of triazine-containing azo dyes, Procion Red MX-5B and Reactive Brilliant Red K-2G, in aqueous TiO_2 dispersions revealed that decolouration and desulfuration occurred at the almost same rate in the first step of the photo-oxidation process. Dyes molecule containing naphthalene groups are hydroxylated more easily compared to the dyes having a triazine group thus, the rate of mineralization is slow for the latter. Most of the intermediates are aromatic and aliphatic carboxylic acids and oxidized slowly to CO_2. For both the dyes, the rate of mineralization is much slower in comparison to the rate of decolourization and ultimately converted to cyanuric acid a stable structure for photocatalysis (Hu et al. 2003).

Methyl substituents have an impact on photocatalytic efficiency in the degradation process. For example, between Acid Orange 7 (AO7) and Acid Orange 8 (AO8), the reactivity of AO8 is lower due to the presence of methyl ($-CH_3$) substituents (Neppolian et al. 2002). In the case of AO8, AO10, AO12, the order of the photocatalytic decolourization kinetics is AO10 > AO12 > AO8. This is attributed to the presence of two sulfonic acid groups in AO10 which helps binding the catalyst surface (Khataee and Kasiri. 2010). In the case of Acid Red 29 (AR29) and chromotrope 2B, the latter one contains a nitrite group in the *para* to the azo bond and showed a slightly higher reaction rate. The nitrite group can interact with the phenyl ring by delocalization of the π-electrons of the ring and unpaired electrons of the heteroatom which leads to favour the electrophilic attack (Qamar et al. 2004). In presence of chlorine atoms, the inductive effect (−I effect) largely exceeds the mesomeric effect (+M effect) and the ring is deactivated thus $OH^{\bullet}$ can easily generate Cl^- in the solution. The electronic properties of a −OH group are −I and +M effects. That is why the rate of photocatalytic decolourization of Acid Red 29 (containing two hydroxyl substituents) is higher in comparison to that of Orange G (containing one hydroxyl substituent) (Lachheb et al. 2002). Similarly, a carboxylic group also accelerates the rate of decolourization. For example, amongst Alizarin S (AS), Orange G (OG), Methyl Red (MR), and Congo Red (CR) the photocatalytic rate constant order is MR > OG $\approx$ AS > CR. This is because of the higher degradability of Methyl Red due to the presence of carboxylic group which can readily react with H^+ via a photo-Kolbe reaction (Khataee and Kasiri 2010; Lachheb et al. 2002).

2.6.4 Comparison of Cationic and Anionic Dye

In the presence of UV light and TiO_2, most of the cationic and anionic dyes are prone to photocatalytic degradation. As evaluated by colour index (C.I.) it was found that only cationic dyes are adsorbed by TiO_2, except for Quinizarin (adsorption efficiency of 21.8%) and Basic Orange 2 (no adsorption is displayed). This was explained concerning the surface structure of TiO_2. The oxygen atoms present on the surface of the unmodified crystal of TiO_2 contain a negative charge and, therefore, the cationic molecules are more readily absorbed. Thus, for cationic dyes, the highest values of photocatalytic degradation rate constant are observed and these results suggest that the adsorption hypothesis of dye has a significant effect on its susceptibility to photocatalytic degradation. However, no correlation was found between adsorption efficiency and the rate constant values of photodegradation. A linear correlation was observed between k^{-1} (inverse value of photocatalytic degradation rate constant) and the absorbance of the illuminated solution when measured at 366 nm. The surface of the photocatalyst can only adsorb cationic dyes and their photocatalytic degradation is faster in comparison to that of anionic dyes (Baran et al. 2008).

2.6.5 Correlation of Dye Degradation with Its Type

Dyes are classified mainly based on the conformational structures, sources, colours, and method of applications in the colour index, which has continuously been studied. Further, based on the chromophore, different categories of dyes have also been considered. Dyes are differentiated with respect to their core skeletonviz. acridine, azo, arylmethane, anthraquinone, nitro, xanthene, quinone-amine, etc. However, in literature, there are no examples of any correlation between percentage degradation and structure or class of dyes for PD. However, from various studies, the order of degradation is observed as follows: Auramine O > Safranin O > Malachite Green > Amido Black > Rhodamine B > Carmine, which was established using the UV/H_2O_2 photolytic AOP. The same order is also observed for other AOPs as hydroxyl radicals ($OH^{\bullet}$), responsible for the dye degradation. In general, it was revealed that diarylmethane class dyes are most effectively degraded, whereas anthraquinone class dyes are least degraded (Table 2.1).

2.6.6 Effect of Doping and Mixed Semiconductors

Doping refers to the introduction of impurities/foreign atoms into the basic structure of a parent material that can improve photocatalytic performance for dye degradation. Several metals including Pt, Li^+, Zn^{2+}, Cd^{2+}, Ag^+, Co^{3+}, Cr^{3+}, Fe^{3+}, Al^{3+}, etc. have been used as dopants (Pelaez et al. 2012; Li and Li. 2002; Su et al. 2007). Doping with metals and non-metals is one of the promising strategies to alter the optical response of TiO_2 photocatalyst. Doping changes the bandgap energy as well as enhances the characteristics and activity of photocatalytic species under visible light. The incorporation of transition metals in TiO_2 crystal lattice may lead to form a new energy level between VB and CB, inducing a shift of light absorption towards the visible light region (Pelaez et al. 2012). Various techniques to make TiO_2 absorb photons of lower energy have been employed, including surface modification (via organic materials and semiconductor coupling), bandgap modification (by creating oxygen vacancies), and oxygen sub-stoichiometric adjustments by co-doping of non-metals and metals. The transition metal doping increases the photocatalytic activity by scavenging electrons that reduce the recombination of charges and therefore favour the $OH^{•}$ formation. Doping modifies the surface properties of the material active sites due to induced defects, thus increase the adsorption and favour the interfacial reactions. Non-metals such as nitrogen, sulfur, phosphorous, fluorine, and carbon are also used as dopants (Garg et al. 2019). Nitrogen is the most promising dopant that can easily substitute oxygen in TiO_2 lattice. Due to the versatility in the structure-electronic properties and biocompatibility, metal oxides such as TiO_2, ZnO, and WO_3 are characterized by their identical bandgap excitation mechanism and the potential of VB holes to generate hydroxyl radicals. A wide range of pollutants in gas and liquid regime can be completely mineralized under light illumination conditions using these metal oxides. Co/N co-doped TiO_2 with variable dopant composition could be synthesized by wet impregnation method and used as a dopant by increasing the cobalt composition for the degradation of bisphenol-A [2,2-bis (4-hydroxyphenyl) propane]. The physical properties of the nanoparticles are disturbed by dopants like Co and N, which altered the crystal structure, energy bandgap as well as elemental composition (Garg et al. 2019).

2.7 Types of Heterogeneous Photocatalysts

Heterogeneous photocatalysis accelerates a photoreaction by chemical transformation. Semiconductors are the most important photocatalyst materials having a VB and a CB with a defined bandgap. Various kinds of applications viz., purification of water and air, water disinfection and environmental remediation can be accomplished using different semiconducting photocatalytic materials. In this context, in (1972), Fujishima and Honda discovered the photochemical splitting of water into hydrogen and oxygen in the presence of TiO_2 (Fujishima and Honda 1972). The metal oxides such as TiO_2, ZnO, Fe_2O_3, SnO_2, ZrO_2, MgO, GeO_2, Sb_2O_3, V_2O_5, Cu_2O, Nb_2O_5, and perovskites have been studied as semiconductors. The photocatalytic activity of Fe_2O_3, SnO_2, ZrO_2, MgO, CdS, or V_2O_5 was also analysed but their photocatalytic efficiencies were somewhat lesser than the TiO_2. Herein, we will focus on two well-explored photocatalysts TiO_2 and ZnO. According to their efficiencies, photocatalystsis categorized into three generations. The first generation consists of single-components (such as TiO_2, ZnO, CdS), whilst the second generation is multiple components (such as $WO_3/NiWO_4$, $BiOI/ZnTiO_3$, C_3N_4/Ag_3VO_4) whereas, the third generation contains photocatalyst immobilized on solid substrates (e.g., FTO/WO_3-ZnO, Steel/TiO_2-WO_3, Glass/P-TiO_2) (Anwer et al. 2019).

2.7.1 *TiO_2 Catalyst*

Titanium dioxide is one of the most effective photocatalysts utilized in oxidation of organic pollutants due to its chemical and photochemical stability, high photoactivity, low energy consumption and biocompatibility. It can be activated under UV irradiation, forming an active oxygen species e.g., hydroxyl radicals ($OH^•$) on the surfaces of the TiO_2 crystals. The bandgap of TiO_2 is more than 3.0 eV (~3.0 eV for rutile and ~3.2 eV for anatase), which makes it primarily active for UV light. Most of the organic dyes could be decomposed into CO_2 and H_2O by the attack of these radicals and a variety of intermediates are formed in the aqueous solution. TiO_2 can only be excited by ultraviolet light having a wavelength (λ) less than 390 nm, therefore the light utilization efficiency to solar irradiation is quite low. The photocatalytic activities of TiO_2 are being explored by upgrading the morphology of TiO_2 such as surface area, crystal composition, particle size, bandgap, porosity, and hydroxyl density on the surface (Chen et al. 2020; Luttrell et al. 2015).

The crystal structure of TiO_2 consists of three phases, anatase, rutile, and brookite (Fig. 2.6). The most common commercial photocatalyst Degussa P-25 contains both rutile and anatase crystalline forms. Due to the synergistic effect of phase mixture of different polymorphs, photocatalytic activity increases compared to pure phases. However, the pure phase anatase exhibits higher photocatalytic activity compared to rutile TiO_2. The bandgap of the anatase form of TiO_2 is larger compared to the rutile form. This reduces the light absorption and increases the valence band to a higher energy levels corresponding to redox potentials of adsorbed molecules. As a result, the oxidation power of electrons increases and facilitates electron transfer from the TiO_2 to adsorbed molecules. In addition, anatase exhibits an indirect bandgap that is smaller compared to its direct bandgap, whilst for rutile structure, the fundamental

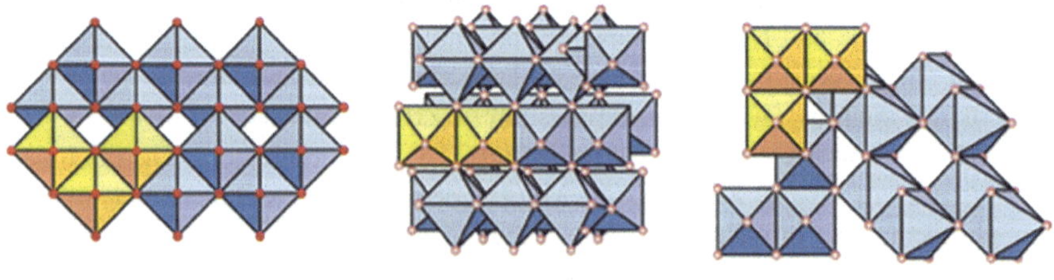

Fig. 2.6 Polymorphs of TiO_2. **a** Anatase, **b** brookite and **c** rutile (Khataee and Kasiri 2010)

bandgap is either direct or indirect bandgap which is similar to its direct bandgap. In general, the semiconductors with indirect bandgap exhibit longer charge carrier lifetimes compared to direct gap materials. Therefore, anatase with a longer electron–hole pair life is more likely to participate in surface reactions compared to the rutile (Luttrell et al. 2015).

2.7.1.1 TiO_2 Doped Photocatalyst

Photocatalytic degradation increases with an increase in the amount of dopant which can trap electrons, and initiate the reaction with water to form oxidative radicals responsible for the degradation of organic pollutants. However, too much dopants prevent the TiO_2 from capturing photons resulting in the restricted generation of electron–hole pairs thereby decreasing photocatalytic efficiency. The efficiency of TiO_2 is restricted due to its large bandgap (>3 eV), which impedes its application in photodegradation using solar light. Therefore, considerable modifications have been made to shift the absorption of TiO_2 from UV to visible light. Decrease in the bandgap of TiO_2 is the primary target of doping, i.e., to induce a bathochromic shift and extend its wavelength range correspond to visible regions. The degree of degradation of organic pollutants is caused by the single-doped and co-doped TiO_2 photocatalysts, where co-doped TiO_2 generally have higher photocatalytic activities in comparison to single-doped TiO_2. For modification of the electronic structure of TiO_2, transition metal ions, such as Pd^{2+}, Cu^{2+}, Cd^{2+}, Cr^{3+}, V^{5+}, Fe^{3+}, Ni^{2+} are investigated. This introduces donor and/or acceptor levels in the bandgap, which excites the photocatalyst even with lower energy photons and exhibits higher photocatalytic activity under visible light (Pan et al. 2010). In addition to this, the higher quantum efficiency is also responsible for dopant in TiO_2 photocatalytic reaction, as a result, it prevents from recombination of generated electron–hole pairs. Therefore, the charge transfer of electrons and holes is increased between the rutile and anatase form of TiO_2 by this mixed-phase (Lai et al. 2016). At high dopant concentration, the metal ions can behave as recombination centres for the photoinduced charge carriers, thereby decreasing the quantum efficiency. Vanadium, copper, and cobalt doping on TiO_2 offer a promising photocatalytic approach under visible light. Due to the synergistic effect, Co-doped TiO_2 generally exhibits higher visible light absorption with respect to single-doped TiO_2, as a result increasing the charge carrier lifetime. During irradiation of higher wavelengths of light, the degradation of rutile C, N, S-TiO_2 nanorods shows higher efficiency in comparison to anatase C, N, S-TiO_2 nanorods even though the latter has a higher surface area (Wang et al. 2017). This suggests that the catalyst phase is a more important factor than the surface area, which affects the co-doping catalytic activity.

2.7.1.2 Principle and Mechanistic Study of TiO_2 in Photocatalytic Reaction

The mechanism of photocatalysis can be investigated as a correlation of semiconductor solid catalyst with the light of suitable wavelength. For a photocatalyst, the electrons lie in the valance band at ambient temperature. For a favourable photocatalysed reaction, the recombination of electrons and hole must be prevented. The ultimate goal is to have a reaction between the electrons and oxidant to produce a reduced product and a similar reaction between holes and reductant to give an oxidized product. Under UV light irradiation (photons) of TiO_2, electrons and positive holes are generated in the conduction ($e_{CB}{}^{-}$) and valence band ($h_{VB}{}^{+}$) according to the Eq. 2.21 (Chen et al. 2020; Poulopoulos and Philippopoulos 2004). The holes can either react directly with organic molecules (Eq. 2.26) or form hydroxyl radicals (Eqs. 2.22and 2.23) that subsequently oxidize organic molecules (Eq. 2.27) (Ahmed et al. 2011; Akpan and Hameed 2009). The electrons can also react with organic compounds to provide reduced products (Eq. 2.28). The oxygen (Eq. 2.24) can react with electrons and reduce the dye or react with electron acceptors such as adsorbed O_2 on the surface of Ti(III) or dissolved in water, reducing it to $O_2{}^{-\bullet}$. The photogenerated holes can also oxidize the organic molecule to form R^+, or react with OH^-or H_2O and oxidize them to $OH^\bullet$. The resulting $OH^\bullet$, being a very strong oxidizing agent (standard redox potential +2.8 V) can oxidize most azo dyes to mineral end-products (Akpan and Hameed 2009). According to this, the appropriate reactions at the semiconductor surface causing the degradation of dyes can be expressed as follows:

$$TiO_2 + h\nu(UV) \rightarrow TiO_2\left(e_{CB}^- + h_{VB}^+\right) \tag{2.21}$$

$$TiO_2\left(h_{VB}^+\right) + H_2O_{absorbed} \rightarrow TiO_2 + H^+ + OH^\bullet \tag{2.22}$$

$$TiO_2\left(h_{VB}^+\right) + OH^-_{surface} \rightarrow TiO_2 + OH^\bullet \tag{2.23}$$

$$TiO_2\left(e_{CB}^-\right) + O_{2\ absorbed} \rightarrow TiO_2 + O_2^{-\bullet} \tag{2.24}$$

$$O_2^{-\bullet} + H^+ \rightarrow HO_2 \tag{2.25}$$

$$h_{VB}^+ + \text{Organic} \rightarrow \text{Oxidation products} \tag{2.26}$$

$$OH^\bullet + \text{Organic} \rightarrow \text{Degradation products} \tag{2.27}$$

$$e_{CB}^- + \text{Organic} \rightarrow \text{Reduction products} \tag{2.28}$$

where *hv* is the energy of photon required to excite the semiconductor electron from the VB region to CB region.

2.7.2 ZnO Catalyst

ZnO is another photocatalyst that is getting increasing attention owing to its high photocatalytic activity, good stability, eco-friendly, and cost-effective nature. It is preferred over other semiconducting metal oxides as it can absorb a large fraction of the solar spectrum. Thus, ZnO has emerged as an attractive PC to remove persistent organic pollutants from wastewater. ZnO is a well-defined naturally occurring oxide having three different crystalline structures which are commonly known as rocksalt, wurtzite, cubic (zinc blende) and their structure is shown in Fig. 2.7. The rocksalt structure of ZnO is quite rare because it can be obtained under high pressure. Amongst the three structures, wurtzite form is the most common structure of ZnO and has the highest thermodynamic stability. At ambient pressure and temperature, ZnO exists in hexagonal wurtzite structure with two lattice parameters, a and c, values of 0.3296 nm and 0.52065 nm respectively (Baruah and Dutta 2009). The hexagonal wurtzite structure of ZnO belongs to the P63mc space group and exhibits a non-centrosymmetric structure, which causes ZnO to be piezoelectric and pyroelectric (Lee et al. 2016).

The wider bandgap energy (3.2–3.7 eV) of ZnO restricts its absorption in the visible region and enhances recombination of charge carriers resulting in low PD efficiency. ZnO photo-corrodes in acidic aqueous suspensions under UV irradiation and suffers dissolution to form $Zn(OH)_2$ on its surface. Photo-corrosion suppresses the dye removal efficiency of ZnO. However, the optical, structural, and magnetic

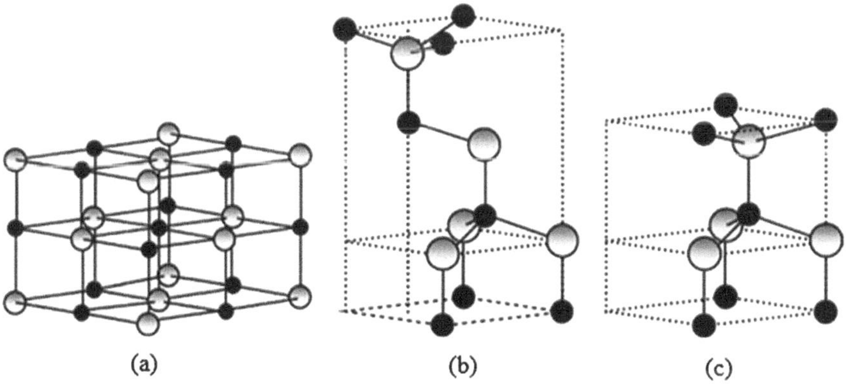

Fig. 2.7 **a** Rocksalt (cubic), **b** zinc blende (cubic) and **c** wurtzite (hexagonal) structures model of ZnO (Lee et al. 2016)

properties of ZnO can be improved and altered by surface modification of the catalysts. The greatest advantage of ZnO is that it can absorb a large portion of the solar spectrum in comparison to TiO_2 (Lee et al. 2016).

2.7.2.1 ZnO Doped Photocatalyst

For improvement of the photocatalytic activity of ZnO, it is modified with the metals that can transfer the photo-response of ZnO in the visible region to maximize photocatalytic efficiency. To counter the recombination of electrons and holes, metal doping is considered as an effective process as it improves the lifetime of charge carriers by enhancing charge separation.

Dopants may also trap the electrons and thus reduce the chances of $e^- - h^+$ recombination and eventually enhancing the PD. The photoactivity of metal-doped semiconductor mostly depends on the concentration and ion nature of the dopant, operating conditions and synthesis method. The incorporation of other elements for the preparation of ZnO photocatalyst is an important challenge. The dopants may be metal or non-metal. The doped ZnO catalyst surpasses its photocatalytic degradation performances. The cationic or metal dopants of ZnO oxide include Li, Na, Ag, W, Cd, Mn, Ni, Pd, Co, Cu. The dopant site of ZnO semiconductors is either a Zn-site or an O-site. Also, it is noticed that the synthesis or substitution of doped catalysts is dependent on the nature of dopants. For example, group-I elements are appropriate for the substitution on Zn-site whilst group-V elements are favourable for substitution on an O-site. Group-I metals like Li^+ have a radius close to Zn^{2+} which favours its incorporation into the ZnO lattice. Transition metals such as Cu^{2+}, Ni^{2+}, Co^{2+}, and Mn^{2+} which are isomorphic to zinc ions are used for the synthesis of doped catalysts. Kumar et al. explained that even 1.0 mol% Fe doped ZnO show better photocatalytic activity towards MB dye degradation (Kumar et al. 2014). Rare earth metals are another important dopant for enhancing photocatalytic degradation. ZnO can also be doped with Er, Hf, where Er-doped ZnO can be prepared through a simple chemical solution route in the absence of surfactant whilst Hf-doped ZnO is prepared by a simple sol–gel method. The Er-doped ZnO showed a better performance in phenol degradation than undoped ZnO which was reported through photoluminescence results (Lam et al. 2012; Din et al. 2021).

Also, non-metal ion doping can improve the photocatalytic degradation performance of ZnO in the visible region compared to pure ZnO. Non-metals such as C, N, F, and S are used as dopants for ZnO. Amongst these non-metals, carbon has a potential advantage due to introducing a new energy level near the ZnO VB edge. As a result, the bandgap energy is reduced and VB is moved to higher energy compared to undoped ZnO and therefore, charge transfer efficiency is increased which enhances the PD (Ullah and Dutta 2008). For N-doped ZnO, incorporation of N occurs into the O site rather than Zn-sites. This modifies the bandgap energy of the catalyst and shifts the absorption efficiency towards visible regions leading to greater PD efficiency (Rajbongshi et al. 2014).

2.7.2.2 Principle of ZnO Photocatalysis and Mechanistic Pathways

Upon irradiation of ZnO in the Ultraviolet (UV) light, the particles absorb photons with energies equal or greater than its bandgap energy. Because of this excitation, the photoinduced electron is promoted from the VB to the CB, forming a positive hole (h_{VB}^+) and electron (e_{CB}^-) on the surface of the ZnO particle (Eq. 2.29). The wavelength for UV light energy typically corresponds to $\lambda < 387$ nm. The bandgap energy depends not only on the crystal structure of the semiconductor but also on its morphology, defects, etc. The reactions of the photocatalytic process are as follows:

$$\text{ZnO} + hv(\text{UV}) \rightarrow \text{ZnO}\left(e_{\text{CB}}^- + h_{\text{VB}}^+\right) \tag{2.29}$$

$$\text{ZnO}\left(e_{\text{CB}}^- + h_{\text{VB}}^+\right) \rightarrow \text{ZnO} + \Delta \tag{2.30}$$

$$\text{ZnO}\left(h_{\text{VB}}^+\right) + \text{H}_2\text{O} \rightarrow \text{ZnO} + \text{H}^+ + \text{OH}^\bullet \tag{2.31}$$

$$\text{ZnO}\left(h_{\text{VB}}^+\right) + \text{OH}^- \rightarrow \text{ZnO} + \text{OH}^\bullet \tag{2.32}$$

$$\text{ZnO}\left(e_{\text{CB}}^-\right) + \text{O}_2 \rightarrow \text{ZnO} + \text{O}_2^{\bullet -} \tag{2.33}$$

$$\text{O}_2^{\bullet -} + \text{H}^+ \rightarrow \text{HO}_2^\bullet \tag{2.34}$$

$$\text{HO}_2^\bullet + \text{HO}_2^\bullet \rightarrow \text{H}_2\text{O}_2 + \text{O}_2 \tag{2.35}$$

$$\text{ZnO}\left(e_{\text{CB}}^-\right) + \text{H}_2\text{O}_2 \rightarrow \text{OH}^\bullet + \text{OH}^- \tag{2.36}$$

$$\text{H}_2\text{O}_2 + \text{O}_2^{\bullet -} \rightarrow \text{OH}^\bullet + \text{OH}^- + \text{O}_2 \tag{2.37}$$

$$\text{Organic pollutant} + \text{OH}^\bullet \rightarrow \text{Degradation products} \tag{2.38}$$

$$\text{Organic pollutant} + \text{ZnO}\left(h_{\text{VB}}^+\right) \rightarrow \text{Oxidation products} \tag{2.39}$$

$$\text{Organic pollutant} + \text{ZnO}\left(e_{\text{CB}}^-\right) \rightarrow \text{Reduction products} \tag{2.40}$$

The recombination of photo-generated holes (h_{VB}^+) in the VBand photo-excited electrons (e_{CB}^-) in the CB dissipate as heat (Eq. 2.30). The holes (h_{VB}^+) are then scavenged by oxidizing species {for example, H_2O, OH^-, organic compounds, Eqs. (2.32), (2.33) and (2.39)}, and electrons (e_{CB}^-) by reducing species {for example, O_2, Eq. (2.33)} in the solution. The reaction of h_{VB}^+ with OH^- (Eq. 2.32) may lead to the generation of $OH^{\bullet}$ whilst superoxide radical anions ($O_2^{\bullet-}$) are obtained on the reaction of e_{CB}^- with oxygen (Eq. 2.33). The $O_2^{\bullet-}$ then forms hydroperoxyl radicals ($HO_2^{\bullet}$) and subsequently $H_2O_2$by reacting with a surface charge of ZnO (Eqs. 2.34 and 2.35). These radicals are active oxidizing species for the photocatalytic oxidation reaction, which can degrade a variety of organic compounds (toxic and non-toxic) and biological agents (Ullah and Dutta 2008; Lam et al. 2012).

2.7.3 Other Photocatalysts

In the last few decades, various photocatalysts have been reported that are categorized into metal-oxides, metal-sulfides, metal-nitrides, and metal-free compounds like polymers or graphenes. A large variety of entirely novel photoactive semiconductors have been introduced recently as possible substitutes for TiO_2 and ZnO. Amongst these, mixed oxides of transition metal (Nb, V, or Ta) with main group elements such as Ga, In, Sb or Bi have been frequently used to obtain materials with photoactivity in the visible range. In addition, some solids with high surface area, such as cation interchanged zeolites, have been also assessed as photocatalysts, even though they don't have any semiconductor properties. But the photocatalytic activity of other metal oxides is considered to be relatively weak in comparison to the TiO_2. Nevertheless. metal oxides like Cu_2O, WO_3 have found application in the photodegradation of polyethylene plastic and organic dyes respectively (Bedja et al. 2002). Further, Nb_2O_5 strongly increases TiO_2 activity towards the degradation of dichloro-benzene and In_2O_3 helps in the photodegradation of ethane (Ishchenko et al. 2016; Hernández-Alonso 2009).

Semiconductors such as ZnS, CdS, $SrTiO_3$, MoS_2, GaP are also frequently used as photocatalysts. For example, GaP and CdS were used for the synthesis of NH_3 from water and nitrogen with a better yield compared to TiO_2 or ZnO. This process is favourable when the conduction band energy of the semiconductor becomes more negative. This can be explained by the greater affinity of photoexcited electrons to reduce nitrogen in a more negative conduction band. Larger yields were reached when Pt black was incorporated into the catalyst (Hernández-Alonso et al. 2009). CdS can also absorb visible light owing to its short bandgap (2.42 eV) which makes it a potential catalyst for photocatalysis and solar cells. In (2016), Samadi-Maybodi and Sadeghi-Maleki developed a method for *in-situ* generation of stable CdS quantum dots (CdS-QD) using $Na_2S_2O_3$ as a precursor and thioglycolic acid as a catalyst as well as capping agent. Good photocatalytic activity and recycling stability were shown by CdS-QD under visible light irradiation compared to the bulk CdS for degradation of alizarin, mordant red (MR), acid violet (AV), and thymol blue (TB) (Samadi-Maybodi and Sadeghi-Maleki 2016).

2.8 Application of Heterogeneous Photocatalysis

Heterogeneous photocatalysis has grown rapidly in recent years and experienced various developments especially concerning energy and the environment. Solar water splitting, purification of air and water containing low concentrations of pollutants are the two most important applications of heterogeneous photocatalysis. It has been successfully utilized by different developing countries to eliminate pathogens and algal blooms in freshwater supplies. Heterogeneous photocatalysis is a very cheap technique compared to reverse osmosis and ultra or nanofiltration. Humic substances which help in bacterial growth can be treated by TiO_2 assisted photosensitized disinfection. The multidisciplinary nature of the process is extended semiconductor physics, surface sciences, materials science, photo and physical chemistry, and chemical engineering (Fujishima et al. 2000; Ibhadon and Fitzpatrick 2013). The various applications of photocatalysis are shown in Fig. 2.8.

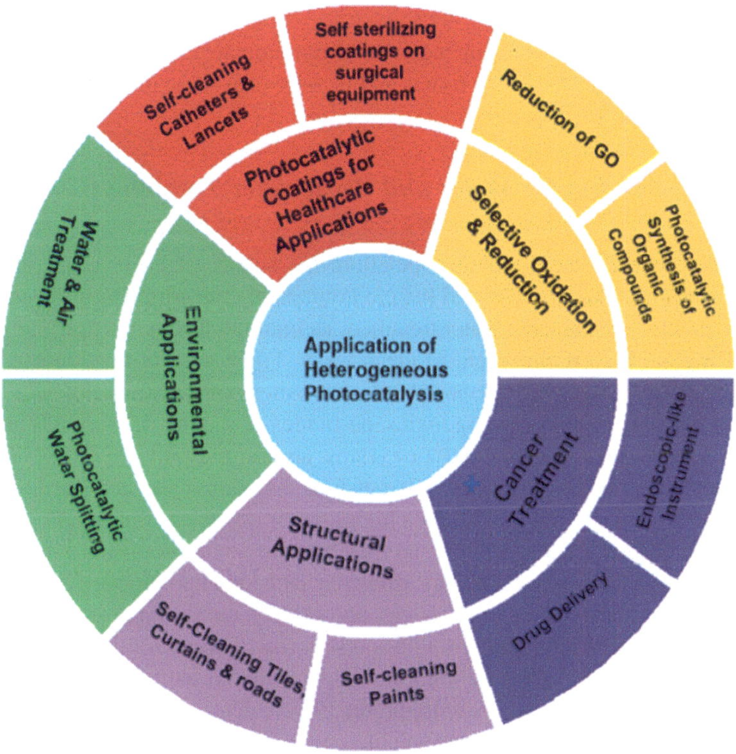

Fig. 2.8 Various applications of heterogeneous photocatalysis

2.8.1 Self-Cleaning

- Heterogeneous photocatalysis has found application in the self-cleaning process of materials for residential and office buildings in exterior tiles, kitchen and bathroom components, interior furnishings, plastic surfaces, aluminum siding, building stone, curtains, and paper window blinds.
- Photocatalysis can also be utilized in indoor and outdoor lamps like translucent paper for indoor lamp covers, coatings for fluorescent lamps, and a cover glass of highway tunnel lamps.
- For other road materials such as tunnel wall, soundproofed wall, traffic signs, and reflectors as well as for tent material, for hospital garments, uniforms as well as spray coatings for cars.

2.8.2 Air Cleaning

- The methods can be used in indoor and outdoor air cleaning. Photocatalyst-equipped air conditioners for rooms and factories are some indoor applications.

2.8.3 Application for Water and Wastewater Treatment

Due to the growth in the global population, the supply of clean water has now been diminished which heightened the environmental concerns. Thus, the employment of effective sustainable water treatment technology has drawn much attention considering the high requirement of clean water. For water purification and treatment, advanced oxidation processes (AOPs) have shown tremendous potential in the elimination of naturally occurring toxins, pesticides, and other deleterious contaminants. UV light-assisted photocatalytic systems are useful for the treatment of water containing trace amounts of pollutants (such as estrogens) and also for industrial wastewater treatment which contains a high load of organic pollutants. Photocatalysis with nanocatalysts and membrane filtration are some promising methods for disinfection with high water cleaning efficiency (Ibhadon and Fitzpatrick 2013).

Different photocatalytic materials can be used for the decomposition of humic substances present in both highly saline and seawater. The rate of decomposition of humic substances was fast in pure water as compared to seawater media and during the process of decomposition, no toxic by-products were detected. Another study found that the decomposition of seawater-soluble crude-oil fractions was achieved using TiO_2 nanoparticles under artificial light irradiation. Photocatalysis techniques are also used to kill bacteria and viruses present in water bodies. Algal blooms found in freshwater supplies are also degraded on an immobilized TiO_2 catalyst. Degradation of the green algae having thick cell walls by photo disinfection sensitized by TiO_2 has been applied greatly to treat water especially in remote and disaster areas of many developed and developing countries (Fukushima et al. 2000; Ahmed and Haider 2018).

2.8.4 Removal of Trace Metals

Heavy metals such as Hg, Cr, Pb, Ni, Cu, etc. are highly hazardous to human health. Removal of these toxic heavy materials from freshwater can be obtained by the application of heterogeneous photocatalysis. Again, photoreduction by photocatalysts has been applied in the extraction of expensive metals, such as Au, Pt, and Ag from industrial effluents (Uyguner and Bekbolet 2008).

2.8.5 Removal of Inorganic Compounds

Several inorganic compounds are also sensitive like organic compounds to photochemical transformation on the catalyst surface. Inorganic species such as BrO_3^-, ClO_4^-, N_3^-, NO, halide ions, Pd- and Rh- species, sulfur species, metal salts like $AgNO_3$, HgCl, organometallic compound (e.g., CH_3HgCl) can be decomposed and removed using heterogeneous photocatalysis (Uyguner and Bekbolet 2008).

2.8.6 Applications in Photodynamic Therapy

Photocatalysis has significant application for the treatment of different cancer cells such as colon or kidney cancer (tumour therapy). This technique has also responded well to the treatment of tumours in rats. Generally, TiO_2 is introduced in cancer sites and photosensitized using an optic fibre cable to introduce the illumination. Thus, reactive oxygen species are generated during activation of photosensitizer on illumination that destroys the tumour cells (Al-Mamun et al. 2019).

Heterogeneous photocatalysis has found applications in hospitals garments worn during operations that have 'doses' of TiO_2 added to the fabric. TiO_2 fabrics control infections in hospital garments, methicillin-resistant *Staphylococcus* (MRSA) as well as antimicrobial photodynamic therapy (APDT) is used to decolonize MRSA from patients (Fujishima et al. 2000).

2.9 Conclusion and Outlook

In this chapter, we attempt to assess and summarize the latest work on heterogeneous photocatalytic degradation of organic effluents using both pure and doped catalysts. The chapter comprises the type of dyes their impact on the environment; advancements to enhance photocatalytic efficiency; the effects of key operational parameters on the photocatalytic performance are detailed. A brief synopsis of the application of heterogeneous photocatalysis in various fields is also given. The principle of photocatalytic activity of metal oxides in particular ZnO and TiO_2 is discussed in detail. The non-toxic ZnO and TiO_2 semiconductors are used to convert effluents to lesser toxic forms. Apart from metal oxides, other low bandgap semiconductors and their modified form should be explored extensively. As per reported literature, the surface morphology of photocatalyst has a great impact on catalytic efficiency. Thus, for designing of novel catalytic system, the optimization of the key parameters is very important for a smooth operation. In the real world, pollutants present in wastewater are a mixture of various contaminants. Thus, future research must be focused on the development of techniques for multi-components rather than a single pollutant. For smooth industrial-scale applications, the economic practicability of a heterogeneous catalyst is another aspect that needs alarming attention.

References

Abdullah FH, Rauf MA, Ashraf SS (2007) Kinetics and optimization of photolytic decolouration of carmine by UV/H2O2. Dyes Pigm 75:194–198

Abid MF, Zablouk MA, Abid-Alameer AM (2012) Experimental study of dye removal from industrial wastewater by membrane technologies of reverse osmosis and nanofiltration. J Environ Health Sci Engineer 9:17

Ahmed SN, Haider W (2018) Heterogeneous photocatalysis and its potential applications in water and wastewater treatment: a review. Nanotechnology 29:342001

Ahmed S, Rasul MG, Brown R, Hashib MA (2011) Influence of parameters on the heterogeneous photocatalytic degradation of pesticides and phenolic contaminants in wastewater: a short review. J Environ Manage 92:311–330

Ajaz M, Rehman A, Khan Z, Nisar MA, Hussain S (2019) Degradation of azo dyes by Alcaligenes aquatilis 3c and its potential use in the wastewater treatment. AMB Expr 9:64

Akpan UG, Hameed BH (2009) Parameters affecting the photocatalytic degradation of dyes using TiO_2-based photocatalysts: a review. J Hazard Mater 170:520–529

Alahiane S, Sennaoui A, Sakr F, Qourzal S, Dinne M, Assabbane A (2017) A study of the photocatalytic degradation of the textile dye reactive yellow 17 in aqueous solution by TiO_2-coated non-woven fibres in a batch photoreactor. J Mater Environ Sci 8:3556–3563

Al-Mamun MR, Kader S, Islam MS, Khan MZH (2019) Photocatalytic activity improvement and application of UV-TiO_2 photocatalysis in textile wastewater treatment: a review. J Environ Chem Eng 7:103248

Alnuaimi MM, Rauf MA, Ashraf SS (2007) Comparative decolouration study of neutral red by different oxidative processes. Dyes Pigm 72:367–371

Andronic L, Duta A (2007) TiO_2 thin films for dyes photodegradation. Thin Solid Films 515:6294–6297

Anwer H, Mahmood A, Lee J, Kim K-H, Park J-W, Yip ACK (2019) Photocatalysts for degradation of dyes in industrial effluents: opportunities and challenges. Nano Res 12:955–972

Atarod M, Nasrollahzadeh M, Mohammad Sajadi S (2016) Euphorbia heterophylla leaf extract mediated green synthesis of Ag/TiO_2 nanocomposite and investigation of its excellent catalytic activity for reduction of variety of dyes in water. J Colloid Interface Sci 462:272–279

Babu SG, Karthik P, John MC, Lakhera SK, Ashokkumar M, Khim J, Neppolian B (2019) Synergistic effect of sono-photocatalytic process for the degradation of organic pollutants using CuO-TiO_2/rGO. Ultrason Sonochem 50:218–223

Bandara WRLN, de Silva RM, de Silva KMN, Dahanayake D, Gunasekara S, Thanabalasingam K (2017) Is nano ZrO_2 a better photocatalyst than nano TiO_2 for degradation of plastics? RSC Adv 7:46155–46163

Baran W, Makowski A, Wardas W (2008) The effect of UV radiation absorption of cationic and anionic dye solutions on their photocatalytic degradation in the presence TiO_2. Dyes Pigm 76:226–230

Baruah S, Dutta J (2009) Hydrothermal growth of ZnO nanostructures. Sci Technol Adv Mater 10:013001

Behnajady MA, Modirshahla N, Shokri M (2004) Photodestruction of Acid Orange 7 (AO7) in aqueous solutions by UV/H_2O_2: influence of operational parameters. Chemosphere 55:129–134

Bedja I, Hotchandani S, Kamat PV (2002) Photoelectrochemistry of quantized WO_3 colloids: electron storage, electrochromic, and photoelectrochromic effects. J Phys Chem 97:11064–11070

Belpaire C, Reyns T, Geeraerts C, Van Loco J (2015) Toxic textile dyes accumulate in wild European eel Anguilla anguilla. Chemosphere 138:784–791

Benatti CT, Tavares CRG, Guedes TA (2006) Optimization of Fenton's oxidation of chemical laboratory wastewaters using the response surface methodology. J Environ Manage 80:66–74

Ben Fradj A, Boubakri A, Amor H, Hamouda S (2019) Removal of azoic dyes from aqueous solutions by chitosan enhanced ultrafiltration. Results Chem 2:100017

Benkhaya S, M' rabet S, El Harfi A, (2020) A review on classifications, recent synthesis and applications of textile dyes. Inorg Chem Commun 115:107891

Blanco-Galvez J, Fernández-Ibáñez P, Malato-Rodríguez S (2007) Solar photocatalytic detoxification and disinfection of water: recent overview. J Sol Energy Eng 129:4–15

Chandanshive VV, Kadam SK, Khandare RV, Kurade MB, Jeon B-H, Jadhav JP, Govindwar SP (2018) In situ phytoremediation of dyes from textile wastewater using garden ornamental plants, effect on soil quality and plant growth. Chemosphere 210:968–976

Chen D, Cheng Y, Zhou N, Chen P, Wang Y, Li K, Huo S, Cheng P, Peng P, Zhang R, Wang L, Liu H, Liu Y, Ruan R (2020) Photocatalytic degradation of organic pollutants using TiO2-based photocatalysts: a review. J Clean Prod 268:121725

Chequer FMD, Oliveira GAR de, Ferraz ERA, Cardoso JC, Zanoni MVB, Oliveira DP de (2013) Textile dyes: dyeing process and environmental impact. IntechOpen, pp 151–176

Che Ramli ZA, Asim N, Isahak WNRW, Emdadi Z, Ahmad-Ludin N, Yarmo MA, Sopian K (2014) Photocatalytic degradation of methylene blue under UV light irradiation on prepared carbonaceous TiO_2. Sci World J 2014:1–8

Chen X, Wu Z, Liu D, Gao Z (2017) Preparation of ZnO photocatalyst for the efficient and rapid photocatalytic degradation of azo dyes. Nanoscale Res Lett 12:143

Crini G, Lichtfouse E (2019) Advantages and disadvantages of techniques used for wastewater treatment. Environ Chem Lett 17:145–155

Dahiya A, Patel BK (2021) Photocatalytic degradation of organic dyes using heterogeneous catalysts. In: Photocatalytic degradation of dyes. Elsevier, pp 43–90

Daneshvar N, Salari D, Khataee AR (2003) Photocatalytic degradation of azo dye acid red 14 in water: investigation of the effect of operational parameters. J Photochem Photobiol A Chem 157:111–116

Din MI, Khalid R, Hussain Z (2021) Recent research on development and modification of nontoxic semiconductor for environmental application. Sep Purif Rev 50:244–261

Dionysiou DD, Khodadoust AP, Kern AM, Suidan MT, Baudin I, Laîné J-M (2000) Continuous-mode photocatalytic degradation of chlorinated phenols and pesticides in water using a bench-scale TiO_2 rotating disk reactor. Appl Catal B24:139–155
Drumond CFM, de Oliveira GAR, Ferraz ERA, Carvalho J, Zanoni MVB, de Oliveir DP (2013) Textile dyes: dyeing process and environmental impact. In: Gunay M (ed) Eco-Friendly Textile Dyeing and Finishing. InTech. 10.5772/53659
Es-sahbany H, Berradi M, Nkhili S, Bassir D, Belfaquir M, Youbi MSE (2018) Valorization of Moroccan clay: application to the adsorption of cobalt ions contained in wastewater synthesized. Mor J Chem 6:173–179
Fenoll J, Martínez-Menchón M, Navarro G, Vela N, Navarro S (2013) Photocatalytic degradation of substituted phenylurea herbicides in aqueous semiconductor suspensions exposed to solar energy. Chemosphere 91:571–578
Ferreira ESB, Hulme AN, McNab H, Quye A (2004) The natural constituents of historical textile dyes. Chem Soc Rev 33:329–336
Franca RDG, Vieira A, Carvalho G, Oehmen A, Pinheiro HM, Barreto Crespo MT, Lourenço ND (2020) Oerskovia paurometabola can efficiently decolourize azo dye Acid Red 14 and remove its recalcitrant metabolite. Ecotoxicol Environ Saf 191:110007
Fujishima A, Honda K (1972) Electrochemical photolysis of water at a semiconductor electrode. Nature 238:37–38
Fujishima A, Rao TN, Tryk DA (2000) Titanium dioxide photocatalysis. J Photochem Photobiol C 1:1–21
Fukushima M, Tatsumi K, Morimoto K (2000) Influence of iron(III) and humic acid on the photodegradation of pentachlorophenol. Environ Toxicol Chem 19:1711–1716
Garg A, Singhania T, Singh A, Sharma S, Rani S, Neogy A, Yadav SR, Sangal VK, Garg N (2019) Photocatalytic degradation of bisphenol-a using N, Co codoped TiO_2catalyst under solar light. Sci Rep 9:765
Gnanaprakasam A, Sivakumar VM, Thirumarimurugan M (2015) Influencing parameters in the photocatalytic degradation of organic effluent via nanometal oxide catalyst: a review. Indian J Mater Sci 2015:1–16
Haibach MC, Kundu S, Brookhart M, Goldman AS (2012) Alkane metathesis by tandem alkane-dehydrogenation–olefin-metathesis catalysis and related chemistry. Acc Chem Res 45:947–958
Hassena H (2016) Photocatalytic degradation of methylene blue by using Al_2O_3/Fe_2O_3nano composite under visible light. Mod Chem Appl 4:176
Haq I, Raj A, Markandeya (2018) Biodegradation of Azure-B dye by Serratia liquefaciens and its validation by phytotoxicity, genotoxicity and cytotoxicity studies. Chemosphere 196:58–68
Hasnat M, Uddin M, Samed A, Alam S, Hossain S (2007) Adsorption and photocatalytic decolourization of a synthetic dye erythrosine on anatase TiO2 and ZnO surfaces. J Hazard Mater 147:471–477
Hernández-Alonso MD, Fresno F, Suárez S, Coronado JM (2009) Development of alternative photocatalysts to TiO2: challenges and opportunities. Energy Environ Sci 2:1231
Holkar CR, Jadhav AJ, Pinjari DV, Mahamuni NM, Pandit AB (2016) A critical review on textile wastewater treatments: possible approaches. J Environ Manage 182:351–366
Hu C, Yu JC, Hao Z, Wong PK (2003) Effects of acidity and inorganic ions on the photocatalytic degradation of different azo dyes. Appl Catal B 46:35–47
Ibhadon AO, Fitzpatrick P (2013) Heterogeneous photocatalysis: recent advances and applications. Catalysts 3:189–218
Ishchenko OM, Rogé V, Lamblin G, Lenoble D (2016) TiO_2- and ZnO-based materials for photocatalysis: material properties, device architecture and emerging concepts. In: Cao W (ed) Semiconductor photocatalysis—materials, mechanisms and applications. InTech
Ito T, Adachi Y, Yamanashi Y, Shimada Y (2016) Long-term natural remediation process in textile dye-polluted river sediment driven by bacterial community changes. Water Res 100:458–465
Jadhav SB, Yedurkar SM, Phugare SS, Jadhav JP (2012) Biodegradation studies on acid violet 19, a triphenylmethane dye, by pseudomonas aeruginosa BCH. Clean Soil Air Water 40:551–558

Javaid R, Qazi UY, Kawasaki S-I (2016) Highly efficient decomposition of Remazol Brilliant Blue R using tubular reactor coated with thin layer of PdO. J Environ Manage 180:551–556

Javaid R, Qazi UY, Ikhlaq A, Zahid M, Alazmi A (2021) Subcritical and supercritical water oxidation for dye decomposition. J Environ Manage 290:112605

Khan S, Malik A (2018) Toxicity evaluation of textile effluents and role of native soil bacterium in biodegradation of a textile dye. Environ Sci Pollut Res Int 25:4446–4458

Khataee AR, Kasiri MB (2010) Photocatalytic degradation of organic dyes in the presence of nanostructured titanium dioxide: influence of the chemical structure of dyes. J Mol Catal A Chem 328:8–26

Kiran Avasarala B, Tirukkovalluri SR (2016) Magnesium doped titania for photocatalytic degradation of dyes in visible light. J Environ Anal Toxicol 6:2

Kong G, Pang J, Tang Y, Fan L, Sun H, Wang R, Feng S, Feng Y, Fan W, Kang W, Guo H, Kang Z, Sun D (2019) Efficient dye nanofiltration of a graphene oxide membrane via combination with a covalent organic framework by hot pressing. J Mater Chem A 7:24301–24310

Kuang S, Yang L, Luo S, Cai Q (2009) Fabrication, characterization and photoelectrochemical properties of Fe_2O_3 modified TiO_2 nanotube arrays. Appl Surf Sci 255:7385–7388

Kumar K, Chitkara M, Sandhu IS, Mehta D, Kumar S (2014) Photocatalytic, optical and magnetic properties of Fe-doped ZnO nanoparticles prepared by chemical route. J Alloys Compd 588:681–689

Kushniarou A, Garrido I, Fenoll J, Vela N, Flores P, Navarro G, Hellín P, Navarro S (2019) Solar photocatalytic reclamation of agro-waste water polluted with twelve pesticides for agricultural reuse. Chemosphere 214:839–845

Lachheb H, Puzenat E, Houas A, Ksibi M, Elaloui E, Guillard C, Herrmann J-M (2002) Photocatalytic degradation of various types of dyes (Alizarin S, Crocein Orange G, Methyl Red, Congo Red, Methylene Blue) in water by UV-irradiated titania. Appl Catal B 39:75–90

Lai C, Wang M-M, Zeng G-M, Liu Y-G, Huang D-L, Zhang C, Wang R-Z, Xu P, Cheng M, Huang C, Wu H-P, Qin L (2016) Synthesis of surface molecular imprinted TiO_2/graphene photocatalyst and its highly efficient photocatalytic degradation of target pollutant under visible light irradiation. Appl Surf Sci 390:368–376

Lakshmi Prasanna V, Rajagopalan V (2016) A new synergetic nanocomposite for dye degradation in dark and light. Sci Rep 6:38606

Lam S-M, Sin J-C, Abdullah AZ, Mohamed AR (2012) Degradation of wastewaters containing organic dyes photocatalysed by zinc oxide: a review. Desalination Water Treat 41:131–169

Lee KM, Lai CW, Ngai KS, Juan JC (2016) Recent developments of zinc oxide based photocatalyst in water treatment technology: a review. Water Res 88:428–448

Lewis DM (2011) The chemistry of reactive dyes and their application processes. In: Handbook of textile and industrial dyeing. Elsevier, pp 303–364

Li FB, Li XZ (2002) The enhancement of photodegradation efficiency using Pt–TiO_2 catalyst. Chemosphere 48:1103–1111

Liu X, Liu Z, Guo X, Liu C, Zhai H, Wang Q, Chang L (2014) Controllable synthesis and photocatalytic activity of spherical, flower-like and nanofibrous bismuth tungstates. Mater Sci Eng B188:35–42

Liu F, Leung YH, Djurišić AB, Ng AMC, Chan WK (2013) Native defects in ZnO: effect on dye adsorption and photocatalytic degradation. J Phys Chem C 117:12218–12228

Li Y, He J, Zhang K, Hong P, Wang C, Kong L, Liu J (2020) Oxidative degradation of sulfamethoxazole antibiotic catalyzed by porous magnetic manganese ferrite nanoparticles: mechanism and by-products identification. J Mater Sci 55:13767–13784

López-Ramón MV, Rivera-Utrilla J, Sánchez-Polo M, Polo AMS, Mota AJ, Orellana-García F, Álvarez MA (2019) Photocatalytic oxidation of diuron using nickel organic xerogel under simulated solar irradiation. Sci Total Environ 650:1207–1215

Luttrell T, Halpegamage S, Tao J, Kramer A, Sutter E, Batzill M (2015) Why is anatase a better photocatalyst than rutile?—model studies on epitaxial TiO_2 films. Sci Rep 4:4043

Lü W, Chen J, Wu Y, Duan L, Yang Y, Ge X (2014) Graphene-enhanced visible-light photocatalysis of large-sized CdS particles for wastewater treatment. Nanoscale Res Lett 9:148

Moura DC de, Quiroz MA, Silva DR da, Salazar R, Martínez-Huitle CA (2016) Electrochemical degradation of Acid Blue 113 dye using TiO_2-nanotubes decorated with PbO_2 as anode. Environ Nanotechnol Monit Manag 13–20

Mozia S, Tomaszewska M, Morawski AW (2005) Photocatalytic degradation of azo-dye acid red 18. Desalination 185:449–456

Muruganandham M, Swaminathan M (2006) TiO_2–UV photocatalytic oxidation of reactive yellow 14: effect of operational parameters. J Hazard Mater 135:78–86

Neppolian B, Choi HC, Sakthivel S, Arabindoo B, Murugesan V (2002) Solar/UV-induced photocatalytic degradation of three commercial textile dyes. J Hazard Mater 89:303–317

Olama N, Dehghani M, Malakootian M (2018) The removal of amoxicillin from aquatic solutions using the TiO_2/UV-C nanophotocatalytic method doped with trivalent iron. Appl Water Sci 8:97

Ollis DF, Pelizzetti E, Serpone N (1991) Destruction of water contaminants. Environ Sci Technol 25:1522–1529

Pan L, Zou J-J, Zhang X, Wang L (2010) Photoisomerization of norbornadiene to quadricyclane using transition metal doped TiO_2. Ind Eng Chem Res 49:8526–8531

Pelaez M, Nolan NT, Pillai SC, Seery MK, Falaras P, Kontos AG, Dunlop PSM, Hamilton JWJ, Byrne JA, O'Shea K, Entezari MH, Dionysiou DD (2012) A review on the visible light active titanium dioxide photocatalysts for environmental applications. Appl Catal B 125:331–349

Pingmuang K, Chen J, Kangwansupamonkon W, Wallace GG, Phanichphant S, Nattestad A (2017) Composite photocatalysts containing $BiVO_4$ for degradation of cationic dyes. Sci Rep 7:8929

Poulopoulos SG, Philippopoulos CJ (2004) Photo-assistedo of chlorophenols in aqueous Solutions using hydrogen peroxide and titanium dioxide. J Environ Sci Health A 39:1385–1397

Prihod'ko RV, Soboleva NM (2013) Photocatalysis: oxidative processes in water treatment. J Chem e168701

Qamar M, Saquib M, Muneer M (2004) Photocatalytic degradation of two selected dye derivatives, chromotrope 2B and amido black 10B, in aqueous suspensions of titanium dioxide. Dyes Pigm 65:1–9

Rahmat M, Rehman A, Rahmat S, Bhatti HN, Iqbal M, Khan WS, Bajwa SZ, Rahmat R, Nazir A (2019) Highly efficient removal of crystal violet dye from water by MnO_2 based nanofibrous mesh/photocatalytic process. J Mater Res Technol 8:5149–5159

Rajabi HR, Khani O, Shamsipur M, Vatanpour V (2013) High-performance pure and Fe^{3+}-ion doped ZnS quantum dots as green nanophotocatalysts for the removal of malachite green under UV-light irradiation. J Hazard Mater 250–251:370–378

Rajbongshi BM, Ramchiary A, Samdarshi S (2014) Influence of N-doping on photocatalytic activity of ZnO nanoparticles under visible light irradiation. Mater Lett 134:111–114

Rauf MA, Ashraf SS (2009) Fundamental principles and application of heterogeneous photocatalytic degradation of dyes in solution. Chem Eng Technol 151:10–18

Ray MB (2000) Photodegradation of the volatile organic compounds in the gas phase: a review. Chem Eng Process 8:405–439

Reddy ChV, Babu B, Reddy IN, Shim J (2018) Synthesis and characterization of pure tetragonal ZrO_2 nanoparticles with enhanced photocatalytic activity. Ceram Int 44:6940–6948

Rehman A, Usman M, Bokhari TH, Haq A, Saeed M, Rahman HMA, Siddiq M, Rasheed A, Nisa M (2020) The application of cationic-nonionic mixed micellar media for enhanced solubilization of Direct Brown 2 dye. Jmol Liq 301:112408

Rehman R, Uz-Zaman W, Abbas A, Mitu L (2019) Rapid photocatalytic degradation of methylene blue, tartrazine and brilliant green dyes by high-flux UV irradiation photolysis reactor. BCC 51:337–341

Saeed K, Khan I, Park S-Y (2015) TiO_2/amidoxime-modified polyacrylonitrile nanofibres and its application for the photodegradation of methyl blue in aqueous medium. Desalin Water Treat 54:3146–3151

Sahel K, Perol N, Chermette H, Bordes C, Derriche Z, Guillard C (2007) Photocatalytic decolourization of Remazol Black 5 (RB5) and Procion Red MX-5B—isotherm of adsorption, kinetic of decolourization and mineralization. Appl Catal B 77:100–109
Sahoo C, Gupta AK, Pillai IMS (2012) Photocatalytic degradation of methylene blue dye from aqueous solution using silver ion-doped TiO2 and its application to the degradation of real textile wastewater. J Environ Sci Health Part A 47:1428–1438
Sakthivel S, Neppolian B, Shankar MV, Arabindoo B, Palanichamy M, Murugesan V (2003) Solar photocatalytic degradation of azo dye: comparison of photocatalytic efficiency of ZnO and TiO_2. Sol Energy Mater Sol Cells 77:65–82
Salem IA, El-Ghamry HA, El-Ghobashy MA (2014) Catalytic decolourization of acid blue 29 dye by H_2O_2 and a heterogeneous catalyst. Beni-Seuf Univ J Basic Appl Sci 3:186–192
Samadi-Maybodi A, Sadeghi-Maleki M-R (2016) In-situ synthesis of high stable CdS quantum dots and their application for photocatalytic degradation of dyes. Spectrochimica Acta Part A Spectrochim Acta A Mol Biomol Spectrosc 152:156–164
Serpone N, Emeline AV (2002) Suggested terms and definitions in photocatalysis and radiocatalysis. Int J Photoenergy 4:91–131
Sharma S, Hasan A, Kumar N, Pandey LM (2018) Removal of methylene blue dye from aqueous solution using immobilized Agrobacterium fabrum biomass along with iron oxide nanoparticles as biosorbent. Environ Sci Pollut Res Int 25:21605–21615
Silva PM dos S, Fiaschitello TR, Queiroz RS de, Freeman HS, Costa SA da, Leo P, Montemor AF, Costa SM da (2020) Natural dye from Croton urucurana Baill. bark: extraction, physicochemical characterization, textile dyeing and colour fastness properties. Dyes Pigm 173:107953
Solís M, Solís A, Pérez HI, Manjarrez N, Flores M (2012) Microbial decolouration of azo dyes: a review. Process Biochem 47:1723–1748
Sripiboon S, Suwannahong K (2018) Colour removal by ozonation process in biological wastewater treatment from the breweries. IOP Conf Ser Earth Environ Sci 167:012010
Štengl V, Bakardjieva S, Murafa N (2009) Preparation and photocatalytic activity of rare earth doped TiO_2 nanoparticles. Mater Chem Phys 114:217–226
Stylidi M (2003) Pathways of solar light-induced photocatalytic degradation of azo dyes in aqueous TiO_2 suspensions. Appl Catal B 40:271–286
Su B, Wang K, Bai J, Mu H, Tong Y, Min S, She S, Lei Z (2007) Photocatalytic degradation of methylene blue on Fe^{3+}-doped TiO_2 nanoparticles under visible light irradiation. Front Chem China 2:364–368
Tanaka K, Padermpole K, Hisanaga T (2000) Photocatalytic degradation of commercial azo dyes. Water Res 34:327–333
Tayade RJ, Natarajan TS, Bajaj HC (2009) Photocatalytic degradation of methylene blue dye using ultraviolet light emitting diodes. Ind Eng Chem Res 48:10262–10267
Tkaczyk A, Mitrowska K, Posyniak A (2020) Synthetic organic dyes as contaminants of the aquatic environment and their implications for ecosystems: a review. Sci Total Environ 717:137222
Touati A, Hammedi T, Najjar W, Ksibi Z, Sayadi S (2016) Photocatalytic degradation of textile wastewater in presence of hydrogen peroxide: effect of cerium doping Titania. J Ind Eng Chem J 35:36–44
Ullah R, Dutta J (2008) Photocatalytic degradation of organic dyes with manganese-doped ZnO nanoparticles. J Hazard Mater 156:194–200
Uyguner CS, Bekbolet M (2008) Aqueous photocatalysis, natural organic matter characterization and removal: a case study of the photacatalytic oxidation of fulvic acid. Dangerous pollutants (xenobiotics) in urban water cycle. Springer, Netherlands, Dordrecht, pp 247–256
Vázquez-Ortega F, Lagunes I, Trigos Á (2020) Cosmetic dyes as potential photosensitizers of singlet oxygen generation. Dyes Pigm 176:108248
Vinu R, Akki SU, Madras G (2010) Investigation of dye functional group on the photocatalytic degradation of dyes by nano-TiO2. J Hazard Mater 176:765–773
Wang F, Ma Z, Ban P, Xu X (2017) C, N and S codoped rutile TiO_2 nanorods for enhanced visible-light photocatalytic activity. Mater Lett 195:143–146

Wang N, Li J, Zhu L, Dong Y, Tang H (2008) Highly photocatalytic activity of metallic hydroxide/titanium dioxide nanoparticles prepared via a modified wet precipitation process. J Photochem Photobiol A 198:282–287
Wang Z, Gao M, Li X, Ning J, Zhou Z, Li G (2020) Efficient adsorption of methylene blue from aqueous solution by graphene oxide modified persimmon tannins. Mater Sci Eng C Mater Biol Appl 108:110196
Wu C-H, Chang H-W, Chern J-M (2006) Basic dye decomposition kinetics in a photocatalytic slurry reactor. J Hazard Mater 137:336–343
Yan-fen F, Ying-ping H, De-fu L, Yang H, Wei G, Johnson D (2006) Photocatalytic degradation of the dye sulforhodamine-B: a comparative study of different light sources. Res J Environ Sci 19:97–102
Yoon J, Lee Y, Kim S (2001) Investigation of the reaction pathway of OH radicals produced by Fenton oxidation in the conditions of wastewater treatment. Water Sci Technol 44:15–21
Yuan H, Chen L, Cao Z, Hong F (2020) Enhanced decolourization efficiency of textile dye Reactive Blue 19 in a horizontal rotating reactor using strips of BNC-immobilized laccase: Optimization of conditions and comparison of decolourization efficiency. Biochem Eng J 156:107501
Zada N, Khan I, Saeed K (2017) Synthesis of multiwalled carbon nanotubes supported manganese and cobalt zinc oxides nanoparticles for the photodegradation of malachite green. Sep Sci Technol 52:1477–1485
Zahrim AY, Hilal N (2013) Treatment of highly concentrated dye solution by coagulation/flocculation–sand filtration and nanofiltration. Water Resour Ind 3:23–34
Zangeneh H, Zinatizadeh AAL, Habibi M, Akia M, Hasnain Isa M (2015) Photocatalytic oxidation of organic dyes and pollutants in wastewater using different modified titanium dioxides: a comparative review. J Ind Eng Chem 26:1–36
Zazo JA, Pliego G, Blasco S, Casas JA, Rodriguez JJ (2011) Intensification of the Fenton process by increasing the temperature. Ind Eng Chem Res 50:866–870
Zhang G, Zhang S (2020) Quantitative structure-activity relationship in the photodegradation of azo dyes. J Environ Sci (china) 90:41–50
Zhang L, Jaroniec M (2017) Toward designing semiconductor-semiconductor heterojunctions for photocatalytic applications. Appl Surf Sci 430:2–17
Zhang T, Oyama T, Horikoshi S, Hidaka H, Zhao J, Serpone N (2002) Photocatalyzed N-demethylation and degradation of methylene Blue in titania dispersions exposed to concentrated sunlight. Sol Energy Mater Sol Cells 73:287–303
Zhang X, Wu Y, Xiao G, Tang Z, Wang M, Liu F, Zhu X (2017) Simultaneous photocatalytic and microbial degradation of dye-containing wastewater by a novel g-C3N4-P25/photosynthetic bacteria composite. PLoS ONE 12:e0172747
Zhao J, Chen C, Ma W (2005) Photocatalytic degradation of organic pollutants under visible light irradiation. Top Catal 35:269–278
Zheng Y, Cao L, Xing G, Bai Z, Huang J, Zhang Z (2019) Microscale flower-like magnesium oxide for highly efficient photocatalytic degradation of organic dyes in aqueous solution. RSC Adv 9:7338–7348

Chapter 3
Recent Developments in Photocatalytic Techniques of Dye Degradation in Effluents

Barkha Tiwari and Hui Joon Park

Abstract In this chapter, we describe how a new trend of materials "nano" and its composites preferential grown as different sizes of small crystallites of structures using varied synthesis routes for photocatalytic dye degradation. Here, we will discuss about the different synthesis methods "traditional and trending" for the purification of water by degradation of dyes using photocatalytic techniques. To support the main outline of this chapter, we have supported various characterizations done to confirm the presence of nanomaterials like XRD for its crystal structure, SEM/TEM for morphology and XPS for elemental analysis; confirming with some functional groups using FTIR and Raman analysis. The modulation of parameters for the purification of water/effluents is pivotal in such work. So, we have compared the new novel techniques and materials trending all over research community as compared to the conventional ones. Certain nanomaterials synthesized are even doped with metals like rare-earth ions and transition metal ions to bring an effective output when illuminated to UV or visible light. Photocatalytic degradation has grown in last 20 years because of the impurities added to water bodies as source from industries and sewage, mainly comprising of waste like chemical dyes. This has paved way for lot of research in the fields of materials science, chemistry, environmental science and energy. Majorly synthetic dyes used in clothes, plastic, leather and other accessories are biggest motif behind this research. Our aim in this chapter is to summarize the whole scenario in compact but covering all relevant information. Usually, the procedure for photocatalytic dye degradation is step-wise process, involving choice of catalyst, concentration of dyes, light illumination and time taken for degradation of dye. Time taken for degradation of dye and choice of materials are inter-related, however efficiency of catalyst for certain dyes is better than other dyes. To make it clearer we will first give a brief introduction on the dyes, photocatalysis process, materials used as catalysts and purification.

Keywords Photocatalysis · Dye degradation · Advanced oxidation

---

B. Tiwari (✉) · H. J. Park
Department of Organic and Nano Engineering, Hanyang University Seoul, Seoul, South Korea
e-mail: kit152barkha@gmail.com

S. Dave and J. Das (eds.), *Trends and Contemporary Technologies for Photocatalytic Degradation of Dyes*, Environmental Science and Engineering,
https://doi.org/10.1007/978-3-031-08991-6_3

3.1 Introduction

From the last 20 years, the on-going trend in the field of photocatalytic degradation of dyes has increased relatively and gradually showing pace in its extensive usage; synthetic dyes are most prominently used in textile and plastic industries. As, it has become a major issue of textile, leather and plastic industries that the waste "effluents" generated are polluting the water bodies (Sacco et al. 2012; Mukhtish et al. 2013; Khan et al. 2017a; Khataee et al. 2010; Akpan and Hameed; Rauf and Salman). Most of the dyeing process effluents enter our environment in a large amount, approximately 20%; thus, hampering our natural resources (Rauf et al. 2011). So, to easily remove this issue, one such method adapts to tackle it involves the degradation of dyes. Where large molecules of dyes are oxidized down into smaller molecules like water, carbon dioxide, hydrogen and other by-products. But we are still struggling to find way to completely utilize all the dye molecules for degradation, as most of it released in water remains unused; making it difficult for researchers.

New modern methods now revolve around heterogenous photocatalysis for the dye's degradation; hetero here means more than one material. Spintronics of the charge carriers or the charge dipole at different energy levels gives rise to its transfer from the valence band to the conduction band when illuminated by light of particular wavelength (occurs in semiconducting material oxides) (Pandit et al. 2015). The exciting electrons generated at the energy levels react with the water molecules or oxygen molecules thus producing hydroxide radicals and oxide anions. Such high oxidizing species have better efficiency to degrade n number of molecules; effluents of industries and others. Advanced Oxidation Process (AOP) is well known type of Fenton processes in the scientific world which is actually the decontamination process using reactive oxygen species and other molecular species. So, advanced Oxidation Process has become an important research area in the common history of photocatalytic degradation but in this chapter, we will limit our discussions to the semiconducting medium types of catalysts (Elmorsi et al. 2010; Gul and Yildlrln 2009; AlHamedi et al. 2009).

It is compulsory at this stage to detail about the demand of this topic as a review. Some of the cause for this to happen have been listed below (Xie and Li 2006):

1. Several groups in the field of science and technology are working and collaborating in this realm. So, it's good to compile the complete literature and its ongoing trend at uniform recurring.
2. This method which is known for its various advantages over the years has major benefits using it in a bigger realm; as it is economically plausible thus making the photocatalytic procedure of degradation of dyes as demanding research.
3. The by-products evolved after degradation are mostly non-toxic components like oxygen, carbon dioxide and water.
4. The efficiency of degradation of the pollutants could be priorly calculated using the reagents like the hydroxide (OH group) to the potential applied for reduction of the pollutant and numerical counts of the pollutant's oxidation potential (Odling and Robertson 2017).

Variety of chemical dyes are used in coloring things in the industries and markets for our daily necessities. These dye materials have been classified in accordance with its molecular structures, optical properties, color, wavelength and applications. Major difference in dyes is done through its chromophoric group (having colour) in the molecular species such as azo dyes, nitro dyes, quinine-amine dyes, acridine dyes, xanthene dyes, anthraquinone dyes and so on (Rauf et al. 2011). The photocatalytic dye degradation studies reported mainly dealing with the variables such as the number of catalysts used, intensity of the light irradiated, concentration of dye, time taken for effect to occur when irradiated by light evolving oxygen and other species. The pseudo first order reaction with kinetic data fitting to the equation $-\ln(C/C0) = kt$ is nothing but the kinetics of photocatalytic degradation of dyes. The importance of kinetic data fit has been shown separately in other sub-section.

Although humongous studies have been reported for degradation of dyes using photocatalytic technique but still there are several areas which have not been attentively checked. Now we will point out those issues lying in this field that need to be tackled. As the Fig. 3.1 shows in the bar graph that there is a sharp increase of research in the field of catalysis in the last 15 years; five times in the 10 years of span thus the publication number have risen double in last 5 years (Akpan and Hameed; Rauf and Salman; Rauf et al. 2011). As we have told in the outline of this chapter that mostly research done in this area are metal oxides or its composites (hybrid) varying the crystallites size to tune the overall performance. Before we start with the ongoing trends in the research of photocatalysis and its limitations, we would briefly illustrate about the literature (comprehensive history).

We have tried to summarize some of the important research done in the field of photocatalysis in the last few years in Table 3.1. Most of the research works that have been reported in literature state the degradation effect of dyes that depends on variety of stuff like concentration of dye, amount of catalyst taken, pH of solution, time taken for light illumination, source of illumination etc. The first order system is followed in the study of the kinetics of photocatalytic dye degradation (Khan et al. 2017b).

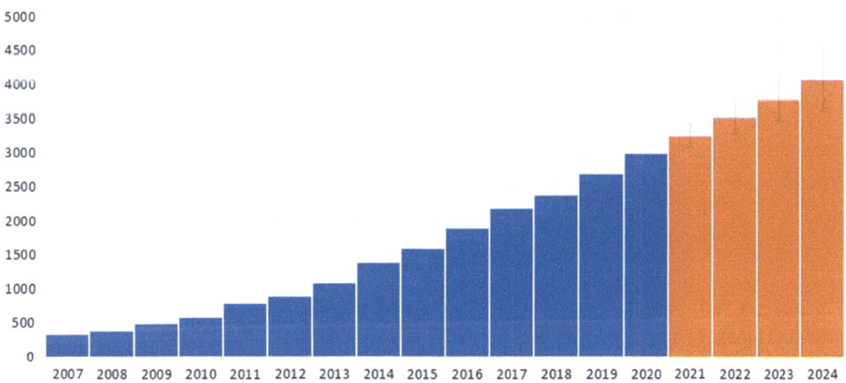

Fig. 3.1 Number of publications versus year plot (*Source* web of science)

Table 3.1 Brief comparative research history data on the degradation of dyes

Catalyst systems studied	Dyes employed	Comments	References
N-doped TiO_2	Methylene blue, Methyl orange	Depends on illuminated visible light and nitrogen content in catalyst	Sacco et al. (2012)
High surface area TiO_2	Methylene blue, Congo red	Sol–gel method preparation of TiO_2	Mukhtish et al. (2013)
Nanostructured TiO_2	Mono, di and tri-azo class of dyes	Degradation depends on the chemical structure of dye	Khan et al. (2017a)
Graphene-nanogold composite	Methyl blue, Rhodamine B	Methylene blue has better efficiency as compared to Rhodamine B when illuminated by visible light	Khataee et al. (2010)
Nano TiO_2 (anatase and rutile)	Methylene blue, Nitrobenzene, Acetophenone	Anatase shows better performance as compared to rutile	Akpan and Hameed
TiO_2, ZnO, SnO_2	Crystal violet, Methyl red	ZnO shows better performance	Rauf and Salman
Mg-TiO_2	Methyl orange	This catalyst has better efficiency in comparison to un-doped TiO_2	Rauf et al. (2011)
ZnO nanoflowers	Congo red, Methyl orange	New photocatalyst synthesized of ZnO showing better results for methyl orange	Pandit et al. (2015)
ZnO nanopowder	Rhodamine B	95% dye degraded under sun light	Elmorsi et al. (2010)
TiO_2	Congo red, Methyl orange, Methylene blue	Anatase phase of TiO_2 was dependent on its size and adsorption of catalyst	AlHamedi et al. (2009)
TiO_2	Indigo carmine, Indigo	Visible light was used as source of light	Xie and Li (2006)
Immobilized TiO_2 on polyvinyl alcohol and polyacrylamide	Reactive orange, Methylene blue	On PVA trapped TiO_2 showed better result than polyvinyl alcohol	Odling and Robertson (2017)

(continued)

Table 3.1 (continued)

Catalyst systems studied	Dyes employed	Comments	References
Nano-TiO_2	Acidic red, Acidic orange	Sulphonate and azo groups shows degradation	Khan et al. (2017b)
Ag, Au doped SiO_2 nanoparticles	Methyl red	Hydroxide radicals help in dye degradation	Xiong et al. (2010)
TiO_2	Emerald green	pH effects the dye degradation	Rajesh et al. (2007)
(Solar, pH, UV) TiO_2	Procyon yellow	Solar light initiates dye degradation in TiO_2	Avasarala et al. (2016)
TiO_2	Reactive red	H_2O_2 and sulphates were studied	Mahadwad et al. (2011)
Thermally activated ZnO	Congo red	Second order kinetics	Mohammed et al. (2016)
Sol–gel TiO_2	Lissamine green B	Thin film of TiO_2 showed better result	Nagaraja et al. (2012)
Ag-TiO_2 core–shell structures	Reactive blue 220	Core–shell shows good dye degradation under solar light	Mo and Ching (1995)
Nano-TiO_2 Anatase	Reactive blue 4	In the presence of H_2O_2, dye degradation is increased	Iakandar et al. (2007)
TiO_2/ZnO	Methylene blue	ZnO shows better performance	Bavykin et al. (2006)
P160-TiO_2	Yellow-28	Presence of carbonate ions increased the performance	Koelsch et al. (2002)
Orthorhombic WO_3	AO7 dye	Oxalic acid initiates reaction	Morgan and Watson (2010)
Ni doped TiO_2	Malachite green	Oxidizing species used are hydroxyl ions	Qamar et al. (2008)
TiO_2	Solo phenyl red 3BL	Concentration of hydroxide and oxygen ions determine rate of degradation	Sclafani and Herrmann (1996)
TiO_2	Reactive green 19, Azo orange	Azo orange has better efficiency as compared to reactive green under sunlight	Bakardiieva et al. (2005)
TiO_2	Disperse and azo dyes	Tuning parameters	Ohno et al. (2004)

(continued)

Table 3.1 (continued)

Catalyst systems studied	Dyes employed	Comments	References
TiO_2	Carmine indigo	UV light, pH = 4	Li (2011)
ZnO	Methylene blue,	Basic solution has better performed	Ohtani et al. (1997)
Carbon-TiO_2	Amido black	Activated O_2 ions are responsible for degradation	Singh et al. (2013)
ZnO	Direct red-31	UV irradiation and annealing temperature (500–800 °C)	Hoffman et al. (1995)
Sol–gel TiO_2 thin films	Congo red, methyl orange	Dip coated TiO_2 showed better performance	Meng and Juan (2008)
Undoped and Fe doped CeO_2	Methyl orange	1.5% doping of Fe in Titanium dioxide showed better dye degradation	Hernandez-Alonso et al. (2009)
Immobilized TiO_2	Methylene blue	Deposition of photosensitive layer decreases the performance	Mills et al. (1993)
Ni-$MgFe_2O_4$	Malachite green	Depends on illuminated visible light	Hashimoto et al. (2005)
UV-TiO_2	Methylene blue	Mineralization of sulphur, carbon and nitrogen	Ye et al. (2010)
Cu doped P-25	Azo dye orange	It is better than H_2O_2/UV reaction	Colmenares et al. (2009)
Ag–Ni TiO_2	Methyl red	Doped is better than undoped TiO_2	Anpo et al. (1987)
Cr doped TiO_2	Congo red, Methylene blue	Anatase to rutile phase transition	Lin et al. (2006)
Au and Ag doped ZnS quantum dots	Methylene blue	Loading of metals favors better performance	Sarkhanpour et al. (2017)
Mesoporous CeO_2	Rhodamine B	Hydroxyl groups are active species	Ajmal et al. (2014)
ZnS	Rose Bengal	OH^- groups show the result	Tesfay et al. (2015)

(continued)

Table 3.1 (continued)

Catalyst systems studied	Dyes employed	Comments	References
Films of C-TiO_2	Azorubine	Photocatalytic reaction and absorption of catalyst effect performance	Paola et al. (2013)
La-Y/TiO_2	Methylene blue	4 g/L catalyst	Nguyen-Phan et al. (2011)
Ag-TiO_2	Direct red 23	3 g/L catalyst	Ioannis and Triatafyllos (2004)
ZnO	RBB dye	1st order kinetics	Kurny and Fahmida (2017)
ZnO	Crystal violet	Better optical property, high crystallinity, high surface area is responsible for good dye degradation	Fox and Dulay (1993)
In/ZnO nanoparticles	Methylene blue	Indium is well dispersed on ZnO	Tunesi and Anderson (1991)
TiO_2-P25-ZnO	Methylene blue	ZnO has better result in visible light as compared to TiO_2	Tang and An (1995)
TiO_2 nanoparticles	Methylene blue	Basic medium is better for degradation	Guillard et al. (2003)
ZnO	Indigo carmine, Indigo	Visible light was used as source of light	Reutergarth and Iangpashuk (1997)
BiOI	Reactive orange, Methylene blue	On PVA trapped TiO_2 showed better result than polyvinyl alcohol	Baran et al. (2008)
TiO_2	Acidic red, Acidic orange	Sulphonate and azo groups show degradation	Wang et al. (2000)
ZnO, TiO_2	Methyl red	Hydroxide radicals help in dye degradation	Baran et al. (2003)
ZnS doped with Mn	Emerald green	pH effects the dye degradation	Tang and An (1995)
Undoped and Cu doped TiO_2	Methylene orange, Rhodamine B	ZnO shows lower performance	Alaton and Balcioglu (2001)
Mn_3O_4 nanoparticles	Procyon yellow	Solar light initiates dye degradation in TiO_2	Poulios and Aetopoulou (1999)

(continued)

Table 3.1 (continued)

Catalyst systems studied	Dyes employed	Comments	References
TiO_2 with Pt	Reactive red	H_2O_2 and sulphates were studied	Poulios et al. (2000)
TiO_2	Congo red	Second order kinetics	Zhang et al. (2001)
TiO_2 coated cotton fabric	Lissamine green B	Thin film of TiO_2 showed better result	Zhiyong et al. (2007)
Silver phosphate	Methylene blue	Industrial waste water is used	Bauer et al. (2001)
$CeCrO_3$	Reactive blue 220	Core–shell shows good dye degradation under solar light	Xiaoqing et al. (2017)
ZnO	Reactive blue 4	In the presence of H_2O_2, dye degradation is increased	Abo-Farha (2010)
CeO_2-ZnO	Yellow-28	Presence of carbonate ions increased the performance	Saber et al. (2011)
MgO	Methylene blue	90% degradation	Kirupavasam and Allen (2012)
TiO_2	Acid orange	UV light source as better effect than visible light	Quang et al. (2017)
TiO_2 on polyethylene film	Crystal violet, Methylene blue	Solar light dye decolorization	Mahmoud et al. (2009)
Mo doped TiO_2	Toluidine blue-o	Pseudo first kinetics	Sun et al. (2006)
Copper ferrite	Glycerol, Methylene blue	Glycerol degradation is not good until H_2O_2 is added	Huang et al. (2008)
TiO_2	Tartrazine	Influence of other salts	Sun et al. (2008)
Zn-TiO_2	Direct blue 71	Zn doped has better performance over undoped	Wei et al. (2007)
Ag-ZnO	Reactive orange 16	Ag doped is better than undoped	Guettai and Amar (2005)
TiO_2	Reactive orange 16	Shows good performance	Sharma et al. (2013)
ZnO-CuO	Reactive black 5	It is good for reduction of effluents in environment	Bhati et al. (2010)

(continued)

Table 3.1 (continued)

Catalyst systems studied	Dyes employed	Comments	References
TiO_2 on polyethylene glycol	Congo red, Methylene orange	UV light high efficiency	Min et al. (2015)
g-C_3N_4 and CaCl	Rhodamine B	Modified system is 50 times better than bare system	Dharmarajan et al. (2013)
CdO/TiO_2	Reactive orange 4	UV irradiation and annealing temperature (500–800 °C)	Nosaka and Nosaka (2017)
ZnO	Reactive blue	Photocatalytic performance good in coupled TiO_2-ZnO	Salem et al. (2015)
BiOCl	Methyl orange	Visible light degradation	Bubacz et al. (2010)
ZnS-Cr doped	Methyl orange	Visible light is better than UV	Salem et al. (2015)
Nano-TiO_2 (C-fe doped)	Basic blue 9	Waste water usage	Bubacz et al. (2010)
CeO_2-SnO_2	Direct black 32	Activity is comparable with TiO_2-P25	Raheem and Hameed (2015)
TiO_2-ZnO	RB 21 dye	UV photoreactor and TiO_2	Mehta and Surana (2013)
CaO	Indigo carmine dye	pH 9 was suitable	Ameta et al. (2015)
$SrTiO_3$	Methyl orange, Methylene blue	Non-selective process	Rupa et al. (2015)
CuO/Ag_3AsO_4/GO	Phenol	Photosensitivity	Narde et al. (2017)
TiO_2/diatomite	Rhodamine B, Methylene blue, Methyl orange	Waste water treatment	Zangi et al. (2017)
$PbCrO_4$	Rhodamine B	Visible light sensitive	Simovic et al. (2017)
ZnO	Acid red 27	Photocatalytic reaction and absorption of catalyst effect performance	Mezughi et al. (2014)

(continued)

Table 3.1 (continued)

Catalyst systems studied	Dyes employed	Comments	References
CuS	Methylene blue, Rhodamine B, Congo red	Pseudo first order kinetics	Kamal et al. (2014)
N-doped ZnO	Azure A	Effective for number of reactions	Marinovic et al. (2017)
TiO_2 (brookite)	Rhodamine	1st order kinetics	Long et al. (2017)
TiO_2 (brookite + rutile)	Orange dye	Annealed at varied temperatures for good result	Dhatshabamurthi et al. (2015)
g-CN	Methylene blue	g-CN under UV is much more efficient	Tiwari and Ram (2019)

Traditional wet chemical, physical and bio-friendly synthesis routes have been used extensively for dye degradation present in effluents. These synthesis processes have shown certain disadvantages like high energy, high cost, by-products formed end up in other types of secondary pollutants. Therefore, Advanced Oxidation Process i.e., AOP has gained considerable notice for the degradation of such harmful dyes in recent trends (Xiong et al. 2010).

3.2 Literature

This is one of the research fields that is done all over the world when compared with other research areas. To validate more about our above point that the research in this particular area "photocatalytic degradation" has always been in trend since more than 10 years as compared to other realms of science and technology; we have shown it in Table 3.1. The pictorial representation of dye degradation phenomena happening in the spintronic level of molecules is clearly shown in Fig. 3.2. The most common of all materials used for the degradation of dyes as photocatalysts is titanium dioxide. But unfortunately, the main flaws of using titanium dioxide (TiO_2) are its wide bandgap i.e., 3.2 eV. Out of all the crystal phases of TiO_2, in reports it is researched that the anatase gives better performance in dye degradation in photocatalysis due to its high absorption intensity for photons. This shows that the crystal phase in TiO_2 plays a major role for dyes degradation (Rajesh et al. 2007; Dnyaneshwar 2017; Avasarala et al. 2016; Mahadwad et al. 2011; Mohammed et al. 2016; Nagaraja et al. 2012; Mehra and Sharma 2012).

Out of the three most prominent phases of titanium dioxide such as anatase, rutile and brookite; anatase and rutile phases have been studied more as shown in Table 3.1. The oxygen ions positioned in the anatase crystal structure of TiO_2 is in such an

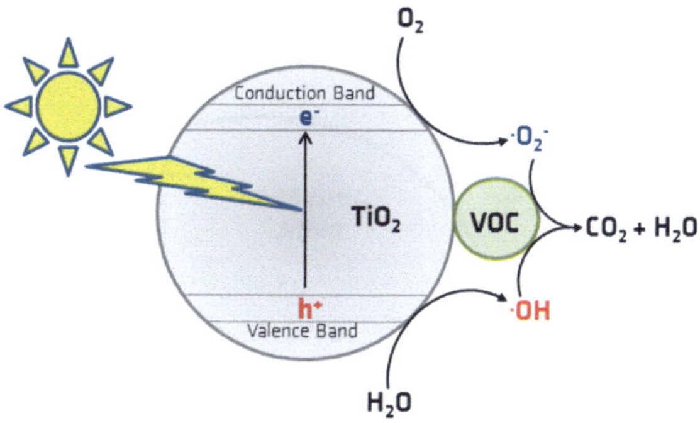

Fig. 3.2 Photocatalytic degradation of dyes in semiconducting materials (ref. Birte et al. (2017))

arrangement that allows maximum absorption of organic pollutants, even the orientation of titanium ions creates a condition favorable for reaction condition with those absorbed dyes (Mo and Ching 1995; Iakandar et al. 2007; Bavykin et al. 2006). But unfortunately, such beneficial crystal structural arrangements of molecules are not present in rutile phase of titanium dioxide. As reviewed, the researchers have found out that the pure form of anatase with small percent of rutile phase is more conductive for porosity and favoring better dye degradation. Spintronically, the charge carriers are excited by illumination of light thus hovering from lower energy level "valence band" to higher energy levels "conduction band" thus creating electron–hole pair, this generates radical ions which are an important reason for the degradation of dyes into water, carbon dioxide and some other species (Koelsch et al. 2002; Morgan and Watson 2010; Qamar et al. 2008; Sclafani and Herrmann 1996; Bakardiieva et al. 2005; Ohno et al. 2004; Colon et al. 2006; Li 2011; Ohtani et al. 1997).

For the dyes degradation using photocatalytic effect needs a material "catalyst" with larger surface area to have good adsorption of the effluents waste thus increasing the kinetics of the reaction. It will be considered slowly. In the standard P-25 TiO_2, which is the mixture of 20% rutile and 80% anatase phase, making the overall material as a core–shell system with high productivity and active in most of the cases, that's the reason it is used as standard material in comparison. For the treatment of effluents, various ways have been tried in history but dye removal using photodegradation has implied a pivotal role. Due to environmental and aesthetic concerns, the dye degradation from the waste in water bodies of industries like textiles, leather, plastic, etc. played a significant role (Elahee 2010). Metal oxides like ZnO and TiO_2 have been used in nanospheres, nanorods, thin films, nanowires and nanofibers on a polymeric matrix; they have high performance, low price and eco-friendly. There are other semiconducting materials too like CdS, WO_3, ZrO_2, etc. for the dye degradation under light irradiation; these materials are already included in Table 3.1. The cons of such materials like titanium dioxide are that it has got high band gap, and need UV light illumination to degrade the dye in effluents. To give a clear picture of photodegradation of dyes, Fig. 3.3. Typically, some other materials and their composites show such photodegradation which is shown in Fig. 3.4a, b using pictorial presentation. Brookite has also been served for photocatalysis phenomena in dye degradation as

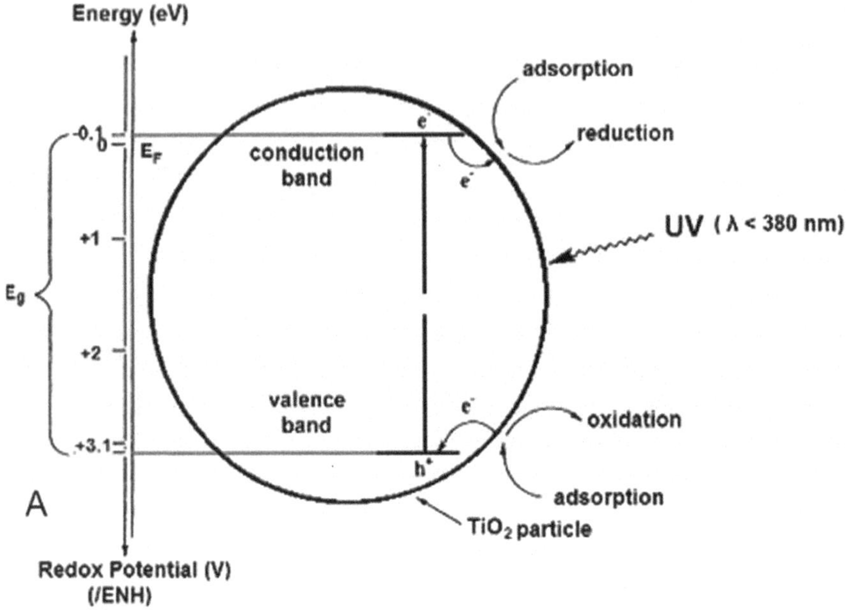

Fig. 3.3 Energy band diagram of photocatalytic degradation of dyes in semiconducting materials (ref. Elena et al. (2017))

one of the phases of titanium dioxide. Commonly, the studies carried out on photocatalytic dye degradation as checked by chemical oxygen demand measurements. Still, it is said to check the product concentration while formation and clear various routes through which these products are formed. There has always been a limitation to using such studies in large scale as it is important to see if the degradation material products are non-toxic to living beings and environment (Singh et al. 2013; Hoffman et al. 1995; Meng and Juan 2008; Hernandez-Alonso et al. 2009; Mills et al. 1993; Hashimoto et al. 2005). Optimization of parameters is pivotal as like the choice of photocatalyst material for successful technique implementation in purifying effluents in this new trending technology. So, the trend in this field has shown that with increase in light intensity of irradiating source, the degradation of dyes percent can also be increased. High value of irradiation light, more will be the recombination (electron–hole pairs) and vice-versa.

The photocatalytic response mainly depends on the chemical nature and annealing temperature of the semiconductor. Most of our choices of semiconducting materials and its crystal structures are based on several parameters like the physical form of the semiconductor and its stability under adverse the conditions of chemical reactions (Neppolian et al. 1999; Ye et al. 2010; Colmenares et al. 2009; Anpo et al. 1987; Lin et al. 2006; Sarkhanpour et al. 2017; Ajmal et al. 2014). Environmentally friendly, low cost or economically viable, less toxic are pivotal characteristics for good photocatalytic materials and out of this titanium dioxide has been reported and

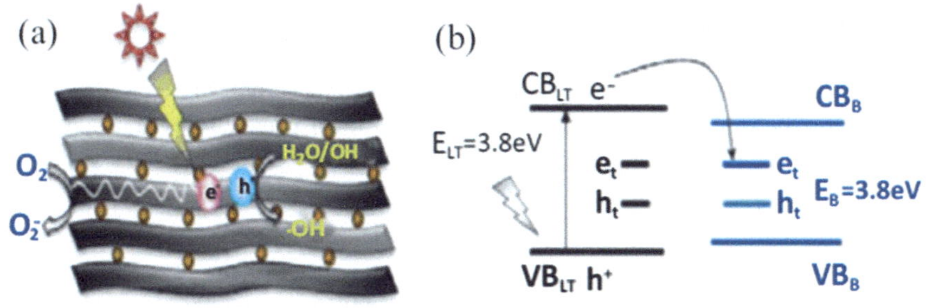

Fig. 3.4 **a** Diffusion behavior of charge carriers in layered nanosheets and **b** transport pathway of the excited electron in the mixed photocatalyst (ref. Thuy-Duong Nguyen et al. (2011))

most trending material of all in this area of research. As the last research has been drafted and reviewed, it is seen that the standard material for comparison of photocatalytic performance of any new material is compared with P-25 TiO_2 (anatase has performance better than rutile but less than standard material). Although, various parameters are optimized to get best result but the amount of photocatalyst material has been used mainly depends on the chemical behavior of the semiconducting property of it; thus, photocatalytic reaction has been easily tuned with certain modifications in literature (Tesfay et al. 2015; Paola et al. 2013; Nguyen-Phan et al. 2011). The changes have been done with various levels like modulating wavelengths of light, and spin-energy level coupling in semiconduction for efficient usage of the charge pair or e^- hole pair. Recently a new trend has emerged in materials used as catalysts for dye degradation using photocatalysis phenomena i.e., use of graphitic carbon nitride (g-C_3N_4) as shown in Fig. 3.5. The V.B. level (valence band level) in the system could be shifted to better value which enhances the oxidation ability.

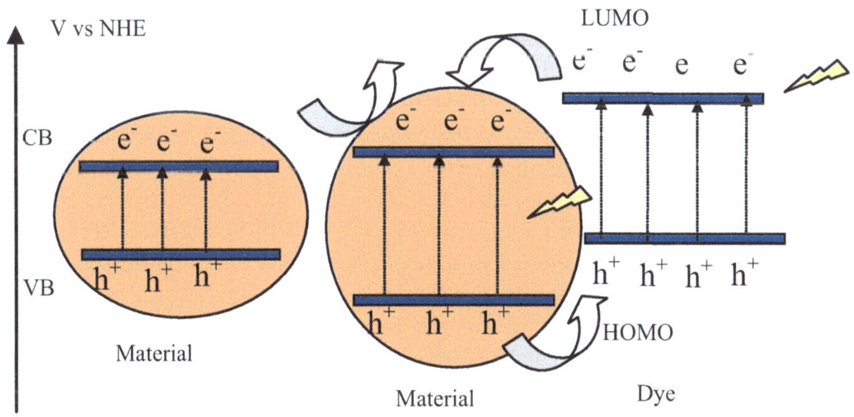

Fig. 3.5 Band diagram (discrete energy levels) of carbon nitride (CN) and CN-Ca for degrading dye

Simultaneously, using the photoexcitation in the dye degradation, the transfer of electrons to the higher energy levels i.e., conduction band. This is predominant in the presence of visible light. Other than this many researchers have tried doping the semiconductor catalysts for better in large realm of wavelengths and degradation performance (Ioannis and Triatafyllos 2004; Kurny and Fahmida 2017; Fox and Dulay 1993).

3.3 Photocatalytic Dye Degradation Chemical Phenomena in Core–Shell

Due to progressively increasing textiles, paper and other coloring industries around the world, the residues sank out of them are a major issue of chemical pollutants/hazards of damaging our routine life-cycle by contaminating water bodies and badly affecting the living beings (both aquatic and non-aquatic). Color dyes have been classified into various classes in the literature, which are often non-biodegradable of a source of badly affecting our eco-system of a green nature. Out of all this, a crystal violet (CV) dye is one of the cationic dyes being used in inks that is majorly polluting our green nature throughout the globe. To control an environmental pollution, it is thus highly required to degrade it (before it is damped as a waste in nature) in an ecofriendly method, such as its photocatalytic degradation in a water host, which is the mostly easily available in plenty in the nature. Usually, a CV dye strongly absorbs visible light with absorption maximum lying at 570 or 590 nm, which can be explored in degrading it using a suitable photocatalyst, for example, C-$CaIn_2O_4$ we synthesized in small core–shell crystallites in this work. As a result, in this investigation, we analyze how a CV dye can be degraded using a core–shell C-$CaIn_2O_4$ as a photocatalyst as follows (Tiwari et al. 2018).

A photocatalytic experiment was carried out under UV irradiation (24 W power, 400 lx intensity, 300–420 nm range with its average peak position lying at λ_{max} = 365 nm) of a CV dye (5×10^{-5} ML^{-1} concentration in water) charged on a sample C-$CaIn_2O_4$ in water while stirring. Typically, a 1.0 gL^{-1} of a C-$CaIn_2O_4$ sample was dispersed in each solution to let the pollutant adsorb over its surfaces to the equilibrium in a deep dark for a sufficient time-period of 30 min. Then, the sample was degraded in parts by irradiating for equal intervals of 15 min followed by centrifuged at 6000 rpm for 5 min to separate the dye from the C-$CaIn_2O_4$ catalyst. Recovered solutions in this way were studied in terms of UV–visible absorption spectra to determine any residual dye left-out after degradation in the solutions. An average of three consecutive measurements was used in analyzing the degradation process and its kinetic parameters. A good catalyst offers its photocatalytic features in four major factors (i) how it uploads a given pollutant in a liquid carrier, (ii) how the uploaded species absorb light in a wide band, (iii) how the "$e^- - h^+$" ion pairs decouple in transient interfaces formed on the catalyst by irradiating so obtained sample and (iv) how degraded species after the reaction ultimately get separated from the dye and disperse back into the solution of a reservoir. Small C-$CaIn_2O_4$ core-shells and tiny pores of a sample we used here promptly adsorb a dye over large surfaces having large charge carriers. The whole reaction process occurs in successive steps that can be put in a simple model reaction scheme as follows. Model molecular diagrams given in Fig. 3.6 describe how (a) core–shell C-$CaIn_2O_4$ crystallites uploads (b) a crystal violet dye (when adding in a solution) on (c) its reactive C-sp^2 modified surfaces so that the dye degrades thereon in a light induced reaction and (d) ultimately the degraded dye species segregate and disperse into the solution as a separate phase (Tiwari and Ram 2019).

As a result, as soon as the dye degrades in this process it proportionally loses its absorbance of its characteristic light absorption band as shown in Fig. 3.7. In this figure, only selective absorption bands, which were recorded in a regular interval of 30 min of UV-irradiating (over 300–420 nm wavelengths) an as-prepared C-$CaIn_2O_4$ sample-1 dispersed along with a CV dye in water, are compiled in order to determine how the irradiation causing the dye to be progressively degrading over a time scale, which is prolonged up to 240 min in a continuous UV irradiation. No any measurable degradation appears in absence of any UV–visible light in a complete dark medium ensuring that it is only the UV-irradiation that it leads the dye to degrade on the $CaIn_2O_4$ sample-1 in the photocatalytic reactions with its reactive surfaces. The reaction is mostly complete in a 240 min of a regular irradiation after which only a

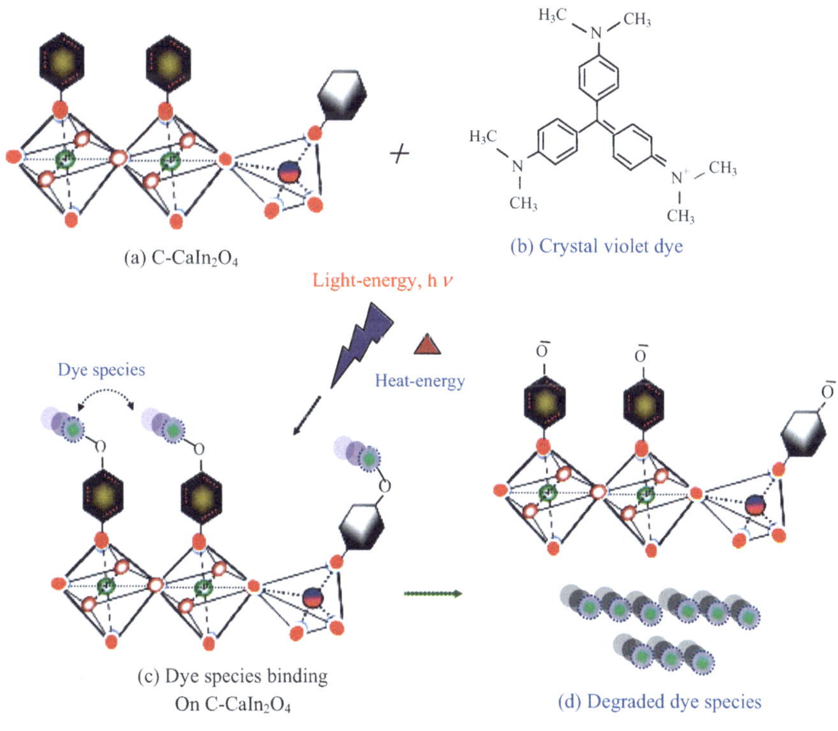

Fig. 3.6 Catalytic performance of **a** C-$CaIn_2O_4$ in (**b, c**) adsorbed dye degradation and **d** removing a crystal violet dye on irradiating UV-light (115)

Fig. 3.7 Light-absorption by the dye adsorbed over C-$CaIn_2O_4$ core-shells in a catalytic reaction when irradiated by a UV-light

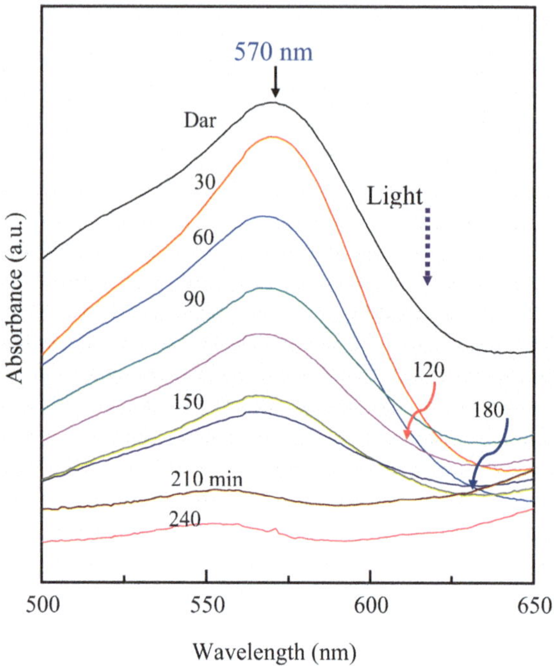

small residual absorption band lasts and it does not change any more (Mills et al. 1993; Hashimoto et al. 2005; Neppolian et al. 1999).

$$\text{Semiconductor} + \text{photons}\ (h\nu) \rightarrow e^-(CB)$$
$$+ h^+(VB)\ (1)\ h^+ + H_2O \rightarrow H^+ + OH^-$$
$$h^+ + OH^- + OH$$
$$e^- + O_2 \rightarrow O_2$$
$$2e^- + O_2 + H_2O \rightarrow H_2O_2$$
$$e^- + H_2O_2 \rightarrow OH + OH^-$$
$$\text{Organic pollutant} + OH + O_2 \rightarrow CO_2 + H_2O + \text{other degradation products}$$

As usual, excited photons in the UV-irradiation lead to excite an electron from the valence band to the conduction band, leaving behind a hole "h^+" in the valence band in a decoupled "$e^- - h^+$" ion-pair. These holes could readily oxidize the donor dye molecules so as to promptly react with H_2O forming refined hydroxyl species (Ye et al. 2010; Colmenares et al. 2009; Anpo et al. 1987; Lin et al. 2006). We optimized adsorption of the dye on the different C-$CaIn_2O_4$ samples-1, 2, 3 dispersed in water in a fixed volume in a completely dark room in terms of its absorbance in its characteristic light-absorption band. Both as-prepared and annealed C-$CaIn_2O_4$ samples readily adsorb the CV dye on its multiple surfaces of small core–shell crystallites. A dispersion of the dye on C-$CaIn_2O_4$ in water has been equilibrated in its maximum adsorption on its surfaces over an extended exposure time to nearly 90 min

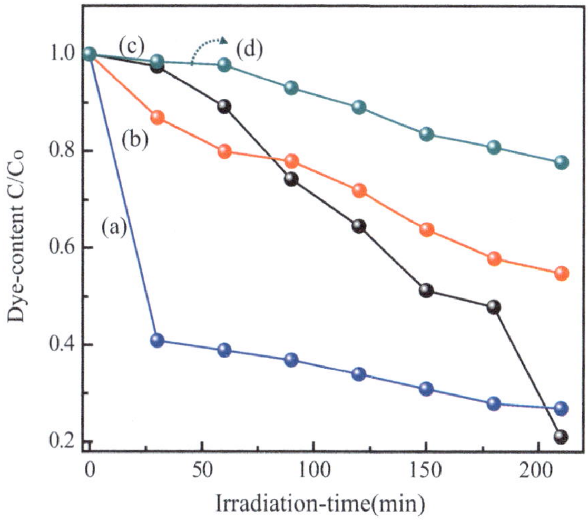

Fig. 3.8 Photocatalytic degradation of the CV dye using **a** the as-prepared and annealed C-$CaIn_2O_4$ powders at **b** 400 °C and **c** 600 °C for 2 h in air, and **d** a standard P25-TiO_2 powder dispersed in water by irradiating over UV–light

until it becomes stable. As soon as the adsorption completed and the final sample becomes stable in absorbing light, we started measuring its degradation by irradiating it by UV-light over predetermined intervals of time of 30 min as shown in the so obtained results, which are plotted in Fig. 3.8. Here, C_0 (which is proportional to the maximum absorbance in the dye in the sample) describes an initial dye concentration before any irradiation and C/C_0 is its normalized residual value after a photocatalytic degradation performed in this experiment. To find out a critical extent of a recurring degradation achieved in this experiment, the intermediate samples were centrifuged at 6000 rpm for 10 min in which so far degraded dye desorbs off the catalyst and disperses in the water.

In the plot in Fig. 3.8a, as the UV-light irradiating time $t \rightarrow 25$ min, the CV dye in sample-1 is decayed rapidly till a minimum $C/C_o \cong 40\%$. A nearly four times larger dye lasts in Fig. 3.8d in decaying much slowly in a standard P25-TiO_2. A reasonably large porosity of ultrafine pores and a reasonably thick C-sp^2 surface layer in this sample-1 promptly favor its catalytic features. A reasonably larger degradation rate constant $k = 0.065\ \text{min}^{-1}$ (against $0.013\ \text{min}^{-1}$ in the P25-TiO_2) thus prolongs in a pseudo first order reaction in a non-linear regression of $\ln(C/C_o) = -kt$ over t as plotted in Fig. 3.9a for this sample. Other samples involve much lower k-values as can be determined assuming linear plots in Fig. 5.9b, c. Further, the degradation plots in Fig. 3.10 describe a non-linear degradation of the CV dye (a) before and (b) after dispersing in a photocatalytic C-$CaIn_2O_4$ sample-3 in water and irradiating over UV-light. In a degradation plot made in Fig. 3.10a, a bare CV dye hardly degrades up to 12% in a very slowly reaction with H_2O molecules by UV irradiation when dispersed in water. This result clearly shows a crucial role a photocatalyst is playing in degrading it reasonably faster as observed in this experiment in the presence of a photocatalyst of a C-$CaIn_2O_4$ sample.

Further, in absence of an irradiation, the dye dispersed in a catalytic sample-3 decays much slowly (hardly by nearly up to 25%) in a marked microscopic effect of a light induced photochemical reaction in this example. A maximum of 81% degradation efficiency, $\eta = (C_0 - C_\infty)\, C_0^{-1}$, of degrading a CV-dye thus has been observed in our photocatalytic C-$CaIn_2O_4$ sample-3 in water after a long exposure of UV irradiation up to 210 min. Here, $C = C_\infty$ as $t \rightarrow \infty$ (taken as 210 min of the degradation reaction) has been used according to the degradation plots given in Fig. 3.8. Nevertheless, an as-prepared C-$CaIn_2O_4$ sample-1 serves to be a faster photocatalyst in degrading the CV-dye over early periods, $t \leq 60$ min, of the UV irradiation in the present experiments. An annealed C-$CaIn_2O_4$ sample-2 at 400 °C for 2 h in air exhibits a reasonably lower $\eta = 40\%$ in accounts of its enhanced electrical conductivity on forming highly regular and stable conductive C-sp^2 surface channels (of a critical $\delta = 1$–2 nm thickness of a conjoint network of a surface layer) as will be described in the following section in this chapter. In this case, a decoupled "e^- – h^+" ion-pair mostly takes part in the electrical conduction rather than in a desired photocatalytic degradation in this experiment. A standard P25-TiO_2 photocatalyst as used over here yields a still inferior $\eta = 20\%$ in degrading a CV-dye in water in the present experimental conductions.

A bar chart given in Fig. 3.11 compares η-values observed in selective samples studied in this work. A few other authors have studied photocatalytic properties of pure and C-sp^2 modified $CaIn_2O_4$ in degrading crystal violet dye dispersed in water and in visible irradiations, with similar degradation results of $\eta \leq 80$ values. Also in these experiments, a surface C-sp^2 modified $CaIn_2O_4$ had duly improved photocatalytic properties according to its local structure.

3.4 Optimization of Variables in Photocatalysis of Dye Degradation

In this we have explained that while optimization of dye degradation chemical phenomena, we not just check chemical synthesis methods for the materials employed but even the wavelength of light used as source for the energy bandgap of semiconducting materials. So, the dye degradation has effects of several variables used in experiment that can be varied and optimized for best result. We have tried to give collective bandgap information on the last 20 years and ongoing semiconducting materials in Table 3.2. Unfortunately, most of the time researchers don't report the light source intensity and it's monitoring; LEDs having good energy efficiency have also been employed, temperature of chemical reaction, light intensity, source of light and the catalyst particles size, surface area of catalyst and varied forms of materials affecting the dye degradation rate.

Table 3.2 Trending material semiconductors (Table 3.1 as reference) that are used for dye degradation with band gap values in eV

Semiconducting materials for dye degradation	Band gap (eV)
TiO_2–Anatase	3.2
TiO_2–Rutile	3.0
TiO_2–Brookite	3.14
ZnO	3.36
WO_3	2.76
CdS	2.42
CuO	1.2
Cu_2O	2.2
MgO	5.9
Mn_3O_4	3.28
CeO_2	3.6
Fe_2O_3	3.19
Fe_3O_4	2.3
ZrO_2	2.25
g-C_3N_4	2.661
Ag_2O	1.4
$SrTiO_3$	3.25
Bi_2WO_6	3.13
$BaTiO_3$	3.302
Bi_2O_3	2.8
CdO	2.2
CoO	2.01
Cr_2O_3	3.5
HgO	1.9
In_2O_3	2.8
MnO	3.6
Nb_2O_3	3.4
NiO	3.5
PbO_2	2.8
PbO	1
Sb_2O_3	3
SnO	4.2
SnO_2	3.5
V_2O_5	2.8
$K_6Ta_{10.8}O_{30}$	3.76

3.4.1 pH Variable Effect on the Dye's Degradation Chemical Phenomena

As shown in Table 3.1, that we have mentioned so beautifully about materials and its dye degradation phenomena is dependent on pH value. The change in oxidation potential as per pH values for the molecular species used in the chemical solution for dye degradation is on pivotal observation. As the hole oxidation potential and electron reduction power that is generated when light is irradiated on the semiconducting catalysts for dye degradation are even dependent on the positions of the high energy level of the valence band and low energy level of conduction band. Thus, pH has greater effect even when adsorption of dye is done on the photocatalyst materials, a change in pH and surface characteristics is noticed. OH^- ions (hydroxyl groups) could be easily formed in these reactions by hydroxide ions having positive holes. At neutral of high pH values, the majorly considered species are hydroxyl radicals whereas the holes (positive charges) are predominantly considered at low values of pH. OH^- ions are produced easier in the alkaline solution by oxidizing the hydroxide species available at the surface of semiconducting materials, thus count to increase in efficiency of dye degradation; mostly dyes have similar behavior like this. According to trends in this research field, dyes have been considered as a function of pH like bromo-cresol purple dye degrade better in acidic medium than in alkaline medium. The synthesis methods for the photo-degradation depends on the formation of products on the way molecules of dyes get adsorbed to the surface of catalysts, thus altering electronic properties.

It has been researched by the scientific teams that in dye degradation using photocatalytic technique the advent of adsorption on unmodified TiO_2 shows better for cationic dyes with positive charges than anionic dyes with negative charges. Therefore, explicitly the behavior of a dye and its pH give a huge impact on photocatalytic reaction efficiency. Azo dyes are cationic dyes having positive charge with low pH of around 6.8 and at higher pH value it gives anionic charges (negatively charged), hence effecting the adsorption of dye on the surface of materials. Lucidly, the effluents treatment should take into consideration two important points that are:

1. pH of the waste should not be neutral
2. Surface behavior of semiconducting materials gets influenced by the substances that will be mixed inside the water.

The dyes adsorbed on the semiconducting materials surface and the electron and hole charge pair separation state which will happen when electrical double layer forms due to charged species. The rate at which dye degrades in the photocatalysis phenomena mostly depends on pH of that dye and the reaction rate for that particular dye is maximum on the nature of pH, whether its acid or alkaline. Like OH^- radicals are formed in alkaline medium that increases the reaction rate of catalysis for dye degradation. We have given trends in materials for dye degradation showing influence of pH value of solution in Table 3.3. So, it is pivotal for us to understand the chemical behavior of these dyes and pH value for proper degradation using photocatalysis. As

Table 3.3 Collective data of the dye's degradation on varied effect of pH

Type of dye	Source of irradiation	Photo-catalyst	pH range	Optimum pH	Ref
Amido black 10B	UV	TiO_2	–	9.0	Sacco et al. (2012)
Acid Yellow	UV	TiO_2	–	3.0	Sacco et al. (2012)
Acid Orange	UV	WO_3-TiO_2	1.0–9.0	3.0	Mukhtish et al. (2013)
Methyl Orange	UV	TiO_2	2.0–10.0	8.0	Khan et al. (2017a)
Rhodamine B	UV	ZnO	2.0–12.0	12.0	Khataee et al. (2010)
Methyl Orange	Visible	Mg doped TiO_2	3.0–8.0	– -	Akpan and Hameed
Bromo-cresol Purple	UV	TiO_2	4.5 & 8.0	4.5	Rauf and Salman
Orange G	UV	Ac/TiO_2/ Sn	1.0–2.0	2.0	Rauf et al. (2011)
Methyl Orange	UV	Pt/TiO_2	2.5–11.0	2.5	Pandit et al. (2015)
Orange G	Visible	N-TiO_2	1.5–6.5	2.0	Elmorsi et al. (2010)
Acid Red B	UV	Ce-TiO_2	1.5–7.0	1.5	Gul and Yildlrln (2009)
Orange H	Solar	Zn-TiO_2	3.0–10.0	3.0	AlHamedi et al. (2009)
Methyl Red	Visible	1.5%Ni-3.0%Ag-TiO_2	3.0–10.0	4.0	Xie and Li (2006)
Methylene B	UV	N-TiO_2	3.0–9.0	Alkaline	Odling and Robertson (2017)
Congo Red	Visible	Chitosan/CdS	6.0–12.0	6.0	Khan et al. (2017b)
Malachite Green	Sun light	Ni/$MgFe_2O_4$	2.5–10.0	4.0	Xiong et al. (2010)
Indigo Carmine	UV	TiO_2	4.0–11.0	4.0	Rajesh et al. (2007)
Textile dye	UV	TiO_2	3.0–7.0	5.0	Dnyaneshwar (2017)
Solophenyl Red 3BI	UV	TiO_2	2.0–14.0	7.0	Avasarala et al. (2016)

(continued)

Table 3.3 (continued)

Type of dye	Source of irradiation	Photo-catalyst	pH range	Optimum pH	Ref
Methyl Orange	UV	Fe2/C/S dopedTiO_2	2.0–12.0	Acidic medium	Mahadwad et al. (2011)
Acid Orange 7	UV	WO_3	3.0–9.0	3.0	Mohammed et al. (2016)
Basic Yellow 28	Sun light	TiO_2	3.0–9.0	5.0	Nagaraja et al. (2012)
Methylene blue	UV	ZnO-TiO_2	1.0–6.0	2.0	Mehra and Sharma (2012)
Reactive Blue 4	UV	Anatase-TiO_2	3.0–13.0	3.0–7.0	Mo and Ching (1995)
Methylene blue	UV	ZnO	2.0–11.0	7.0	Iakandar et al. (2007)
Congo Red	UV	ZnO	5.0–10.0	8.0	Bavykin et al. (2006)
Reactive Red 2	UV	TiO_2	4.0–12.0	4.0–6.0	Koelsch et al. (2002)
Procion Yellow	UV	TiO_2	2.0–10.0	7.8	Morgan and Watson (2010)
Acid Orange 10	UV	TiO_2	1.0–11.0	3.0	Qamar et al. (2008)
Methyl Orange	UV	ZnO	2.0–10.0	Basic medium	Sclafani and Herrmann (1996)

we change the annealing temperatures, pH value varies thus changing the phase or crystal structure of the semiconducting materials. As compared with few examples given in Table 3.3 that TiO_2 degrades faster with good efficiency when at acidic medium. Hence, it degrades the chemically toxic dyes by absorbing H^+ ions on the surface of semiconducting materials; ultimately giving a new trend for wide range of dyes degraded alone with a single material. Based on the mechanism of chemical reaction, effects of pH on the dye degradation should be rationalized. Possibly, there are only 3 ways through which dyes can photocatalytically degraded in relation to pH value:

i. Involvement of positive charges "holes" directly in the oxidation reaction
ii. Hydroxyl species are directly attacked
iii. When electrons are excited to the higher energy bands (conduction bands) get directly reduced.

A detailed collection of the optimized values of pH for different photocatalysts and varied dyes is given in Table 3.2 with references.

3.4.2 Issues Still to Be Handled with New Trends

In the research literature most of the covered portions of this field as variables; source of light for dye degradation, its light intensity, the quantity and concentration of catalyst used, pH of the solution medium, initial dye concentration used for the detailed study of its degradation, time taken for irradiation in the whole dye degradation along with other cations present in the chemical reaction medium. Since ages, literature shows that research community has been following these variables avidly. So, it is considered that it is important to study these variables for getting the correct information for the effluents treatment using photocatalytic phenomenon in the dye industries like textile, paint and plastic. The main purpose of this chapter is to briefly compile and present all the past and ongoing trends in photocatalytic dye degradation for the waste treatment "effluents" and cleaning water bodies that are getting polluted by the industries, considering all parameters.

3.4.2.1 Kinetics of Dye Degradation

The kinetics of dye degradation using photocatalytic phenomenon by these semi-conducting catalysts have been mostly using the first order kinetics. In Table 3.4, we have clearly given detailed information of so many catalysts used in past and present that are trending for dye degradation with references along with its kinetic value "first order rate constant value". The purpose of this analysis is to determine some of the outcomes of treating these first order kinetic data in the degradation of the dye "pollutants" and other toxic effluents in the form of molecular species under the irradiation of light and semiconducting catalysts. The equation used for the first order kinetic in these chemical reaction phenomena could be written as

$$-\ln(C/C_o) = kt;$$

where C is quoted as the concentration, at t time in seconds, C_0 is the initial dye concentration and k represents the rate constant value, and adsorption constant, so on. Typical data of kinetic analysis is given for C-$CaIn_2O_4$ in Fig. 3.9 for dye degradation using photocatalytic phenomenon. This study could be concluded with pivotal points that the innate rate of catalytic reaction is source dependent and self-degradation in first order kinetics (Gouvea et al. 2000; Rong et al. 2015; Rochkind et al. 2014; Eyasu et al. 2013; Raji and Palanivelu 2016; Foletto et al. 2012; He et al. 2015). The lucid rate constant in the kinetic equation is a lumped parameter mostly consisting of those rates constant value, the other parallel chemical reaction rates that took place over semiconducting catalyst surface and several other unclear rates of dye degradation. So, the rate constant for catalytic reaction cannot be taken as a lucid measure for comparing this process that is taking place between two or more catalysts that are not similar; it will have some very serious misconceptions when doing comparative study. In these examples given in Table 3.4, the researchers have reported the chemical

Table 3.4 First order rate constant values for dyes degraded when adsorbed on photocatalyst

Photocatalyst	Dye	First order rate constant value	References
ZnO	Crystal Violet	$0.079\ min^{-1}$	Rauf and Salman
ZnO	Methyl Red	$0.014\ min^{-1}$	Rauf and Salman
ZnO	Basic Blue	$0.1\ min^{-1}$	Rauf and Salman
TiO_2	Crystal Violet	$0.026\ min^{-1}$	Rauf and Salman
TiO_2	Methyl Red	$0.008\ min^{-1}$	Rauf and Salman
TiO_2	Basic Blue	$0.045\ min^{-1}$	Rauf and Salman
SnO_2	Crystal Violet	$0.010\ min^{-1}$	Rauf and Salman
SnO_2	Methyl Red	$0.004\ min^{-1}$	Rauf and Salman
SnO_2	Basic Blue	$0.017\ min^{-1}$	Rauf and Salman
ZnO nanoflowers	Methyl Orange	$0.05485\ min^{-1}$	Pandit et al. (2015)
ZnO nanoflowers	Congo Red	$0.04611\ min^{-1}$	Pandit et al. (2015)
ZnO nanoflowers	Chicago Sky Blue	$0.003182\ min^{-1}$	Pandit et al. (2015)
ZnO nanoflowers	Eosin B	$0.002884\ min^{-1}$	Pandit et al. (2015)
TiO_2	Congo Red	$0.0102\ min^{-1}$	Mukhtish et al. (2013)
TiO_2	Methylene Blue	$0.0085\ min^{-1}$	Mukhtish et al. (2013)
TiO_2	Acid Orange 10	$0.0326\ min^{-1}$	Quang et al. (2017)
TiO_2	Acid Orange 12	$0.0269\ min^{-1}$	Quang et al. (2017)
TiO_2	Acid Orange 8	$0.0235\ min^{-1}$	Quang et al. (2017)
TiO_2	Amido Black 10B	$0.02083\ min^{-1}$	Singh et al. (2013)
TiO_2	Methyl Red	$0.0019\ min^{-1}$	Anpo et al. (1987)
$1.5Ag–0.75Ni/TiO_2$	Methyl Red	$0.0085\ min^{-1}$	Anpo et al. (1987)
$1.5Ag–1.5Ni/TiO_2$	Methyl Red	$0.0326\ min^{-1}$	Anpo et al. (1987)
$1.5Ag–3.0Ni/TiO_2$	Methyl Red	$0.0269\ min^{-1}$	Anpo et al. (1987)
$3.0Ag–1.5Ni/TiO_2$	Methyl Red	$0.0235\ min^{-1}$	Anpo et al. (1987)
$P25\text{-}TiO_2$	Amido Black 10B	$0.02083\ min^{-1}$	Singh et al. (2013)
ZnO	Rose Bengal	$\sim 4.5 \times 10^{-5}\ s^{-1}$	Abo-Farha (2010)
$CeCrO_3$	Fast Green	$\sim 4.4 \times 10^{-4}\ s^{-1}$	Xiaoqing et al. (2017)
TiO_2	RG 19	$\sim 4.69\ h^{-1}$	Quang et al. (2017)
TiO_2	AO 7	$\sim 2.07\ h^{-1}$	Quang et al. (2017)
P-25 TiO_2	Crystal Violet	$0.060\ min^{-1}$	Rauf and Salman
P-25 TiO_2	Methyl Red	$0.012\ min^{-1}$	Rauf and Salman
P-25 TiO_2	Basic Blue	$0.017\ min^{-1}$	Gul and Yildlrln (2009)
ZnO	Methyl Orange	$0.00029\ min^{-1}$	Dnyaneshwar (2017)

(continued)

Table 3.4 (continued)

Photocatalyst	Dye	First order rate constant value	References
ZnO	Rhodamine 6G	0.00027 min^{-1}	Dnyaneshwar (2017)
WO_3	Acid Orange 7	0.0225 min^{-1}	Morgan and Watson (2010)
TiO_2	Carmine dye	0.1456 min^{-1}	Li (2011)
TiO_2	Lissamine Green B	0.0165 min^{-1}	Zhiyong et al. (2007)
ZnO	Methylene Blue	0.0135 min^{-1}	Abo-Farha (2010)
TiO_2	Reactive Red	0.0325 min^{-1}	Min et al. (2015)
ZnO	Congo Red	0.0586 min^{-1}	Mezughi et al. (2014)

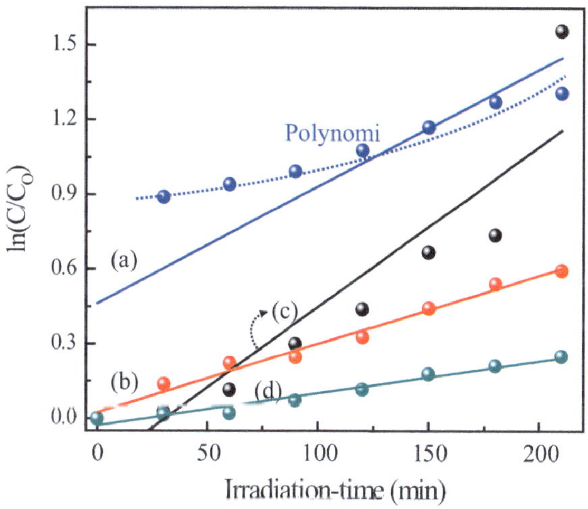

Fig. 3.9 The $\ln(C/C_0)$ versus t plots of a photocatalytic degradation of the CV dye using **a** the as-prepared and annealed $C\text{-}CaIn_2O_4$ at **b** 400 °C and **c** 600 °C for 2 h in air, and **d** a standard P25-TiO_2 powder dispersed in water by irradiating over UV–light

kinetic rate constant for the best catalyst as around 24 min^{-1} whereas the value for the degradation of chlorophenol rate constant of around 3.47 min^{-1} (here photon irradiated degradation is not present) (Li et al. 2017; Paridwala et al. 2017; Liu et al. 2015, 2016; Kumar and Sahare 2013).

Some of the precautions have been exercised when comparing two or more semi-conducting catalytic system phenomenon on the basis of the chemical kinetic constant rate values when used first order for catalysts reactions when dye gets adsorbed on the surface following its degradation. This may apply to certain reactions generally involving multiple steps that may succeed or precede the surface chemical reactions that are sometimes taken as first order reaction kinetics. Degradation of dye can foster several preconditions, out of which one is the adsorption of dyes on the surface of semiconducting catalyst surface and the equilibrium constant must be shown in the rate constant value calculated in data collected. There are so many preconditions that have been reported for the dye degradation, like the surface adsorption

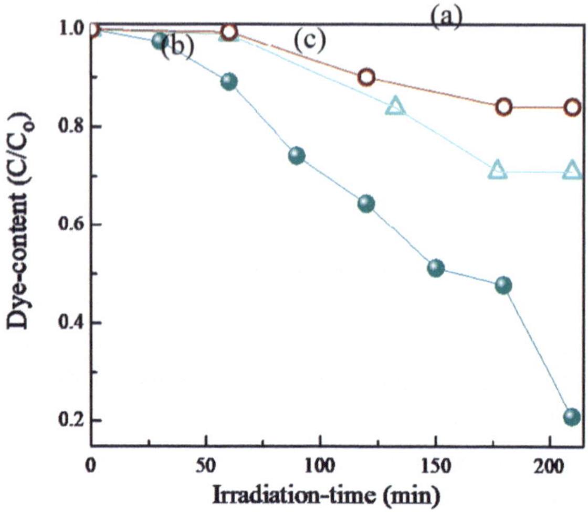

Fig. 3.10 A degradation of the CV-dye **a** before and **b** after dispersing in a photocatalytic C-$CaIn_2O_4$ sample-3 in water by irradiating over UV-light, with **c** degradation of sample (**b**) in absence of any irradiation

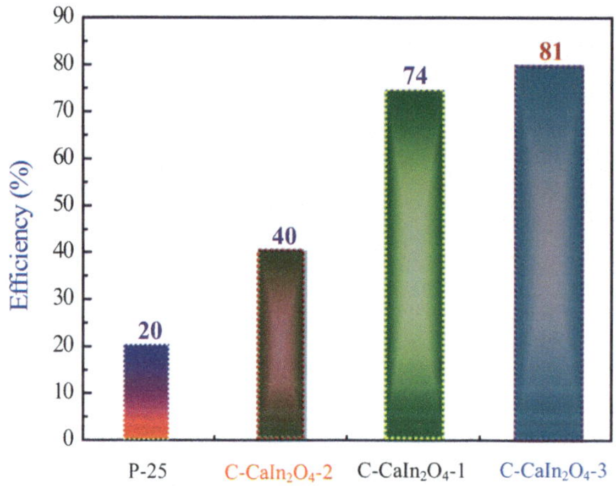

Fig. 3.11 A comparison of efficiencies of photocatalytic degradation of a CV dye in different catalysts dispersed in water and irradiating over UV-light

of organic dyes on catalyst and rate constant value. The surface values are capable of giving different orders of reactivity for adsorbents, and it should be considered when choosing the materials for treatment of effluents. BET adsorption isotherms are very famous Brunauer, Emmett, Teller isotherms as shown in Fig. 3.12 for determining surface area of the catalyst using adsorption curve; Table 3.5 shows the reported surface areas and pore sizes of annealed samples of core–shell C-$CaIn_2O_4$ nanostructures (*This is one of my recent works*). Keeping in mind the surface characteristics of these catalysts and nature of organic dyes have always been considered in this research field, not just this but even pH can change the surface charge carriers thus effecting the adsorption of dyes. Dyes are mostly chosen with surface as acidic or basic characteristics, showing anionic or cationic property in charged dyes. This

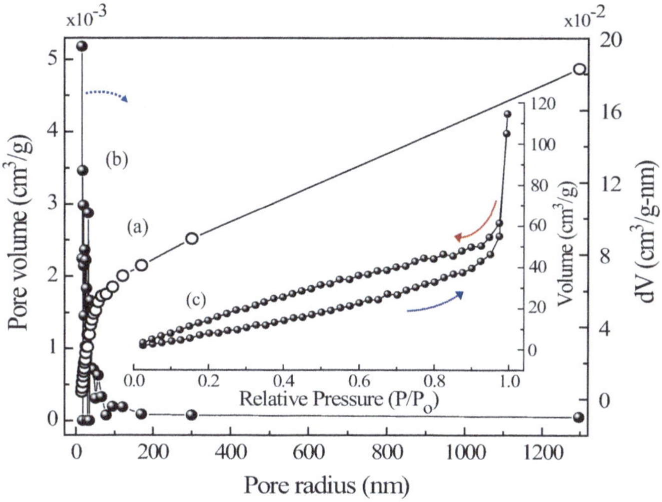

Fig. 3.12 **a** Commutative V_c and **b** its derivative plotted over r-values a C-CaIn$_2$O$_4$ sample, with **c** N_2 adsorption–desorption isotherms

Table 3.5 Surface area and pore size obtained for the as-prepared and annealed C-CaIn$_2$O$_4$ samples using the BET analysis

S. No	Sample	Surface area ($m^{2-}g^{-1}$)	Pore radius (nm)
1	As-prepared	22.312	1.5
2	Annealed at 400 °C for 2 h in air	26.526	1.5
3	Annealed at 600 °C For 2 h in air	47.996	1.5

aspect has not yet been tackled when treating the chemical kinetics of photocatalytic dye degradation on semiconducting materials. In recent studies, the details on adsorption of dyes and its consequences on the kinetic behavior of dye degradation have not been touched or linked yet; this comparison will benefit for future.

3.4.2.2 Loading of Semiconducting Catalyst

Some of the observations that have been checked and recorded in the recent trends of photocatalytic degradation of dyes is that as the weight of dyes load increases, the overall efficiency to degrade dye via adsorption on the catalyst surface decreases. So, it's not an easy one to find out as the surface area exposed to dye adsorption of the catalyst will sometimes not be equivalent to quantity of catalyst present in the solution. The saturation point for dye adsorption on the surface of the semiconducting catalyst is also checked, as beyond certain extent of limit it will stop the photocatalytic phenomenon; most of the research studies report only few kg of around 3–4 g per liter of dye in solution. Industries should now put a control on the effluents that are possibly destroying water bodies and environment, as already mentioned that the weight extent of dye adsorption on the surface of catalyst is still an ongoing limitation. This could be marked as the control of pollution and effluents by the authorities for our environmental safety and protection.

3.4.2.3 Light Wavelength and Its Intensity of Irradiation Effect

The usage of solar light radiation has always been less expensive and less toxic to our lives and environment. The effect of photo light irradiation on the chemical photocatalytic phenomenon has been demonstrated in three ranges:

i. At low intensity, the rate of reaction increases with the light irradiation
ii. At medium and high intensities, the kinetic rate is not dependent on the light intensity.

Overall, the degradation rate for UV–visible light as a source of irradiation is better than the solar light radiation. Secondly, the charge pairs i.e., electron–hole formation & its recombination would be differently done under UV–visible light source. Thus, light wavelength is another major point in affecting the charge pair formation and recombination rate.

3.4.2.4 Dye Degradation Mechanism

Photocatalyst is one of the best methods that can be applied for several applications for example, organic pollutants degradation from wastewater, production of hydrogen gas, purifying of air and other antibacterial activity. In 1972, Fujishima and Honda discovered the beginning of a new era in heterogeneous photocatalysis as a photocatalytic splitting of water on a TiO_2 electrode as noticeable by the phenomena. The wavelength or light (photon) energy and the catalyst are main characteristics on which photocatalytic reaction are dependent. In general, as a catalyst which performs as sensitizer especially, semiconducting materials due to their electronic structure for the irradiation of light stimulated redox process, which is defined by a filled valence band and a vacant conduction band as shown in Fig. 3.13.

When photons in terms of the light energy falls (or interacts) on the surface of a material of energy of incident radiation if equivalent or more than the bandgap energy of the semiconducting material, the electrons present in valence band are excited to the conduction band of the semiconductor. Holes would be charges left in the valence band; these holes can oxidize donor molecules and react with water molecules to generate hydroxyl radicals. The conduction band electrons react with dissolved oxygen species to form superoxide ions. These electrons induce the redox reactions. These holes and electrons could undergo successive oxidation and reduction reactions with any species, which might be adsorbed on the surface of the semiconductor to give

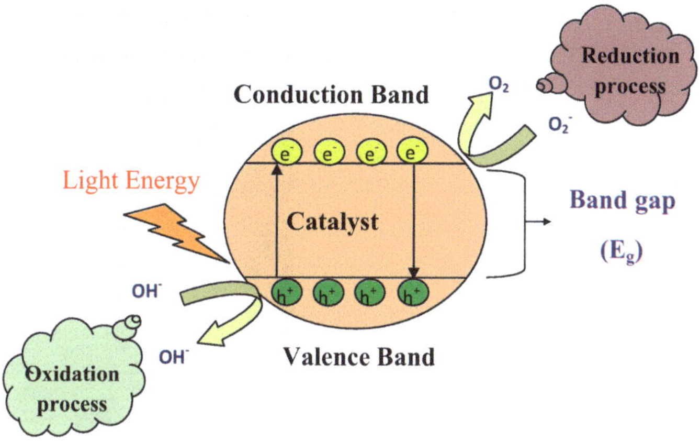

Fig. 3.13 Schematic representation of semiconductor photocatalytic mechanism

the necessary products. The photocatalytic performance of C-$CaIn_2O_4$ powders (as-prepared and annealed at 400 and 600 °C) were studied along with the benchmark Degussa P25 TiO_2 powder by considering crystal violet (CV) as the target dye. Photocatalytic studies were done under UV (24 W, intensity 400 lx, $\lambda < 400$ nm) irradiation at the sample surface under stirring condition.

Aqueous solutions of crystal violet with a concentration 5×10^{-5} M with a photocatalyst loading of 1 g/L were taken. The dispersions were equilibrated for dye adsorption on the photocatalyst surface for 30 min. During the degradation period, the dispersions were collected at regular intervals of thirty minutes in case of UV light irradiation as shown in Fig. 3.14. To note down the extent of the degradation, the samples were centrifuged at 6000 rpm for 10 min to collect the so far degraded dye solution. Absorption spectra of all the collected dye solutions were taken using Perkin-Elmer UV–visible spectrophotometer Lambda 750.

3.5 Summary

In this chapter, we clearly explained the recent developments in the photocatalytic performance of different semiconducting materials for varied dyes that have been thoroughly studied by researchers in terms of optimized parameters like, pH, concentration of solution, quantity of catalyst employed, cationic and anionic form of dye, bandgap of semiconducting materials and source of light. The research in this field has increased immensely in recent years and is only going to grow as trend says. The reports of these studies mentioned in our literature mainly deal with some common variables as given above. In end the summarization of the results in this detailed study, this work deals with several aspects which we can explore in following way:

1. Insufficiently treating the chemical kinetic information collected in terms of the pseudo first order reaction rate. Hence, we have to priorly take in consideration the adsorption characteristics in this phenomenon.

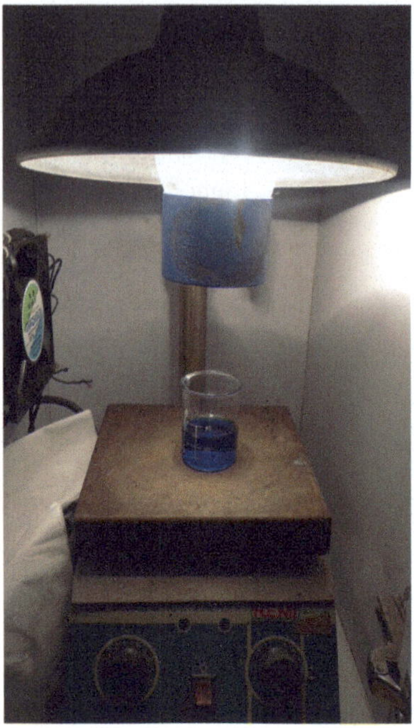

Fig. 3.14 Setup of photocatalytic mechanism

2. The natural effect in this research is the adequate quantity of catalyst, and identification of those active sites and surface area number density that will be adsorbed on the surface of semiconductor.
3. The dye degradation is partially or sometimes completely dependent on the behavior of oxidizing agent that has been incorporated; choice of oxidizing agent is under our control for the redox potential of the species.
4. The dye adsorption on the semiconducting surface decides that whether the pH of the solution is acidic or basic; thus, the process of degradation gets clear. This variation in pH was been demonstrated for the semiconducting surface in response to the pH value.

We can conclude in the end of this book chapter, that all studies done in this photocatalytic field of dye degradation for waste water treatment or effluents treatment could be appropriately be extended to other detailed studies in future thus covering more economical and global significances.

References

Abo-Farha SA (2010) Photocatalytic degradation of monoazo and diazo dyes in waste water on nanometer sized TiO_2. Researcher 2(7):1–20

Ajmal A, Majeed I, Malik RN, Idriss H, Nadeem MA (2014) Principles and mechanism of photocatalytic dye degradation on TiO_2 based photocatalysts: a comparative overview. RSC Adv 4:370003–437026

Akpan UG, Hameed BH Parameters affecting the photocatalytic degradation of dyes using TiO2-based photocatalysts: a review. J Hazard Mater 170:520–529

Alaton IA, Balcioglu IA (2001) Photochemical and heterogeneous photocatalytic degradation of waste vinylsulphonate dyes: a case study with hydrolyzed reactive black 5. J Photochem Photobiol A Chem 141:247–254

AlHamedi FH, Rauf MA, Asraf SS (2009) Degradation of Rhodamine B in the presence of UV/H_2O_2. Desalination 239:159–166

Ameta R, Sharma S, Sharma S, Gorana Y (2015) Visible light induced photocatalytic degradation of toluidine blue-O using molybdenum doped titanium dioxide. Eur J Adv Engineer Technol 2:95–99

Anpo M, Shima T, Kodama S, Kubokawa Y (1987) Photocatalytic hydrogenation of propyne with water on small particle titania: size quantization effects and reaction intermediates. J Phys Chem 91:4305–4310

Apostolescu GA, Cernatescu C, Cobzaru C, Tataru-Farmus RE, Apostolescu N (2015) Studies of the photocatalytic degradation of organic dyes using CeO_2-ZnO mixed oxides, 14:415–420

Avasarala BK, Tirukkovalluri SR, Bojja S (2016) Magnesium doped titania for photocatalytic degradation of dyes in visible light. J Environ Anal Toxicol 6:1–8

Bakardiieva S, Subrt J, Stengl V, Dianez MJ, Sayagues MJ (2005) Photoactivity of anatase-rutile TiO_2 nanocrystalline mixtures obtained by heat treatment of homogeneously precipitated anatase. Appl Catal B 58:193–202

Baran W, Makowski A, Wardas W (2003) The influence of $FeCl_3$ on the photocatalytic degradation of dissolved azo dyes in aqueous TiO_2 suspensions. Chemosphere 53:87–95

Baran W, Makowski A, Wardas W (2008) The effect of UV radiation adsorption of cationic and anionic dye solutions on their photocatalytic degradation in the presence of TiO_2. Dyes Pigm 76:226–230

Bauer C, Jacques P, Kalt A (2001) Photooxidation of an azo dye induced by visible light incident on the surface of TiO_2. J Photochem Photobiol A Chem 140:87

Bavykin DV, Friedrich JM, Walsh FC (2006) Protonated titanates and TiO_2 nanostructured materials: synthesis, properties and applications. Adv Mater 18:2807–2824

Bhati I, Pinki BP, Suresh CA (2010) Photocatalytic degradation of fast green using nanosized $CeCrO_3$. Maced J Chem Chem Eng 29:195–202

Birte M, Lennart M, Olaf W (2017) Photocatalytic degradation of toluene, butyl acetate a limonene under UV and visible light with titanium dioxide-graphene oxide as photocatalyst. Environs 4:1–9

Bubacz K, Choina J, Dolat D, Morawski AW (2010) Methylene blue and phenol photocatalytic degradation on nanoparticles of anatase TiO_2 Pol. J Environ Stud 19:685–691

Colmenares JC, Luque R, Campelo JM, Colmenares F, Karpinski Z, Romero AA (2009) Nanostructured photocatalysts and their applications in the photocatalytic transformation of lignocellulosic biomass: an overview. Materials (basel) 2:2228–2258

Colon G, Hidalgo MC, Munuera G, Ferino I, Cutrufello MG, Navio JA (2006) Structural and surface approach to the enhanced photocatalytic activity of sulfated TiO_2 photocatalyst. Appl Catal B 63:45–59

Dharmarajan P, Sabastiyan A, K< SM, Titus S, Muthukumar C (2013) Photocatalytic degradation of reactive dyes in effluents employing copper doped titanium dioxide nanocrystals and direct sunlight. Chem Sci Trans 2:1450–1458

Dhatshabamurthi P, Subhash B, Shanthi M (2015) Investigation on UVA light photocatalytic degradation of azo dye in the presence of CdO/TiO_2 coupled semiconductor. Mater Sci Semicond Process 35:22–29

Di Paola A, Bellardita M, Palmisano L (2013) Brookite, the least known TiO_2 photo-catalyst. Catalysts 3:36–73

Dnyaneshwar R (2017) Photocatalytic degradation of dyes in water by analytical reagent grade photocatalysts—a comparative study. Drink Water Eng Sci 10:109
Elahee K (2010) Heat recovery in the textile dyeing and finishing industry: lessons from developing economies. J Energy Southern Africa 21:9–15
Elena CP, Felix E, Fernando V (2017) Computational nanotechnology to predict photocatalysis of titania nanoparticles in cement-based materials. IEEE Conf Proc 17:208–123
Elmorsi TM, Riyad YM, Mohamed ZH, Abd El Bary HM, Bary E (2010) Decolorization of mordant red 73 azo dye in water using H_2O_2/UV and photo-Fenton treatment. J Hazard Mater 174:352–358
Eyasu A, Yadav OP, Bachheti RK (2013) Photocatalytic degradation of methyl orange dye using Cr-doped ZnS nanoparticles under visible radiation. Int J Chemtech Res 5:1452–1461
Foletto EL, Battiston S, Collazzo GC, Bassaco MM, Mazutti MA (2012) Degradation of leather dye using CeO_2-SnO_2 nanocomposite as photocatalyst under sunlight. Water Air Soil Pollution 223:5773–5779
Fox MA, Dulay MT (1993) Heterogeneous photocatalysis. Chem Rev 93:341
Gouvea CA, Wypych F, Moreaes SG, Duran N, Nagat N, Peralta-Zamora P (2000) Semiconductor-assisted photo-catalytic degradation of reactive dyes in aqueous solution. Chemosphere 40:433–440
Guettai N, Amar HA (2005) Photocatalytic oxidation of methyl orange in presence of titanium dioxide in aqueous suspension Part II Kinetic study. Desalination 185:439–448
Guillard C, Diadier J, Monnet C, Dussaud J, Malato S, Blanco J, Maldonano MI, Herrmann JM (2003) Solar efficiency of new deposited titania photo-catalyst: Chlorophenol, pesticide and dye removal applications. Appl Catal b: Environ 46:319–332
Gul S, Yildlrln O (2009) Degradation of reactive red 194 and reactive yellow 145 azo dyes by O_3 and H_2O_2/UV-C process. Chem Eng J 155:684–690
Hashimoto K, Irie H, Fujishima A (2005) TiO_2 photocatalysis: a historical overview future prospect. Jpn J Appl Phys 44:8269–8285
He Y, Zhang L, Fan M, Wang X, Walbridge ML, Nong Q, Wu Y, Zhao L (2015) Z-scheme SnO_{2-x}/g-C_3N_4 composite as an efficient photocatalyst for dye degradation and photocatalytic CO_2 reduction. Sol Energy Mater Sol Cells 137:175–184
Hernandez-Alonso MD, Fresno F, Suareza S, Coronado JM (2009) Development of alternative photocatalysts to TiO_2: challenges and opportunities. Energy Environ Sci 2:1231–1257
Hoffman MR, Martin ST, Choi W, Bahnemannt DW (1995) Environmental applications of semiconductor photocatalysis. Chem Rev 95:69–96
Huang M, Xu C, Wu Z, Huang Y, Lin J, Wu J (2008) Photocatalytic decolorization of methyl orange solution by Pt modified TiO_2 loaded on natural zeolite. Dyes Pigm 2:327–334
Iakandar F, Nandiyanto ABD, Yun KM, Hogan CJ, Biswas P (2007) Enhanced photocatalytic performance of brookite TiO2 macroporous particles prepared by spray drying with colloidal templating. Adv Mater 19:1408–1412
Ioannis KK, Triatafyllos A (2004) TiO_2-assisted photocatalytic degradation of azo dyes in aqueous solution: kinetic and mechanistic investigations: a review. Appl Catal B 49:1–14
Kamal S, Afshin M, Behzad S, Borhan M, Fardin G (2014) Investigation of photocatalytic degradation of reactive balck 5 dyes using ZnO-CuO nano-composite. Zanki J Med Sci 15:66–74
Karunakaran C, Magesan P, Gomathisankar P, Vinayagamoorthy P (2013) Photocatalytic degradation of dyes by Al_2O_3-TiO_2 and ZrO_2-TiO_2 nanocomposites. Mater Sci Forum 734:325–333
Khan MR, Kurney ASW, Fahmida G (2017a) Parameters affecting the photocatalytic degradation of dyes using TiO_2: a review. Appl Water Sci 7:1569–1578
Khan MR, Kurny ASW, Eahmida G (2017b) Parameters affecting the photo-catalytic degradation of dyes using TiO_2: a review. Appl Water Sci 7:1569–1578
Khataee AR, Pons MN, Zahraa O (2009) Photocatalytic degradation of three azo dyes using immobilized TiO_2 nanoparticles on glass plates activated by UV light irradiation: influence of dye molecular structure. J Hazard Mater 168:451–457

Khataee AK, Kasiri MB (2010) Photocatalytic degradation of organic dyes in the presence of nanostructured titanium dioxide: influence of the chemical structure dyes. J Mol Catal Chem 328:8–26

Kirupavasam EK, Allen GR (2012) Photocatalytic degradation of amido black-10B catalyzed by carbon doped TiO_2 photocatalyst. Int J Green Chem Bioprocess 2(3):20–25

Koelsch M, Cassaignon S, Guillemoles JF, Jolivet JP (2002) Comparison of optical and electrochemical properties of anatase and brookite TiO_2 synthesized by the sol-gel method. Thin Solid Films 403:312–319

Kumar S, Sahare PD (2013) Photocatalytic activity of bismuth vanadate for the degradation of organic compounds. NANO 8:1350007

Kurny ASW, Fahmida G (2017) Parameters affecting the photocatalytic degradation of dyes using TiO_2: a review. Appl Water Sci 7:1569–1578

Li H (2011) Enhanced photocatalytic activity of electrospun TiO_2 nanofibers with optimal anatase/rutile ratio. J Am Ceramic Soc 94:3184–3187

Li Q, Guan Z, Wu D, Zhao X, Bao S, Tian B, Zhang J (2017) Z scheme BiOCl-Au-CdS heterostructure with enhanced sun light driven photocatalytic activity in degrading water dyes and antibiotics. ACS Sustain Chem Eng 5:6958–6968

Lin H, Huang CP, Li W, Ni C, Shah SI, Tseng Y-H (2006) Size dependency of nanocrystalline TiO_2 on its optical property and photocatalytic reactivity exemplified by 2-cholorophrnol. Appl Catal B 68:1–11

Liu Y, Wang R, Yang Z, Du H, Jiang Y, Shen C, Liang K, Xu A (2015) Enhanced visible light photocatalytic activity of Z-scheme graphitic carbon nitride/oxygen vacancy-rich zinc oxide hybrid photocatalysts. Chin J Catal 36:2135–2144

Liu P, Liu Y, Ye W, Ma J, Gao D (2016) Flower like N-doped MoS_2 for photocatalytic degradation of RhB by visible light irradiation. Nanotechnology 27:225403

Long X, Yan T, Hu T, Gong X, Li H, Chu Z (2017) Enhanced photocatalysis of g-C_3N_4 thermally modified with calcium chloride. Catal Lett 147:1922–1930

Mahadwad OK, Parikh PA, Jasra RV, Patil C (2011) Photocatalytic degradation of reactive black-5 dye using TiO_2 impregnated ZSM-5 Bull. Mater Sci 34:551–556

Mahmoud MA, Poncheri A, Badr Y, Abd El Wahed MG (2009) Photocatalytic degradation of methyl red dye. South Afr. J. Sci. 105:299–303

Marinovic V, Ljubas D, Curkovic L (2017) Effects of concentration and UV radiation wavelengths on photolytic and photocatalytic degradation of azo dyes aqueous solutions by sol-gel TiO_2 films. Holistic Approach Environ 7:3–14

Mehra M, Sharma TR (2012) Photo-catalytic degradation of two commercial dyes in aqueous phase using photo catalyst TiO_2. Adv Appl Sci Res 3:849–853

Mehta R, Surana M (2013) Photodegradation of dye acid orange 67 by titanium dioxide in the presence of visible light and UV light. Res Rev 2:1216

Meng Z, Juan Z (2008) Wastewater treatment by photocatalytic oxidation of nano-ZnO. Global Environ Policy Jpn 12:1–9

Mezughi K, Tizaoui C, Ma'an FA (2014) effect of TiO_2 concentration on photocatalytic degradation of reactive orange 16 dye (RO16). Adv Environ Biol 8:692–695

Mills A, Davies RH, Worsley D (1993) Water purification by semiconductor photocatalysis. Chem Soc Rev 22:417–425

Min OM, Ho LN, Ong SA, Wong YS (2015) Comparison between the photocatalytic degradation of single and binary azo dyes in TiO_2 suspensions under solar light irradiation. J Water Reuse Desalin 5:579–591

Mo S-D, Ching WY (1995) Electronic and optical properties of three phases of titanium dioxide: rutile, anatase and brookite. Phys Rev B Condens Matter 51:13023–13032

Mohammed A, Kapoor K, Mobin SM (2016) Improved photocatalytic degradation of organic dyes by ZnO-nanoflowers. Chem Select 1:3483–3490

Morgan BJ, Watson GW (2010) Intrinsic n-type defect formation in TiO_2: a comparison of rutile and anatase from GGA plus U calculations. J Phys Chem 114:2321–2328

Mukhtish MZB, Najnin F, Rahman MM, Uddin MJ (2013) Photocatalytic degradation of different dyes using TiO_2 with high surface area: a kinetic study. J Sci Res 5(2):301–314
Nagaraja R, Nagaraju KCRG, Nagabhushana BM (2012) Photocatalytic degradation of Rhodamine B dye under UV/solar light using ZnO nano powder synthesized by solution combustion route. Powder Technol 215:91–97
Narde SB, Lanjewar RB, Gadegone SM, Lanjewar MR (2017) Photocatalytic degradation of azo dye congo red using $Ni_{0.6}Co_{0.4}Fe_2O_4$ as photocatalyst. Pharma Chem 9:115–120
Neppolian B, Sakthivel S, Arabindo B, Palanichamy M, Murugesan V (1999) Degradation of textile dye by solar light using TiO_2 and ZnO photocatalysts. J Environ Sci Health Part A Tox Hazard Subst Environ Eng 34:1829–1838
Nguyen-Phan TD, Kim EJ, Hahn SH, Kim W-J, Shin EW (2011) Synthesis of hierarchical rose bridal bouquet- and humming-top like TiO2 nanostructures and their shape-dependent degradation efficiency of dye. J Colloid Interface Sci 356:138–144
Nosaka Y, Nosaka AY (2017) Generation and detection of reactive oxygen species in photocatalysis. Chem Rev 117:11302–11336
Odling G, Robertson N (2017) Silar BiOI-sensitized TiO2 films for visible-light photocatalytic degradation of Rhodamine B and 4- cholophenol. Chem Phys Chem 18:728–735
Ohno T, Tsubota T, Toyofuku M, Inaba R (2004) Photocatalytic activity of a TiO2 photo-catalyst doped with C^{4++} and S^{4++} ions having a rutile phase under visible light. Catal Lett 98:255–258
Ohtani B, Ogawa Y, Nishimoto S-I (1997) Photocatalytic activity of amorphous-anatase mixture of titanium (IV) oxide particles suspended in aqueous solutions. J Phys Chem 101:19
Pandit VK, Arbuj SS, Pandit YB, Naik SD, Rane SB, Mulik UP, Gosavic SW, Kale BB (2015) Solar light driven dye degradation using novel organic-inorganic (6,13-pentacenequinone/TiO_2) nanocomposite. RSC Adv 5:10326–10331
Paridwala JM, Patel FJ, Patel SS (2017) Photocatalytic degradation of RB21 dye by TiO_2 and ZnO under natural sunlight: microwave irradiation and UV reactor. Int J Adv Res Engineer Technol 8:8–16
Poulios I, Aetopoulou I (1999) Photocatalytic degradation of the textile dye reactive orange 16 in the presence of TiO_2 suspensions. Environ Technol 20:479–487
Poulios I, Avrans A, Rekliti E, Zouboulis A (2000) Photocatalytic oxidation of Auramine O in the presence of semiconducting oxides. J. Chem. Biotechnol. 75:205–212
Qamar M, Yoon CR, Oh HJ, Lee NH, Park K, Kim DH (2008) Preparation and photocatalytic activity of nanotubes obtained from titanium dioxide. Catal Today 131:3–14
Quang MD, Nguyen ND, Nguyen QH (2017) Photocatalytic degradation of azo dye (Methyl Red) in water under visible light using AgNi/TiO_2 synthesized by and γ-irradiation method. Int J Environ Agric Biotechnol 2:529–538
Raheem Z, Hameed AM (2015) Photocatalytic degradation for methylene blue dye using magnesium oxide. Int J Basic Appl Sci 4:81–83
Rajesh JT, Praveen KS, Ramachandra GK, Raksh VJ (2007) Photocatalytic degradation of dyes and organic contaminants in water using nanocrystalline anatase and rutile TiO_2. Sci Technol Adv Mater 8:455–462
Raji J, Palanivelu K (2016) Semiconductor coupled solar photo-Fenton's treatment of dyes and textile effluent. Adv Environ Res 5:61–77
Rauf MA, Meetani MA, Hisaidee S (2011) An overview on the photocatalytic degradation of azo dyes in the presence of TiO_2 doped with selective transition metals. Desalination 276:13–27
Rauf MA, Salman A S Fundamental principles and application of heterogenous photocatalytic degradation of dyes in solution. Chem Engineer J 151:10–18
Reutergarth LB, Iangpashuk M (1997) Photocatalytic decolorization of reactive azo dye: a comparison between TiO_2 and CdS photocatalysts. Chemosphere 35:585–596
Rochkind M, Pasternak S, Paz Y (2014) Using dyes for evaluating photocatalytic properties: a critical review. Molecules 20:88–110

Rong X, Qiu F, Zhao H, Yan J, Zhu X, Yang D (2015) Fabrication of single-layer graphitic carbon nitride and coupled systems for the phot-catalytic degradation of dyes under visible-light irradiation. Eur J Inorg Chem 8:1359–1367

Rupa AV, Manikandan D, Divakar D, Revathi S, Preethi ME, Shanthi K, Sivakumar T (2015) Photocatalytic degradation of tetrazine dye using TiO_2 catalyst: salt effect kinetic studies. Indian J Chem Technol 14:71–78

Saber A, Rasul MG, Wayde M, Richard B, Hashib MA (2011) Advances in heterogeneous photocatalytic degradation of phenols and dyes in waste water: a review. Water Air Soil Pollut 215:3–29

Sacco O, Stoller M, Vincenzo V, Ciambelli P, Chianese A, Sannino D (2012) Photocatalytic degradation of organic dyes under visible light on N-doped TiO_2 photo-catalysts. Int J Photoenergy 2012:1–8

Salem MA, Shaban SY, Ismail SM (2015) Photocatalytic degradation of acid green 25 using ZnO and natural sunlight. Int J Energy Technol Adv Eng 5:439–443

Sarkhanpour R, Tavakoli O, Ghiyasi S, Saeb MR, Borja R (2017) Photocatalytic degradation of a chemical industry wastewater: search for higher efficiency. J Residuals Sci Technol 14:44–58

Sclafani A, Herrmann JM (1996) Comparison of the photoelectronic and photocatalytic activities of various anatase and rutile forms of titania in pure liquid organic phases and in aqueous solutions. J Phys Chem 100:13655–13661

Sharma S, Ameta RRKM, Suresh CA (2013) Photocatalytic degradation of rose Bengal using semiconducting zinc sulphide as the photocatalyst. J Serb Chem Soc 78:897–905

Simovic B, Poleti D, Golubovic A, Matkovic A, Scepanovic M, Babic B, Brankovic G (2017) Enhanced photocatalytic degradation of RO16 dye using Ag modified ZnO nano-powders prepared by the solvothermal method Process. Appl Ceramics 11:27–38

Singh P, Mondal K, Sharma A (2013) Reusable electrospun mesoporous ZnO nanofiber mats for photocatalytic degradation of polycyclic aromatic hydrocarbon dyes in wastewater. J Colloid Interface Sci 394:208–215

Sun J, Wang X, Sun R, Sun S, Qiao L (2006) Photocatalytic degradation and kinetics of orange G using nano-sized Sn (IV)/TiO_2/AC photocatalyst. J Mol Catal Chem 260:241–246

Sun J, Qiao L, Sun S, Wang G (2008) Photocatalytic degradation of orange G on nitrogen-doped TiO_2 catalysts under visible light and sunlight irradiation. J Hazard Mater 155:312–319

Tang WZ, An H (1995) UV/TiO_2 photocatalytic oxidation of commercial dyes in aqueous solutions. Chemosphere 31:4158–4170

Tesfay WG, Manjunatha P, Rani MP (2015) Review on the photocatalytic degradation of dyes and antibacterial activities of pure and doped-ZnO. Int J Sci Res 4:2252–2264

Thuy-Duong Nguyen P, Eui Jung K, Sung Hong H, Woo Jae K, Eun Woo S (2011) Synthesis of hierarchical rose bridal bouquet- and humming-top-like TiO_2 nanostructures and their shape-dependent degradation efficiency of dye. J Colloid Interface Sci 356:138–144

Tiwari B, Ram S (2019) Biogenic synthesis of graphitic carbon nitride for photocatalytic degradation of organic dyes. ACS Omega 6:16298–16307

Tiwari B, Ram S, Banerji P (2018) Biogenic synthesis of tunable core-shell C-$CaIn_2O_4$, interface bonding, conductive network channels and tailored dielectric properties. ACS Sustain Chem Eng 6:16298–16307

Tunesi S, Anderson M (1991) Influence of chemisorption on the photodecomposition of salicylic acid and related compounds using suspended titania ceramic membranes. J Phys Chem 95:3399–3405

Wang K-H, Hsieh Y-H, Wu C-H, Chang C-Y (2000) The pH and anion effects on the heterogeneous photocatalytic degradation of omethylbenzoic acid in TiO_2 aqueous suspension. Chemosphere 40:389–394

Wei C-H, Tang X-H, Liang J-R, Tan SY (2007) Preparation, characterization and photocatalytic activities of boron and cerium codoped TiO_2. J Environ Sci (china) 19:90–96

Xiaoqing C, Zhansheng W, Dandan L, Zhenzhen G (2017) Preparation of ZnO photocatalyst for the efficient and rapid photo-catalytic degradation of azo dyes. Nanoscale Res Lett 12:143

Xie YB, Li XZ (2006) Interactive oxidation of photo-electro catalysis and electro-Fenton for azo dye degradation using TiO_2-Ti mesh and reticulated vitreous carbon electrode. Mater Chem Phys 95:39–50

Xiong Z, Zhang LL, Ma J, Zhao XS (2010) Photocatalytic degradation of dyes over graphene-gold nanocomposites under visible light irradiation. Chem Commun (camb) 46(33):6099–6101

Ye M, Zhang Q, Hu Y, GE j, Lu Z, He L, Chen Z, Yin Y (2010) Magnetically recoverable core-shell nanocomposites with enhanced photocatalytic activity. Chemistry 16:6243–6250

Zangi ZM, Ganjidoust H, Ayati B (2017) Analysis of photocatalytic degradation of zao dyes under sunlight with response surfsce method. Desalination Water Treat 63:262–274

Zhang T, Oyama T, Horikoshi S, Hidaka H, Zhao J, Serpone N (2001) Photocatalysed N-demethylation and degradation of methylene blue in titania dispersions exposed to concentrated sunlight, solar energy. Mater Solar Cells 73:287–303

Zhiyong Y, Bensimon M, Sarria V, Stolitchnov I, Jardim W, Laub D, Mielczarski E, Mielczarski J, Kiwi-Minsker L, Kiwi J (2007) $ZnSO_4$-TiO_2 doped catalyst with higher activity in photocatalytic processes. Appl Catal B 76:185–195

Chapter 4
Role of Doped Semiconductors in the Catalytic Activity

Ashish Gaurav, Ananta Paul, and Sushma Dave

Abstract There are several substances in today's world that are neither good conductors (metals) nor insulators (glass). At normal temperature, semiconductors are described as materials with a crystalline structure and very few free electrons. Examples of semiconductor include Silicon, carbon, germanium etc. At room temperature, the semiconductors act like an insulator with its resistivity which lies between that of conductor and insulator and most importantly, the conductivity can be controlled by addition of various impurities. Mostly, the semiconducting materials are crystalline in nature with rarely found semiconductor in form of amorphous and liquid. The electronic industries across the globe includes transistors, solar cells, light-emitting diodes (LEDs), and digital and analog integrated circuits are dependent on semiconductor technology. The diverse property of the semiconductor materials lies at the quantum level which includes the motions of the building blocks like electrons and holes in the crystal and lattice. Electrical conductivity being opposite to that of a metal and its conductivity can be tailored by adding impurities commonly known as doping or by applying different mechanism like electrical fields or lights. Conductivities in semiconductors are due to movement of free electron (n-type) and holes (p-type) and are sensitive to temperature, illumination, magnetic field and impurity atoms introduced. Controlling the concentration and location of p- and n-type dopant under a precise condition is utmost important for electronic devices.

A. Gaurav (✉)
Graduate Institute of Photonics and Optoelectronics, College of Electrical Engineering and Computer Science, National Taiwan University, Taipei City, Taiwan
e-mail: itsgaurav.bitm@gmail.com

A. Paul
Department of MEMS, IIT Mumbai, Mumbai, India

S. Dave
Department of Chemistry, JIET, Jodhpur, Rajasthan, India

S. Dave and J. Das (eds.), *Trends and Contemporary Technologies for Photocatalytic Degradation of Dyes*, Environmental Science and Engineering,
https://doi.org/10.1007/978-3-031-08991-6_4

4.1 Introduction

Nowadays semiconductor materials are used in every sector of modern technology. In technical purpose the high temperature materials are widely used (Rahman 2014).

Once compared to organic dyes and fluorescent proteins, semiconductor nanocrystals have unique physical and photoelectrical properties, such as high emission quantum yields, a narrow and symmetric emission peak with a broad excited wavelength, size-controlled fluorescence properties, and high optical stabilities (Khlyustova et al. 2020; Ilkme and Soylu 2021). But with the development of semiconductor nanocrystals, there is existence of many problems and challenges. In recent years, researchers found that doping impurities into these semiconductor materials could impact the various electrical, optical, and magnetic properties of the materials (Khlyustova et al. 2020; Yang and Lee 2017). So, intense studies have been going around the globe on impurity atoms such as Mn, Cu, Co, and many more doped semiconductor materials for their new enhanced properties and its application in various fields. Numerous techniques have been adopted till date to synthesize high quality doped semiconductor nanocrystals. Basically, doped semiconductor nanocrystals are usually synthesized by forming particular molecular precursors containing the specific constituent elements.

The introduction of impurities or defects into semiconductor lattices is a key method for influencing electrical conductivity, and it can also have a significant impact on the semiconductor's optical, luminescent, magnetic, and other physical properties. For example, whereas pure stoichiometric ZnO is an insulator, the conductivity of ZnO can be increased by a factor of ten with just minor changes in defect concentrations such as interstitial zinc or aluminium addition. Magnetic ions are one of the most intriguing types of dopants in semiconductors. Semiconductors with magnetic impurities, also known as diluted magnetic semiconductors or semimagnetic semiconductors, have been explored for decades (Lee et al. 2018; Zhang et al. 2019). "Giant Zeeman effects" observed in the excitonic levels drew the attention and toward diluted magnetic semiconductors (Li et al. 2021, p. 4; Chen and Pan 2021). The excitonic Zeeman parting of DMSs regularly surpass the parting of the relating nonmagnetic semiconductors by more than two significant degrees, leading to potential applications in optical gating (Ambigadevi et al. 2020). All the more as of late, consideration in DMSs has moved to their utilization in spintronics, or turn based hardware advances (Chen et al. 2020; González-Rodríguez et al. 2020). The gigantic Zeeman parting is used to make turn settled conductivity directs in semiconductors in this field. The subsequent twist energized flows might give different take based on levels of opportunity to semiconductor gadgets, possibly expanding the data content of a charge beat and presenting new capacities not found in current-based semiconductors. Nanoscale DMSs have been utilized as major functional parts in a large number of the gadgets proposed by theoreticians or tried in model forms by experimentalists (Bi et al. 2020; Fu et al. 2020; Sheng et al. 2020). Luminescence Glow activators are one more significant class of dopants for semiconductor nanocrystals. For as long as decade, a large part of the examination into unadulterated semiconductor nanocrystals has been driven by interest in their glowing properties, and controlling these nanocrystals' properties by doping with particles like Mn_2 or Eu_2 can possibly widen the scope of valuable spectroscopic properties that can be accomplished from this class of materials. These glowing luminous colloids are alluring possibilities for optical imaging applications in light of their high quantum yields, slight discharge line morphologies, and expansive excitation profiles. Electronic dopants, what work as shallow donors or acceptors inside the semiconductor band structure to present transporters, are a third intriguing gathering of dopants (Shah et al. 2021; Dave et al. 2021a). Although electronic doping of nanocrystals has not yet been extensively investigated, it is evident that as self-assembled device structures become more accessible, this field will play a significant role in the future of nanotechnology (Bryan and Gamelin 2005). As a result of the exponential expansion in industrialization, both energy and environmental challenges have arisen. The primary source of energy is fossil fuels, such as coal, oil, and natural gas. Fossil fuels, which are used as a primary source of energy, are also responsible for many types of environmental pollution, as air and water pollution have expanded dramatically in recent years, affecting a vast number of cultures around the world (Zhao et al. 2019). In 1972, Fujishima and Honda first founded that the TiO_2 electrode can break down hydrogen in aquatic production under ultraviolet light (UV) (Zhang et al. 2019). Since then, photocatalysis activities have attracted intense attention and researcher from across the globe due to its several advantages such as low cost and attractive efficiency with feature of direct conversion of solar energy to easily stored hydrogen, with no harm to environment as it is a unique process for rectifying energy and environmental issues like degradation of various organic pollutants in wastewater, production of hydrogen, purification of air, and antibacterial activity (Bao et al. 2008). Photocatalysis is a combination of photochemistry and catalysis in which light and catalysts are employed in tandem to speed up chemical reactions. Photocatalysis is defined as "catalysis-driven acceleration of a light-induced process" in simple terms. Photocatalysts are classified as either homogeneous or heterogeneous processes. Furthermore, Frank and Bard successfully oxidized CN- to OCN- using TiO_2 as a photocatalyst, an experiment that offered optimism for the use of photocatalysts in wastewater treatment (Tong et al. 2012). Since then, TiO_2 has been widely accepted as a photocatalyst for a wide range of applications in the field of environmental management. The broader research area currently at this stage includes photocatalyst disinfection (Guha et al. 2003, p. 2), photocatalytic hydrogen production (You et al. 2010; Giahi et al. 2019), photocatalytic reduction of CO_2 (Rahman et al. 2021; Moballegh et al. 2007), photocatalyst wastewater treatment (Szabó-Bárdos et al. 2003; Goto et al. 2004; Li et al. 2004; Kwon et al. 2000), and air purification (Subramanian et al. 2003). Xing et al. (2006) presented the application of photocatalysts in air treatment, including catalyst development and reactor design. Masakazu and Takahit (1987) gave an update on photocatalytic sterilization's recent developments and problems. The majority of existing reviews concentrate on only one or a few characteristics of photocatalysts (Zhang et al. 2019).

4.2 Photocatalytic Mechanism and Influencing Factor

4.2.1 Reaction Mechanism

Photocatalytic response is a synthetic cycle that happens under the coupled activities of light and the photo impetus with different benefits like protection of environment and remarkable toxin destruction with no supplementary products. The fundamental component of a photocatalytic response is portrayed in Fig. 4.1. Photocatalytic activity is characterized as the light-prompted redox response of semiconductors from the point of view of semiconductor photochemistry. Because of their discrete electronic states, semiconducting materials are ordinarily utilized as an impetus that can go about as a sensitizer for the illumination of light reforming redox processes. A semiconductor structure has an energy band structure that comprises of a low-energy valence band (VB) and a high-energy conduction band (CB), with a prohibited band between the conduction and valence groups. At the point when the occurrence light's energy is more noteworthy than the semiconductor's band hole, the electrons in the VB of the semiconductor are unsettled and travel toward the conduction band, making openings structure in the VB. Photogenerated electrons and openings are isolated by an electric field and move to the outer layer of semiconductor particles (Satapathy et al. 2021; Dave et al. 2021c).

Photogenerated pores have huge oxidizing qualities and can oxidize atoms adsorbed on the semiconductor's surface or in its answer. The itemized course of a photocatalytic response is as per the following. Under a specific energy of light,

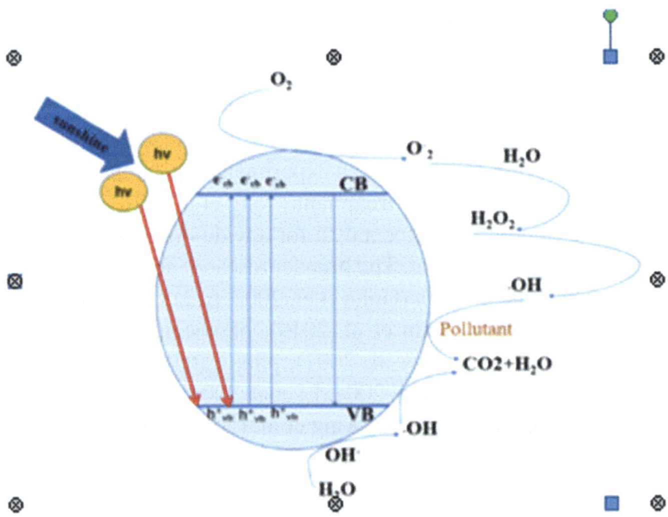

Fig. 4.1 Principle mechanism of a photocatalytic reaction [Reproduced from Zhang et al. (2019) open access]

electrons on a VB are invigorated and leap to a CB, and the openings stay on the VB. The openings on the VB diffuse to the photocatalyst surface and participate in an oxidation interaction, while the electrons on the CB travel to the impetus surface and partake in a decrease response. The electrons can make H_2O_2 or a superoxide revolutionary O_2-with H+ and disintegrated O_2 in the fluid arrangement during the interaction, and the openings can oxidize OH– to create hydroxyl extremists OH, bringing about the degradation of pollutants (Santra and Kamat 2012; Li et al. 2014; Ertis and Boz 2017; Ranjith et al. 2018).

Many factors influence photocatalytic activity, including the catalyst and its properties, surface conditions (charge, adsorbed materials, defect, and composition), reaction media (pH, solvent), light source, and varied reactant concentrations, adsorption, and so on (Dave et al. 2021b).

4.3 Metal-Doped Semiconductor

4.3.1 Silver and Gold Doped Semiconductor

4.3.1.1 Effect of Doping on XRD Pattern

To modify the photocatalytic activity of different semiconductors (TiO_2, ZnO, and CdS), various metal ions (Au, Ag, Pt, and Pd) are doped in the semiconductor materials. Doping with two atoms (co-doping) into semiconductors also has a significant interest for a superior photocatalytic activity.

Figure 4.2 has shown the XRD characterization and pattern of precipitation-decomposition method, Ag doped Au–ZnO shows that it has both hexagonal and wurtzite structure (Senthilraja et al. 2014).

For the bare ZnO all the diffraction peaks matched with standard JCPDS 89-0511. It was found that with the incorporation of 3% Ag, the crystallite size of bare ZnO and co-doped Ag–Au–ZnO semiconductor are found to be almost equal to 4.15 and 4.14 respectively. The crystallite size was determined by the Debye-Scherrer equation given by:

$$D = \frac{K\lambda}{\beta \cos\theta}$$

where D is the crystal size, K is the dimensionless constant, λ is the wavelength of X-ray, β is the full width at half-maximum of the diffraction peak and θ is the diffraction angle.

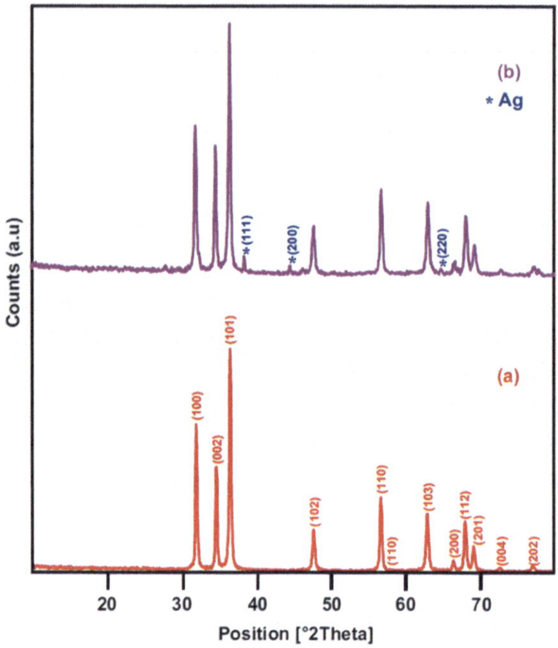

Fig. 4.2 XRD patterns of **a** bare ZnO and **b** Ag–Au–ZnO [Reproduced with permissions from Senthilraja et al. (2014)]

4.3.1.2 Effect of Doping on Morphology

The photocatalytic activity of the crystal is influenced by the structure and morphology of the catalyst. The FESEM image were used to examine the surface morphology of Ag–Au–ZnO. Figure 4.3 shows that the particles are both spherical and hexagonal in shape.

The particle size varies from 4 to 100 nm and these changes of the morphology increase the photocatalytic activity of Ag–Au–ZnO.

4.3.1.3 Effect of Doping on Surface Area

The most critical element influencing photocatalytic activity is the surface area of the catalyst. The surface area was measured by nitrogen gas absorption method. The surface area of Ag–Au–ZnO (20.3 m^2g^{-1}) is higher compared to bare ZnO (14.9 m^2g^{-1}). The increased surface area increases the number of active sites on the surface of the catalyst, resulting in increased photocatalytic activity. The single point total pore volume also increased for Ag–Au–ZnO as per the results.

Fig. 4.3 FE-SEM images of **a** Ag–Au–ZnO (100 nm) and **b** Ag–Au–ZnO (100 nm) [Reproduced with permissions from Senthilraja et al. (2014)]

4.3.2 Silver Doped Semiconductor

4.3.2.1 Effect of Doping on XRD

The XRD pattern of Ag doped ZnO shown in Fig. 4.4 exhibits a minor change in intensity of the (101) plane but preserved the wurtzite ZnO structure. The change in the intensity was associated to the changes in the grain size (Vallejo et al. 2020).

4.3.2.2 Effect of Doping on Morphology

The effect of Ag doped ZnO was shown in the morphology. The roughness and grain size of the pristine ZnO and Ag–ZnO varies from 18.4 nm to 36.6 nm and 190 nm to

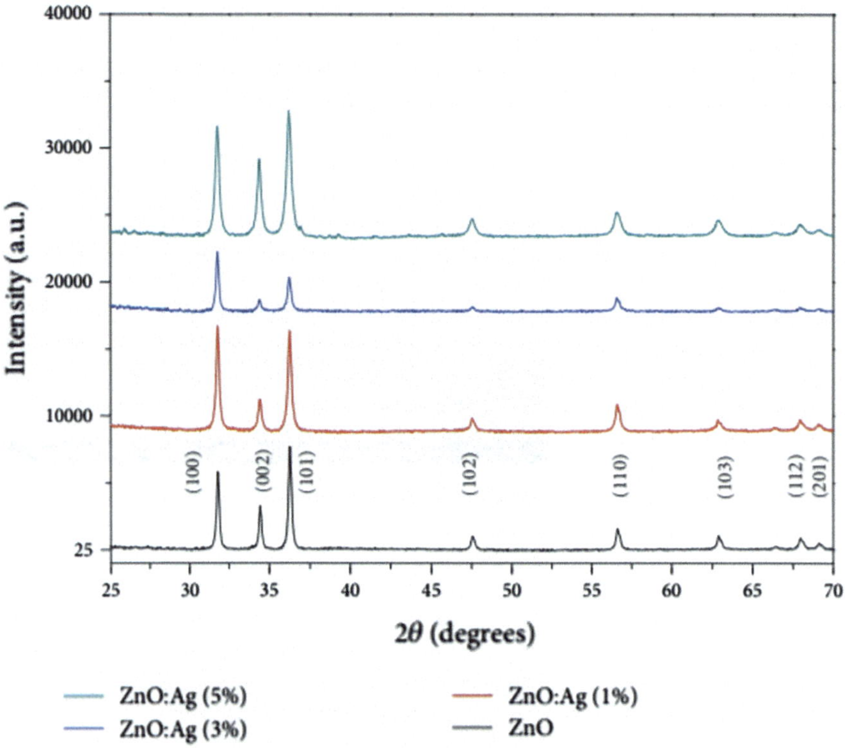

Fig. 4.4 X-ray diffraction patterns for both ZnO and ZnO: Ag thin films [Reproduced with permissions from Vallejo et al. (2020)]

120 nm respectively. The decrease in the grain size is attributed to the incorporation of Ag in the ZnO lattice.

4.3.2.3 Effect of Doping on Optical Properties

The effect of Ag doped ZnO was also realized in optical properties. For pure ZnO, the reflectance was very high (70%) after 350 nm. Figure 4.5 shows that with incorporation of Ag the reflectance spectrum decreases continuously in the visible range. It was also observed that due to oxygen vacancy in the ZnO structure, the bandgap of doped semiconductor decreased from 3.22 eV (ZnO) to 3.10 eV (5% Ag–ZnO) with increasing Ag concentration. Oxygen vacancies are the most favorable defects procedure during ZnO synthesis.

The bandgap E_g of all the samples determined using the Kubelka–Munk function and tauc plot is calculated using following equations:

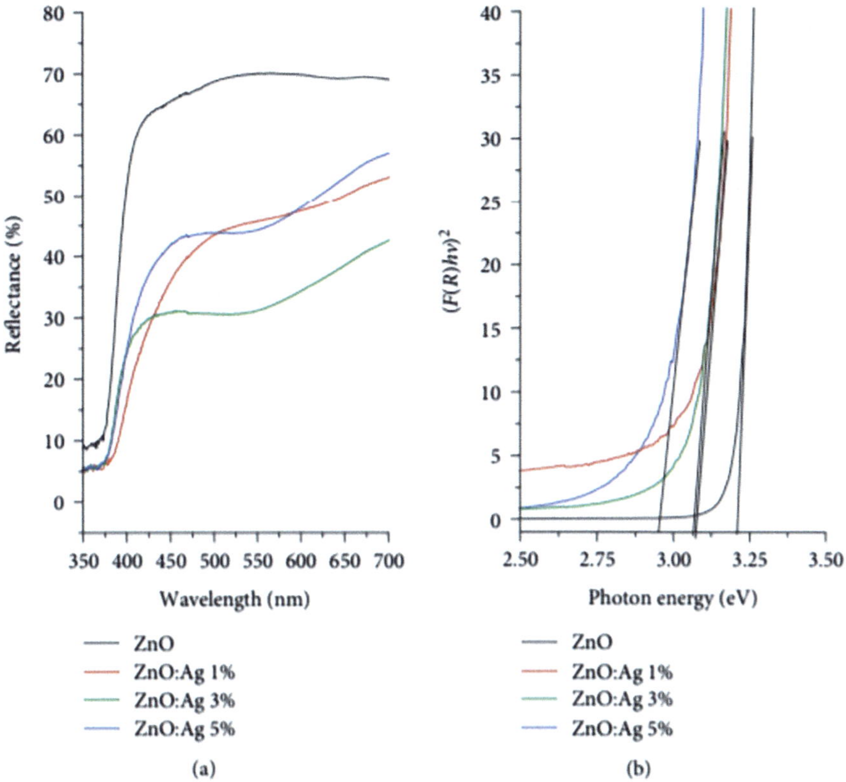

Fig. 4.5 **a** Reflectance diffuse spectra for both ZnO and ZnO:Ag thin films. **b** Kubelka–Munk plots and band gap energy estimation for both ZnO and ZnO:Ag thin films [Reproduced with permissions from Vallejo et al. (2020)]

$$F(R_\infty) = \frac{(1 - R_\infty)^2}{2R_\infty}$$

$$(F(R_\infty)h\upsilon)^{1/2} = A\left(h\upsilon - E_g\right)$$

where $F(R_\infty)$ proportional to the absorption coefficient, λ is the wavelength, E_g is the bandgap, A is the constant.

4.3.2.4 Effect of Doping on Photocatalysis

The Ag doped ZnO exhibit better photocatalytic performance compared to the pure ZnO. With Ag doping, the photocatalytic performance was enhanced because of the reduction in grain size and intra-band transition. The photo degradation performance also increased by 16.5 times with 5% Ag doping.

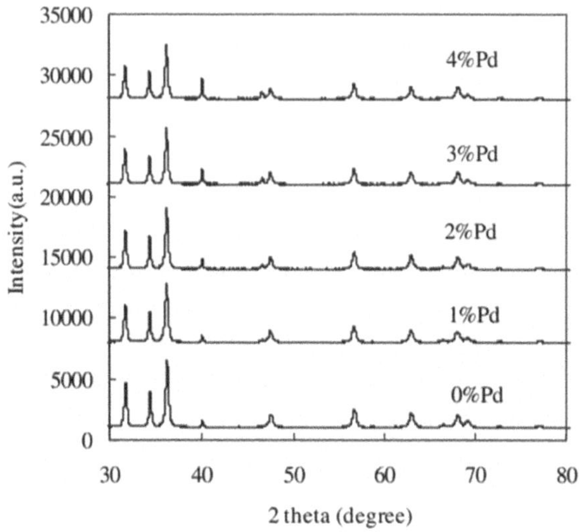

Fig. 4.6 XRD patterns of photocatalysts [Reproduced with permissions from Zhong et al. (2012)]

4.3.3 Palladium Doped Semiconductor

4.3.3.1 Effect of Doping on XRD Patterns

The XRD pattern of Pd doped ZnO (Fig. 4.6) shows that it exhibits both hexagonal wurtzite structure. For the bare ZnO, all the diffraction peaks matched with standard JCPDS 89-0511. There are no traces of PdO which depicts that Pd enters the ZnO lattice.

The crystallite size was determined by the Debye–Scherrer equation and it was observed that the crystallite size increased from 25 to 33 nm with increased Pd concentrations (Zhong et al. 2012).

4.3.3.2 Effect of Doping on Optical Properties

The effect of Pd doped ZnO was also shown in optical properties. For pure ZnO, the reflectance was very high after 350 nm but with incorporation Pd concentration the reflectance decreases continuously as shown in Fig. 4.7.

Pd doped ZnO samples shows higher absorption in the visible range which indicates that with incorporation of Pd in ZnO crystal there might be some trapping states formed (inter-band trap site). By absorbing visible light electrons are excited from valance band of ZnO and captured by the inter-band trap sites. The absorption of the electron by the trap states can promote the photocatalytic activity of the doped samples.

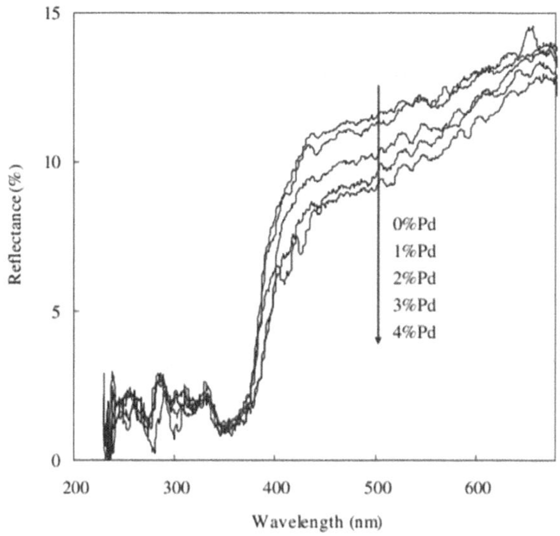

Fig. 4.7 UV/vis diffuse reflectance spectra of photocatalysts [Reproduced with permissions from Zhong et al. (2012)]

4.3.3.3 Effect of Doping on Surface Area

The surface area is measured by nitrogen gas absorption method. The surface area of ZnO (7.24 m^2g^{-1}) is higher compared to Pd doped ZnO (6.49 m^2g^{-1}) (4%). The decreased surface area of the doped samples is due to the larger crystal size of Pd–ZnO. The single point total pore volume also decreased for Pd–ZnO.

4.3.3.4 Photocatalytic Activity

A better performance is seen in the photo catalytic activity of the Pd doped samples in comparison to pure ZnO. The photocatalysis degradation of ZnO occurs due to the generation of an electron-holepair and various radicals such as hydroxyl (•OH), hydroperoxyl (•OOH) and superoxide (O_2•–) radicals. Out of these radicals, hydroxyl (•OH) radical known as most powerful oxidizing species which can attack the organic pollutant near to the surface of the photocatalyst. The ZnO crystal enhanced by Pd doping which was observed that for 3% of Pd doping shows higher photocatalytic activity. The improved photocatalytic activity with doped ZnO is due to increased absorption capability of light and the separation rate of photo induced charge carriers.

In todays' world, a diverse amount of research is going on in the field of doped semiconductors and its role in catalytic activities. Some of the most common doped semiconductor reported includes Ag, Au and Pt/Pd doping and its characteristics effect in the field of catalytic properties and applications. In this book chapter we are going to focus on the some of the doped semiconductor and its significant application and role in catalytic activities.

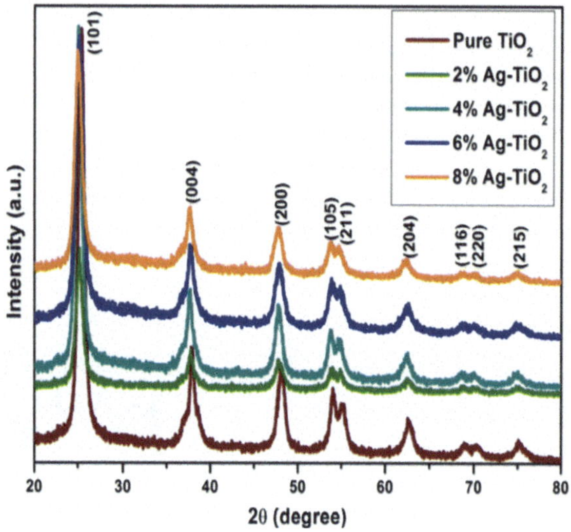

Fig. 4.8 The XRD patterns of the pure and Ag-doped TiO_2 NPs [Reproduced with permissions from Ali et al. (2018)]

4.3.4 Ag-Doped TiO_2

4.3.4.1 XRD Analysis

To explore the crystal structure of pure and Ag–TiO_2 NPs calcined at 400 °C, XRD analysis was carried out in the range of 20°–80°. Figure 4.8 is showing the diffraction peaks in correspondence to the anatase phase of TiO_2 (JCPSD Card: 21e1272).

It emphasizes that Ag doping concentration has no effect on the anatase phase. Furthermore, the absence of an impurity phase demonstrated the successful incorporation of silver ion into the TiO_2 matrix structure. However, as doping ion concentration increases, the primary diffraction peak changes towards the lower 2 value and broadens, which could be attributed to lattice strain in the materials. Using the Williamson Hall relation, the average crystallite size (D) and lattice strain () may be calculated from the diffraction peaks' full-widths at half-maximum (FWHM) (Santra and Kamat 2012). XRD data are used to calculate the average crystallite size (D), lattice strain, lattice parameters (a, b, and c), and lattice volume for all samples. However, when the amount of Ag doped increases (from 0 to 2.0 mol percent), the lattice properties and average crystallite size drop (Ali et al. 2018).

4.3.4.2 Raman Analysis

The Raman peaks for pure TiO_2 and Ag-doped TiO_2 NPs showed (Fig. 4.9) a similar pattern to that of anatase TiO_2.

There is no trace of silver oxide peaks which explains that the Raman spectral results are good agreement with XRD observation. Moreover, Ag-doped TiO_2NPs

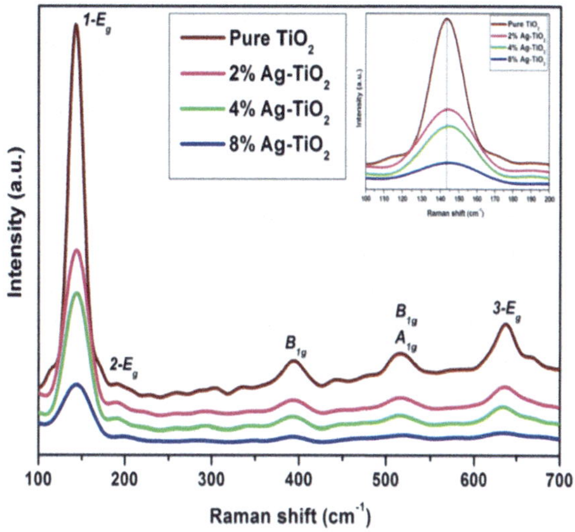

Fig. 4.9 Raman spectra of pure and Ag-doped TiO_2 NPs [Reproduced with permissions from Ali et al. (2018)]

conserved the anatase structure which suggests that Ag dopants are substitutional combined into the structure of TiO_2 framework.

4.3.4.3 Optical Properties

The effect of Ag doped TiO_2 was also shown in Fig. 4.10.

With increasing Ag concentration, it was shown that the bandgap of doped semiconductors drops from 3.12 (TiO_2) to 2.6 (8% Ag–TiO_2) due to oxygen vacancies in

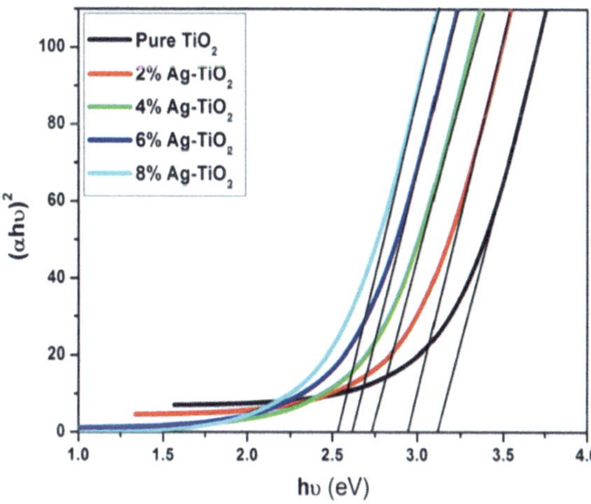

Fig. 4.10 Optical band gap energy plot of pure and Ag-doped TiO_2 NPs [Reproduced with permissions from Ali et al. (2018)]

the TiO_2 structure. During ZnO production, oxygen vacancies are the most desirable flaws.

4.3.4.4 Photocatalytic Properties

The photocatalytic activity of pure and Ag-doped TiO_2 photo-catalysts was tested using methylene blue dye photodegradation. The results showed that when exposed to visible light, an Ag-doped TiO_2 photocatalyst can efficiently degrade MB, with 4.0 mol percent Ag-doped TiO_2 having the highest photocatalytic activity of all the sol-gel samples.

4.3.5 *Au–TiO_2*

4.3.5.1 Experiment

Using 150 mL of hydrochloric acid, 0.75 g of pristine single-walled carbon nanotubes (SWCNT) (quality 98%; D2-10 nm; L 0.3–3 μm) were chemically activated, resulting in the production of oxygen groups on the SWCNT surface. The mixture was ultrasonically sonicated for 10 min at 40 °C, then poured with deionized water, filtered under vacuum, and dried for 12 h at 120 °C. The tertiary TiO_2/Au/SWCNT nanocomposites were made using the sol-gel technique. To begin, 300 mg modified SWNT and 0.4 ml titanium isopropoxide (TTIP) were dissolved in 100 ml deionized water and agitated for 20 min at 30 °C. The pH range was preserved at 4 by adding 4 ml of 1 M nitric acid to the solution. 20 mg of gold chloride (Au/TiO_2 = 10%) was dissolved in 100 ml of deionized water and heated to form the Au solution. Temperature of fusion after being added to an Au solution, the TiO_2 SWNT combination was agitated for two hours. After that, the temperature, coolant, and 20 mL of 0.01 M sodium citrate solution in water were added, and the mixture was continually stirred for another hour. The mixture was heated to 80 °C overnight before incubation at 500 °C for 3 h. To make a pure TiO_2 and TiO_2/Au binary nanocomposite, the same procedures were used without the use of SWCNT (Mohammed 2020).

4.3.5.2 XRD Data

The X-ray diffraction patterns of TiO_2/Au binary and TiO_2/Au/SWCNT ternary nanohybrids show a preference for orientation in the (101) direction at about 25.32° for the anatase phase of TiO_2, with low intense peaks located at 37.37°, 47.44°, 53.851°, and 55.1°, respectively, corresponding to (004), (200), (105), and (211) planes The major XRD peak of the Au is not highlighted due to the overlapping of Au (004) peak and titanium dioxide (111) peak (Fig. 4.11).

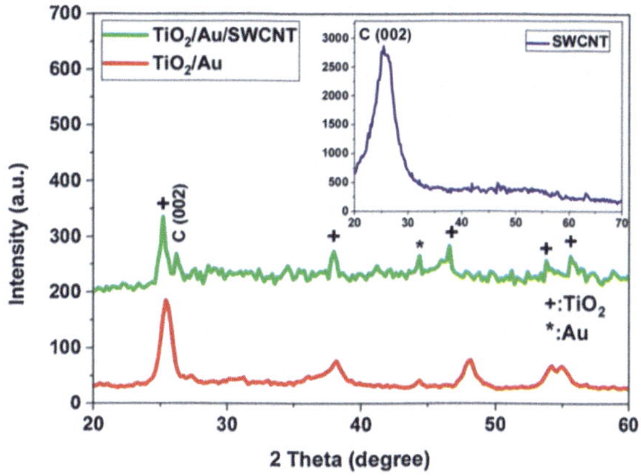

Fig. 4.11 XRD patterns of TiO_2/Au and TiO_2/Au/SWCNT nanohybrid. XRD pattern of raw SWCNT (inset) [Reproduced with permissions from Mohammed (2020)]

The XRD pattern of TiO_2/Au/SWCNT nanohybrid shows the SWCNT (002) diffraction peak at around 26.17°, this increasing in the angle position can be credited to the decrease of interplanar distance between carbon tubes after loading of TiO_2/Au nanoparticles. The intensity of all XRD peaks of the TiO_2/Au unit decreases with the addition of Au dopant and SWCNT because the crystal growth kinetics is affected by the chemical reactivity of the dopant. The particle size of TiO_2/Au and TiO_2/Au/SWCNT are 15 nm and 11 nm respectively.

4.3.5.3 FESEM Data

From the FESEM image (Fig. 4.12), it is clear that the SWCNTs have a tube-like structure with diverse outer diameters and tube lengths. By TiO_2/Au image, the small particles obviously own a spherical shape with different sizes (10–150 nm).

The FESEM image of the TiO_2/Au/SWCNT nanocomposite showed that the TiO_2/Au particles were anchored to the wall of the SWCNTs and the aggregation was also shown in a few locations.

4.3.5.4 Optical Properties

The Tauc relationship was used to determine the band gap of the tertiary hybrids TiO_2, TiO_2/Au, and TiO_2/Au/SWCNT using UV-vis spectra. The band gap of virgin TiO_2 is roughly 3.18 eV, which is similar to the normal band gap of bulk titanium oxide. When Au is included into TiO_2, however, the band gap decreases from 3.18 to 2.93 eV for TiO_2/Au and to 1.95 eV for TiO_2/Au/SWCNT.

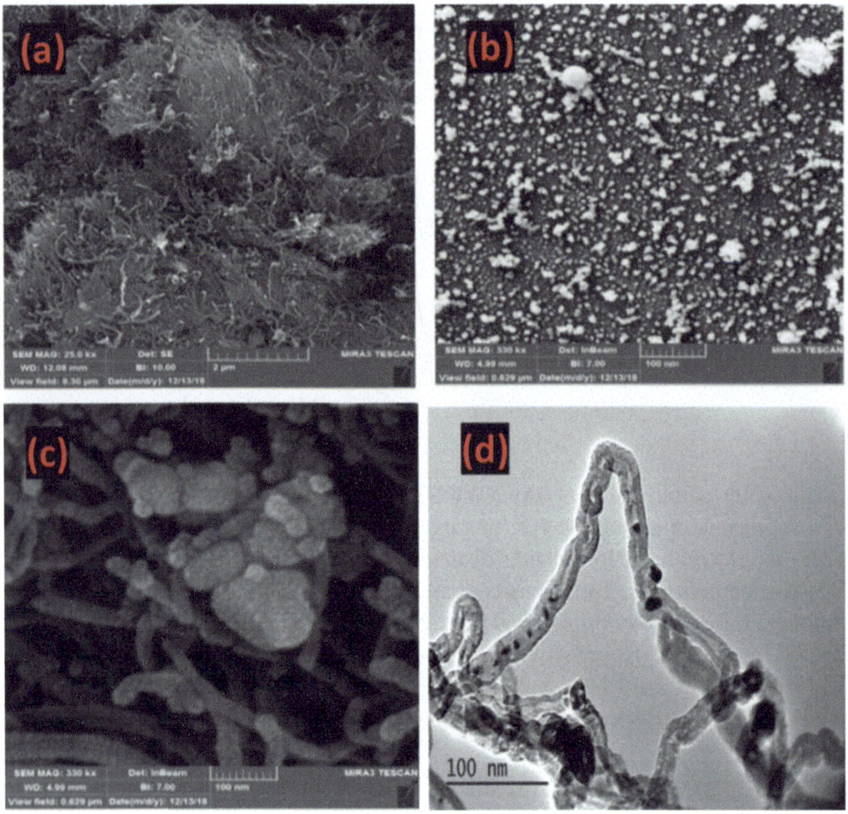

Fig. 4.12 FESEM images of **a** raw SWCNT, **b** TiO_2/Au, **c** TiO_2/Au/SWCNT, and **d** HRTEM of TiO_2/Au/SWCNT nanohybrid [Reproduced with permissions from Mohammed (2020)]

4.3.5.5 Photocatalytic Activity

Ternary nanohybrid estimation shows moderately uniform TiO_2/Au nanoparticles stacking in the SWCNT framework with upgraded UV-vis retention highlights. The TiO_2/Au/SWCNT showed higher effectiveness for the decolorization and mineralization of MB color under sunlight based reenacted light contrasted and unblemished TiO_2 and TiO_2/Au. Since the TiO_2/Au/SWCNT nanohybrids enormously dislodged light assimilation to a more extended frequency, lower energy was required for photocatalytic execution and the MB color corruption effectiveness developed. The current discoveries affirm that the improved photocatalytic action of TiO_2/Au/SWCNT by SWCNTs doing as an adsorbent, framework specialist and charge acceptor to catch photo created electrons from the TiO_2 conduction band because of apparent light illumination and henceforth forestalling the e− to h+ pair recombination.

4.3.5.6 Pt/Pd-TiO_2

The impregnation process was used to create the modified TiO_2. At 373 K, 600 mg TiO_2 (Degussa P25) was agitated in a succession of concentrations of $H_2PtCl_6$and $PdCl_2$ solutions for 60 min. The resulting Pt_4+/TiO_2 and Pd_2+/TiO_2 slurries were then vacuum freeze-dried for 20 h (100–200 mTorr at 57.5 °C). Pt-doped TiO_2 and Pd-doped TiO_2 samples were obtained in a tubular oven at progressively higher temperatures (30, 120, and 400 °C) for 3 h. Before further characterizations, the as-prepared Pt-doped TiO_2 and Pd-doped TiO_2 were exposed to air for 3 h, a period similar to that used in the kinetic studies. Pt-doped TiO_2 and Pd-doped TiO_2 reduced at 400 °C is abbreviated as Pt/TiO_2 and Pd/TiO_2, respectively, on the next pages.

4.3.5.7 Photocatalytic Applications

Photocatalysis is a versatile technique with a lot of room for growth. The applications of various semiconductor photocatalytic shown in Fig. 4.13.

Houas et al. (2001) used methylene blue to imitate environmental wastewater in 2001. The researchers discovered that photocatalysts based on titanium dioxide which could successfully decompose methylene blue. Lachheb et al. (2002) irradiated titanium dioxide photocatalyst with UV light to accomplish dye degradation (alizarin S, crocein orange G, methyl red, Congo red, and methylene blue). The experimental results revealed that the five dyes not only decolorized but also disintegrated completely. Liu et al. (2018) used a conventional electrospinning approach

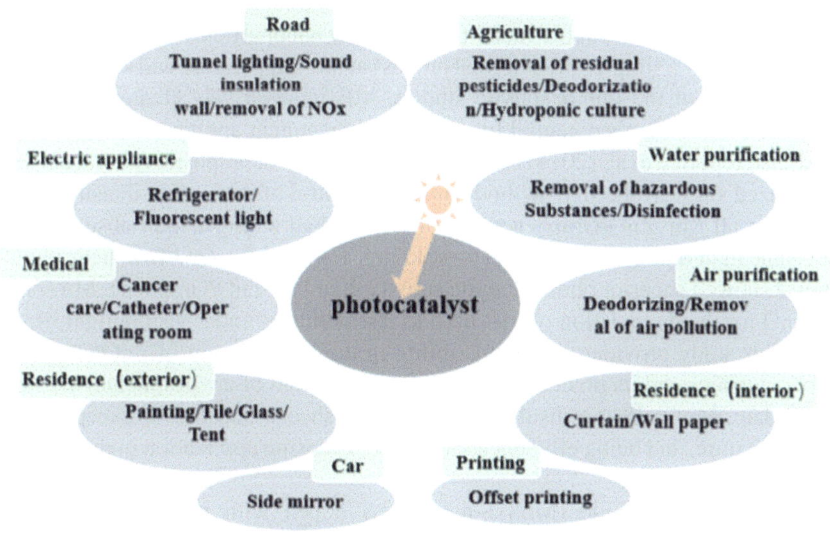

Fig. 4.13 Various applications of Semiconductor photocatalysis [Reproduced from Zhang et al. (2019) open access]

with a calcination procedure to create flexible composite Fe_2O_3/TiO_2 nanofibers, which they effectively used for photocatalytic wastewater treatment and photocurrent monitoring. Nigussie et al. (2018) reported Ag–TiO_2 and Ag–ZnO NPs with high thermal stability and strong antibacterial activity which is expected to serve in applications in the pharmaceutical and nanocomposite fields and also in antibacterial activities.

Demirci et al. (2016) showed that Ag-doped TiO_2 films had a better photocatalytic activity than undoped TiO_2 film. In comparison to undoped TiO_2 films, Ag replacement in TiO_2 matrix increased photocatalytic activity of TiO_2 films under UV light irradiation. This study revealed the effectiveness of using Ag-doped film to remove dissolved organic pollutants in water.

Guillén-Santiago et al. (2010) reported undoped and Ag-doped TiO_2 thin films onto glass substrates deposited by the sol-gel technique. The film thickness is controlled indirectly by the immersions number, and the aging time of the starting solution plays an important role in the degradation performance of MB. A 35% of MB degradation is reached using UV-irradiation. As a result, the photocatalysis processes showed that Ag-doped TiO_2 thin films are attractive for applications in water cleaning. Kumar et al. (2015) reported simple solution technique at low temperature which was employed to make Ag-doped ZnO nanoellipsoids, which were used as an effective photocatalyst and efficient electron mediator in the production of a highly sensitive and robust hydrazine chemical sensor.

Khurshid et al. (2019) have stated artificial hydrothermal method used for the fruitful preparation of ZnO: Ag/rGO nanocomposites. The photo-electrochemical properties of ZnO: Ag/rGO were methodically examined and results were linked with pristine ZnO. The results revealed that the ZnO: Ag/rGO nanocomposite can be employed as a good photoanode in electrochemical applications. According to Muoz-Fernandez et al. (2016), noble metal (Pt, Ag)/ZnO nanoparticles were successfully manufactured by a simple solvothermal process at low temperature and short reaction periods. In all of the photocatalytic studies, MB removal was greater than 70%, confirming the system's potential for wastewater treatment applications.

Ahmad reported et al. (2014) Mn–ZnO/Graphene nanocomposites photocatalysts synthesized via solvothermal method with good control of the coating density. It was found that all Mn–ZnO/Graphenenanocomposites had stronger light absorption in the visible light region than pure and doped ZnO. The 3% Mn–ZnO/Graphene photocatalyst showed superior photocatalytic activity than Mn–ZnO/Graphene, Mn–ZnO, and ZnO for the degradation of MB used as test pollutant under visible light irradiation. This study provides a new possibility in the investigation of ZnO/Graphene nanocomposites and its practical applications in the field of environmental science.

Till date, Molybdenum disulfide (MoS_2) nanosheets, because of exceptionally dynamic nature, and being efficient shows their promising application in the reactant decrease of nitroarenes.

Molybdenum disulfide (MoS_2) nanosheets doped with progress metal is a notable system embraced to upgrade their catalyst effectiveness for the decrease of nitroarenes, nonetheless, the specific and ideal measure of doping is as yet questionable and a subject of future research. Rahman et al. (2021) detailed the creation of couple layered cobalt doped MoS_2 nanosheets with various cobalt content (2, 4, 6, and 8%) by means of the solvothermal technique with appropriate tuning of Co focus as doping in MoS_2 nanosheets with a point of finding a potential alternative of a respectable metal impetus for the reactant decrease of nitroarenes, utilizing sodium molybdate dihydrate ($Na_2MoO_4.2H_2O$), thiourea (CH_4N_2S) and cobalt acetate tetrahydrate [$Co(CH_3COO)_2.4H_2O$] as antecedents and their catalyst execution has been analyzed by monitoring the decrease of p-nitrophenol by NaBH4 continuously utilizing UV-apparent absorption spectroscopy. As per the result, the 6% Co doped MoS_2 nanosheets exhibited predominant reactant movement with a pseudo-first request rate steady of $3.03 \times 103\ s^{-1}$ corresponding to the bountiful deformities in the active edge destinations having a prevailing metallic 1T stage with Co particle initiated inadequate basal planes, sulfur (S) edges, catalytic primary and electronic tweak among MoS_2 and Co particles and enhanced electron move helped through redox cycling in the dynamic locales. The analysis performed showed that the undoped and unique (2, 4, 6, and 8%) nuclear weight rates of cobalt doped MoS_2 nanosheets with the cobalt particle actuated on faulty basal planes, S-edges, Mo locales were incorporated in a solitary advance aqueous response process. The viability of the pre-arranged MoS_2 nanostructures as impetuses for the decrease of 4-nitrophenol to 4-aminophenol in abundance $NaBH_4$ was additionally analyzed. It was found that different arranged morphologies of MoS_2 nanostructures were viewed as powerful and stable impetuses and followed pseudo-first-request energy. Moreover, the catalyst movement property was obviously seen to improve with cobalt doping rates and furthermore gave upgraded outcome contrasted with honorable metal and rGO-based MoS_2 nanocomposites. Moreover, promptly dispersible in polar solvents like water or methanol, the 6% Co-doped MoS_2 nanosheets exhibited exceptional reactant movement toward decrease of destructive nitrophenol from squander waters at encompassing temperature showing high turnover recurrence (4.18×10^{18} particles per g pers). Furthermore, high bad zeta potential was likewise recognized showing that the so shaped examples are steady against accumulation in scattering medium. Results showed some conceivable clarification for the unrivaled catalyst action of the greatest 6% Co doped MoS_2 layers as it could be because of a blend of (a) adjustment of the metallic 1T stage with expanding electrical conductivity (decline in band hole and enactment energy) and (b) better electron catch from the hydride and electron supply to the nitrophenol substrates through reversible redox responses at the Co locales. Ethiraj et al. (2020) disclosed the degradation of phenolic compounds in the presence of pure and Cu doped nickel oxide (Cu–NiO) nano-catalysts. A wet chemical method was opted for the catalyst preparation. Characterization like fourier transform infrared spectroscopy (FTIR) confirms the formation of pure NiO and the existence of copper in doped nano-catalyst samples. Cu–NiO nano-catalysts samples showed a depletion in average crystallite size as compared to pure NiO sample, where it was 24.0, 22.8, and 19.03 nm for 2Cu–NiO and 4Cu–NiO, respectively. Using TEM, the average particle size as determined were about 28.0, 26.6, and 22.8 nm for NiO, 2Cu–NiO, and 4Cu–NiO, respectively with the energy band gap values of 3.26, 3.64, and 3.87 eV for undoped NiO, 2Cu–NiO and 4Cu NiO respectively. It was also found that under UV radiation for 180 min, 4Cu–NiO nano-catalysts showed degradation efficiency of 66.36 and 47.00% for 0.3 and 0.4 mol/L, respectively. After 300 min. of UV radiation, the maximum degradation efficiency was 66.83% in 0.3 mol/L (4Cu–NiO), 75.20 and 72.04% was noted in 0.4 and 0.5 mol/L (2Cu–NiO). The maximum amount of phenol (75.2%) was removed at 300 min. Furthermore, photocatalytic performance of Cu–NiO and pure NiO was systematically examined at various reaction times and Cu doping ratios (2–4 wt.%). Different molar concentrations of phenol were taken into consideration for this experiment. The obtained results showed that the Cu–NiO nano-catalyst exhibited the highest phenol degradation efficiency as compared to their undoped composition. Promising results shows the practical application of these materials can be in efficient removal of phenol from real industrial effluent. The nano-catalyst efficiency for phenol removal was tested in real leather industrial wastewater effluent which could remove about 85.7% within 150 min.

You et al. (2010) presented Ag+ doping TiO_2 nanofibers arranged by in-situ and drenching strategies assigned as SI-TiO_2 and IM TiO_2 nanofibers, separately. The creator attempted to assess the impacts of the two different doping techniques on the material properties. The outcomes showed that the two sorts of nanofibers with the breadth of around 20 nm are micron-sized length, and were covered by nano-wads of Ag_2O with the measurement of 2–5 nm. As compared to IM TiO_2 nanofibers, the SI TiO_2 nanofibers have less Ag_2O particles. Along these lines, the photo reactant movement of IM-TiO_2 nanofibers is higher than SI-TiO_2 nanofibers, which is additionally demonstrated in photo catalysis explore showing the debasement proportion of IM-TiO_2 impetus is 100%, and the corruption proportion of SI-TiO_2 impetus is 95%. To further assess the photo catalyst action of IM-TiO_2 and SI-TiO_2 nanofibers, the decay of MB color in arrangement has been analyzed utilizing the IM-TiO_2 and SI-TiO_2 nanofibers under bright light illumination. Preceding light, the MB arrangement over the impetus is for the most part kept in obscurity for 30 min to acquire the harmony adsorption state. It was observed that the corruption of the MB arrangement was done after 200 min when IM-TiO_2 nanofibers were utilized as the impetus. In a similar photo catalysis condition, the debasement proportion of the MB arrangement was just 95% when SI-TiO_2 nanofibers were utilized as the impetus. The corruption proportion of two nanofibers was higher than that of P_{25} (Degussa TiO_2 nanocrystals with 25 nm measurement). For IM-TiO_2 nanofibers, Ag_2O particles were adsorbed on the outer layer of the TiO_2 nanofibers. By and large, on engrossing a photon, an Ag_2O molecule produces an electron and an opening, and consequently the photo created electron consolidates with an Ag+ particle to frame an Ag iota. The adsorptive O_2 on IM-TiO_2 nanofibers go about as electron acceptors to catch photo generated electrons and diminishes the recombination of electrons and openings, as depicted by the accompanying conditions: $e- + O_2 = O_2-$and $Ag++e-=Ag$. Under a similar photo catalysis condition, IM-TiO_2 nanofibers and SI-TiO_2 nanofibers were reused as photo impetuses. The corruption proportion of IM-TiO_2 nanofibers and SI-TiO_2 nanofibers after 3 reuses for the MB arrangement was 100 and 95% individually, and the photo catalysis time was 200 min. The two nanofibers showed magnificent soundness for its reuse, 90% of starting movement actually stayed after 5 cycle uses. In the wake of rehashing multiple times, the corruption proportion of IM-TiO_2 nanofibers and SI-TiO_2 nanofibers was 85.1% and 84.3% individually. Consequently, IM-TiO_2 nanofibers and SI-TiO_2 nanofibers have high photcatalytic activity action and are not difficult to be reused.

Senthilraja et al. (2014) reported the Ag loaded Au–ZnO synthesized by the precipitation–decomposition method. The formed catalyst was characterized by various sophisticated tools like X-Ray Diffraction (XRD), Field Emission Scanning Electron Microscopy (FE-SEM), Energy Dispersive Spectrum (EDS), Transmission Electron Microscopy (TEM), Diffuse Reflectance Spectra (DRS), Photoluminescence spectra (PL) and BET surface area measurements. The photocatalytic activity of Ag–Au–ZnO was investigated for the degradation of Methylene Blue (MB) in aqueous solution using UV-A light. Interestingly, Ag–Au–ZnO was found to be more effective than Ag–ZnO, Au–ZnO, commercial ZnO, bare ZnO, TiO_2-P_25 and TiO_2 (Aldrich) at pH 7 for the mineralization of Methylene Blue dye. The effects of operational parameters such as the amount of photocatalyst, dye concentration, initial pH on photo mineralization of MB dye was examined. The mineralization of MB was confirmed by Chemical Oxygen Demand (COD) measurements. Additionally, a mechanism was proposed for the degradation of MB under UV-A light. A mechanistic scheme of the charge separation and photocatalytic reaction of Ag–Au–ZnO photocatalyst is shown in Scheme 1 (Fig. 4.14).

For the most part, when metal oxide semiconductors (ZnO) are generally uncovered to UV excitation, the delivered electrons/openings induce a response with appended particles on the material surface and produce receptive revolutionaries. The electrons in the conduction band respond with O_2 to frame superoxide anion revolutionaries (O_2 d–) and the openings in the valence band respond with water

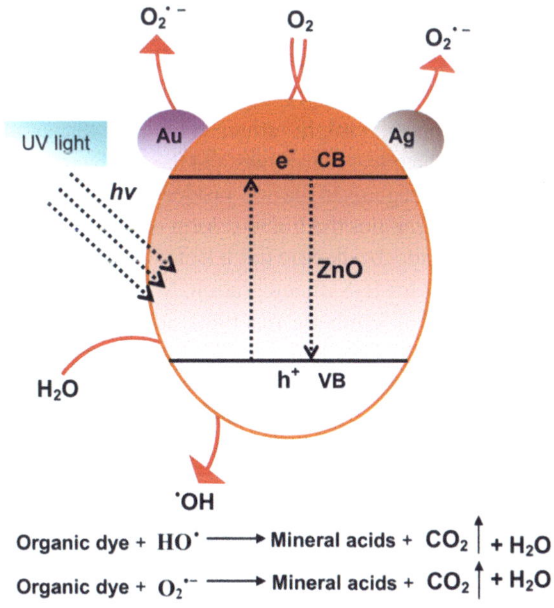

Fig. 4.14 Mechanism of degradation of MB by Ag–Au–ZnO [Reproduced with permissions from Senthilraja et al. (2014)]

to form hydroxyl extremists (dOH) (Lin et al. 2005). Such enacted extremists play the role in degrading the contaminations. It is for the most part proposed that the rate-restricting strides of the photocatalytic response of semiconductors are electron-opening recombination and electron transfer from the ZnO surface to the adsorbed oxygen molecules (Szabó-Bárdos et al. 2003; Goto et al. 2004). "Ag" can trap the electron from conduction band (CB) of the semiconductor to diminish the electron-hole recombination (Subash et al. 2013). "Au" likewise can store the produced electrons by photoexcited electron-opening partition, actuating the Fermi level shift toward more regrettable possibilities. The transfer of electrons to Au go on until the Fermi level approaches the conduction band edge of the semiconductor (Subramanian et al. 2003). Hence Ag and Au enjoy great benefit on moving the photoelectron. Subsequently, the recombination of photoelectron and hole is avoided and the photoinduced age of electron-holepairs will proceed. The diminished electron-opening recombination increases the photocatalytic movement of the impetuses. The superoxide radical anion and hydroxyl revolutionary delivered are used for the corruption of color.

Zhao et al. (2019) reported a progression of Al_2O_3 upheld Cr-doped ZnS semiconductor impetuses with different Cr/Zn molar proportions ready by co-impregnation strategy, and explored in non-warm plasma initiated H_2S disintegration. The as synthesized impetuses were portrayed utilizing XRD, N_2 adsorption/desorption, UV-noticeable spectroscopy, TEM, Raman, ICP and XPS methods. Cr-doped ZnS with various measure of Cr had the essentially upgraded reactant execution, contrasted with undoped ZnS. Varieties in catalyst movement were examined as far as adjustment on the synthetic and actual properties of the Cr-doped ZnS impetuses. Among every one of the impetuses, the Cr0.20–ZnS impetus showed the best reactant conduct with full H_2S transformation at least energy cost. Besides, the drawn out study shows that Cr doped ZnS displays great dependability in H_2S disintegration.

The reactant exhibitions of the different Cr_x–ZnS impetuses were evaluated through H_2S decomposition in non-warm plasma. For comparison, the exhibitions of undoped ZnS and Al_2O_3 support were additionally performed. For the instance of Al_2O_3 support, it might influence both the dispersion of plasma-created dynamic species and the release attributes. On the one hand, Al_2O_3 support has a solid adsorption limit with regards to plasma-produced dynamic species. The adsorption by Al_2O_3 support can drag out the home season of these species. The permeable materials in the hole upgrade the electric field. Both the improved release and the expanded home time are great for H_2S disintegration. More miniature releases, which brought about the inception of synthetic reactions among H_2S atoms, extremists and electrons, happened for Al_2O_3 filled gap. As found in Fig. 4.15, every one of the Cr_x–ZnS impetuses would do well to exercises of H_2S decay than ZnS impetus and Al_2O_3 support, and the H_2S transformation could reach as high as 100%. Examination of the trial results show that the Cr/Zn molar proportion impacted considerably H_2S conversion.

Interestingly, results showed the catalytic activity first increased and then declined with increasing the Cr content with the energy consumption in the decomposition of H_2S and was strongly dependent on the H_2S conversion levels. Among them, the $Cr_{0.20}$–ZnS catalyst exhibited the best catalytic performance, achieving full H_2S

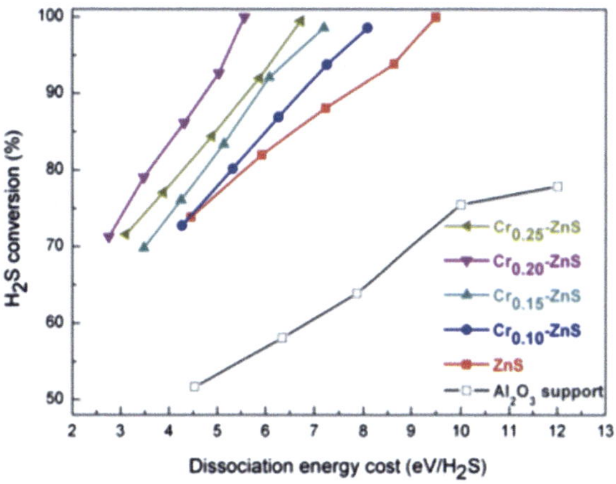

Fig. 4.15 H_2S conversion as a function of dissociation energy cost in the plasma-induced decomposition over the Cr_x–ZnS catalysts with different Cr/Zn molar ratios. Reaction conditions: catalyst bed volume: 15 mL, feed: 20 vol% H_2S in Ar, GHSV: 120 h1 [Reproduced with permissions from Zhao et al. (2019)]

conversion at lowest energy cost. At an energy consumption of 5.57 eV/H_2S, H_2S conversions were 100%, 89.7%, 87.4%, 81.8%, 79.7%, and 55.2% when $Cr_{0.20}$–ZnS, $Cr_{0.25}$–ZnS, $Cr_{0.15}$–ZnS, $Cr_{0.10}$–ZnS, undoped ZnS and Al_2O_3 were filled in the gap, respectively. The characterizations of the Cr_x-ZnS catalysts showed that the chemical and physical properties such as structure, optics, and particle size were dramatically influenced by the Cr/Zn molar ratio. Based on the XRD, TEM, and Raman portrayal, Cr_x–ZnS had the cubic sphalerite structure, and the Cr3+ particles were consistently dispersed in ZnS without isolated debasement stage seen in all the catalysts. Moreover, all the Cr_x–ZnS impetuses showed somewhat high BET surface regions. The reactant exhibitions of the different Cr_x–ZnS catalysts were evaluated through H_2S disintegration in non-warm plasma. Relative exhibitions of undoped ZnS and Al_2O_3 were performed. The high BET surface region is favorable to retain more lights and increment the quantity of the dynamic habitats, as well as shortening the distance for the photo created transporters to come to the surface (Xing et al. 2006; Hu et al. 2005). In the meantime, the molecule size of Cr-doped ZnS was around 8 nm. Little size nanoparticles with unfortunate crystallinity are good for the quick electron transportation from mass to surface, which avoid the recombination of the created electrons and openings in the main part of the catalyst (Bao et al. 2008; Masakazu and Takahit 1987). Along these lines, a reduction in molecule size of Cr doped ZnS likewise added to the higher movement. Fundamentally, the photo physical properties of semiconductor impetus, for example, places of conduction and valence groups and light assimilation, are connected with the synthetic pieces. High reactant exercises of the Cr_x–ZnS impetuses are associated with the low recombination of the openings and electrons and high light assimilation. Moreover, the reasonable debasement energy level could be given through an appropriate doping measure of Cr into the interstitial or cross section of ZnS. This pollutant energy level prompts the simple infusion of the invigorated electrons from valence band (VB) to conduction band (CB) of ZnS (Bodke et al. 2014). The pace of H_2S deterioration over Cr-doped ZnS relies upon the number of inhabitants in opening electron matches. We revealed that upheld metal sulfide semiconductors (like ZnS/Al_2O_3 and CdS/Al_2O_3) functioned admirably along with DBD plasma in the decay of H_2S (Bao et al. 2008). Metal sulfide semiconductors in the terminal hole can be invigorated by both the solid electric field and light illumination, creating profoundly dynamic opening electron matches. The created opening and electron will respond with the adsorbed surface species, like H_2S, HS, and H+, in this way speeding up the development of H_2 and sulfur (Zhao et al. 2013a, b). In this way, since the created opening electron matches are adequately responsive to change H_2S over completely to H_2 and sulfur, a higher populace of opening electron matches can be connected to the better exhibition on H_2S decay. The progressions in molar proportion of Cr/Zn impacted the photo physical properties of Cr doped ZnS. As portrayed by UV-noticeable spectra (found in Fig. 4.16), with expanding the Cr content, a repetitive variety in the assimilation in apparent light area and band hole of Cr_x–ZnS could be found obviously.

For the impetus with smaller band hole, less energy for electrons is expected to hop from VB to CB. Accordingly, a decline in the band hole of impetus can instigate an

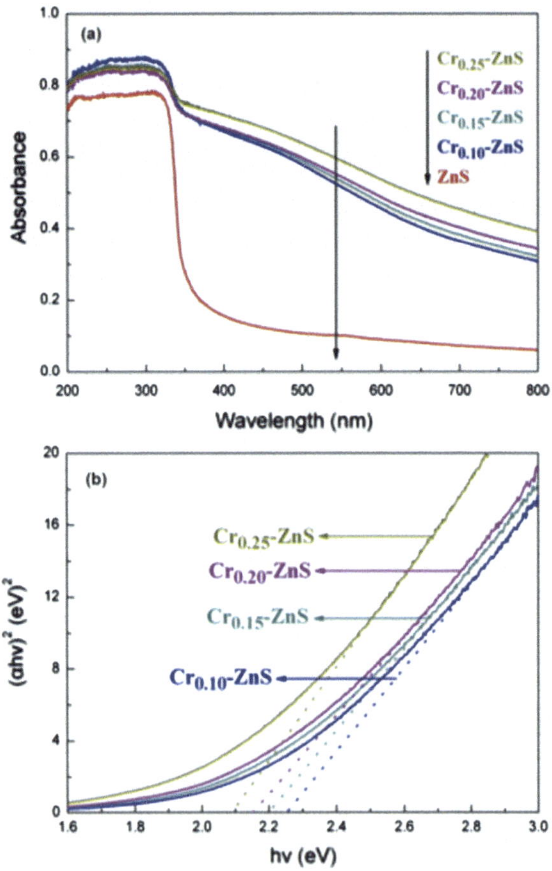

Fig. 4.16 UV-visible diffuse reflection spectra for the ZnS and Cr_x–ZnS catalysts with different Cr/Zn molar ratios [Reproduced with permissions from Zhao et al. (2019)]

increment of opening electron pair populace. Furthermore, the doped Cr3+ particles were approved to be consistently integrated into the interstitial destinations of ZnS grid, which would bring about the outperformed positive charges and the Zn opportunities arrangement. Preferably, one Zn opportunity might be created by the fuse of two Cr3+ particles (Zeng et al. 2012). Subsequently, countless Zn opportunities can be created by the Cr3+ particles consolidation. Zn opportunities advanced the detachment of the energy initiated openings and electrons, which prompted the better catalyst conduct in H_2S deterioration, contrasted and the undoped ZnS impetus. Be that as it may, albeit the $Cr_{0.25}$–ZnS impetus displayed more grounded noticeable light retention than other Cr-doped ZnS impetuses with lower Cr content, the H_2S change was lower. For a diminished catalyst movement of $Cr_{0.25}$ZnS, expanding how much Cr doping would achieve the lopsided dispersion of the Cr3+ particles. The opening electron recombination was considered to become predominant, because of the greater centralization of Cr doping. In this way, a pointless expansion in the Cr content not just given the more recombination habitats to opening electron matches, yet in addition hampered the light assimilation of the Cr-doped ZnS impetuses, hence the catalyst movement was diminished. Creator established that a Cr/Zn molar proportion of 0.20 was the most reasonable for the primary and optical properties regarding H_2S decomposition action. Higher H_2S transformation was gotten for the feed gas with lower H_2S fixation. In the interim, an increment of SIE came about in fundamentally improved H_2S transformation. Reddy et al. (2012a, b) detailed comparable outcomes during the H_2S deterioration in a non-warm plasma. At lower H_2S focus, a bigger piece of the electrons slam into Ar balance gas, demonstrating that Ar balance gas likewise assume fundamental parts in the decay (Zhao et al. 2007). To exhibit the steadiness of the Cr-doped ZnS impetus, $Cr_{0.20}$–ZnS was picked as the delegate impetus to examine the catalyst security at full H_2S conversion and H_2S deterioration response consequence of the $Cr_{0.20}$–ZnS impetus (Fig. 4.17). It uncovered that the reactant action didn't display misfortune in the runs. Besides, the

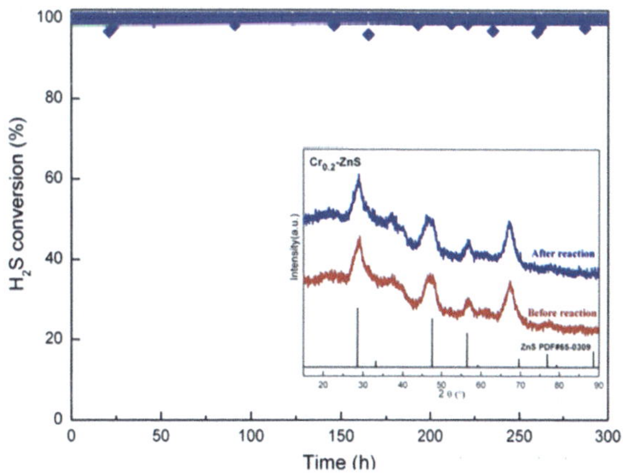

Fig. 4.17 Shows the long-time H_2S decomposition reaction result of the $Cr_{0.20}$ ZnS catalyst [Reproduced with permissions from Zhao et al. (2019)]

XRD profiles of the impetus when the response barely changed demonstrating that the precious stone design went through no variety during the H_2S plasma.

To work on the exhibition of photo electrochemical cells, different metal particles have been integrated into various host semiconductor nanocrystals (Sheng et al. 2020). In any case, such a methodology brings up two new issues that should be paid all due respects to plan and upgrade this new doping technique: (i) how would we precisely assess and rank these dopants? also (ii) how do the dopant related mid hole states tailor the band construction of the host semiconductors and the elements of the photo generated transporters? Creator in this work utilized nanosheet-nanorod ZnO as a detached substrate and CdS as a host for indium, copper, manganese, and cerium doping. Given the uniform actual properties and electronic construction of this climate, the photocurrent thickness and photo electrochemical productivity increment from ZnOeCdS (3.0 mA/cm^2, 0.41%), ZnOeCdSeIn (4.4 mA/cm^2, 0.59%), ZnOeCdSeCu (5.1 mA/cm^2, 0.98%), ZnOeCdSeMn (6.7 mA/cm^2, 1.07%), to ZnOeCdSeCe (8.9 mA/cm^2, 1.9%). With the assistance of electrochemical cyclic voltammetry, open-circuit voltage-rot estimations, and fluorescence spectroscopy, the band edges custom fitted by the mid hole states and the transporter elements not entirely set in stone. Besides, a mind boggling connection between these properties and the material presentation is laid out for precisely interpreting the photo electrochemical action. By observing the OH fixation during light, the powerful photo electrochemical catalyst action is likewise obviously portrayed by involving ZnOeCdSeMn and ZnOCdSeCe as model photocatalysts. Critical work has been done on the photo electrochemical debasement of water-broke down poisons at a lighted semiconductor-electrolyte intersection and plays very much archived the part of hydroxyl revolutionaries (OH). Goodness produced during illumination on the outer layer of a photocatalyst is the fundamental species liable for the corruption of toxin atoms. It is in this manner helpful to screen the convergence of OH to unequivocally assess the catalyst movement (Hisatomi et al. 2014). In this work, the ZnOeCdSeMn and ZnOeCdSeCe photo cathodes are utilized as model photocatalysts. Goodness shaped at these photo cathodes can be distinguished from the photoluminescence strategy utilizing terephthalic corrosive (TA) as a test atom. TA promptly responds with OH to create the emphatically fluorescent item TAOH, and the force of fluorescence is corresponding to how much OH. As displayed in Fig. 4.18a, b, the discharge force increments continuously with expanding light time for both ZnOeCdSeMn and ZnOeCdSeCe. To represent the unique movement of the photocatalysts, the emanation power is gathered each 30 min during illumination. Figure 4.18c shows the relative additions, which demonstrate that the photoelectrochemical catalyst action of ZnOeCdSeCe is better than that of ZnOeCdSeMn. For both model photocatalysts, an enormous grouping of OH shows up during the underlying 90 min of light, trailed by an unmistakable decline in OH fixation. This outcome is predictable with the broadly detailed lessening of the photocatalytic rate, however doesn't totally concur with the clarification given that the contamination beginning focus subordinate rate variety (Sakthivel et al. 2003, p. 2). A seriously convincing clarification is that the impetus deactivation is connected with the photostability. A more cautious investigation of the overall additions in Fig. 18c prompts the end that the photocatalytic movement of ZnOeCdSeCe

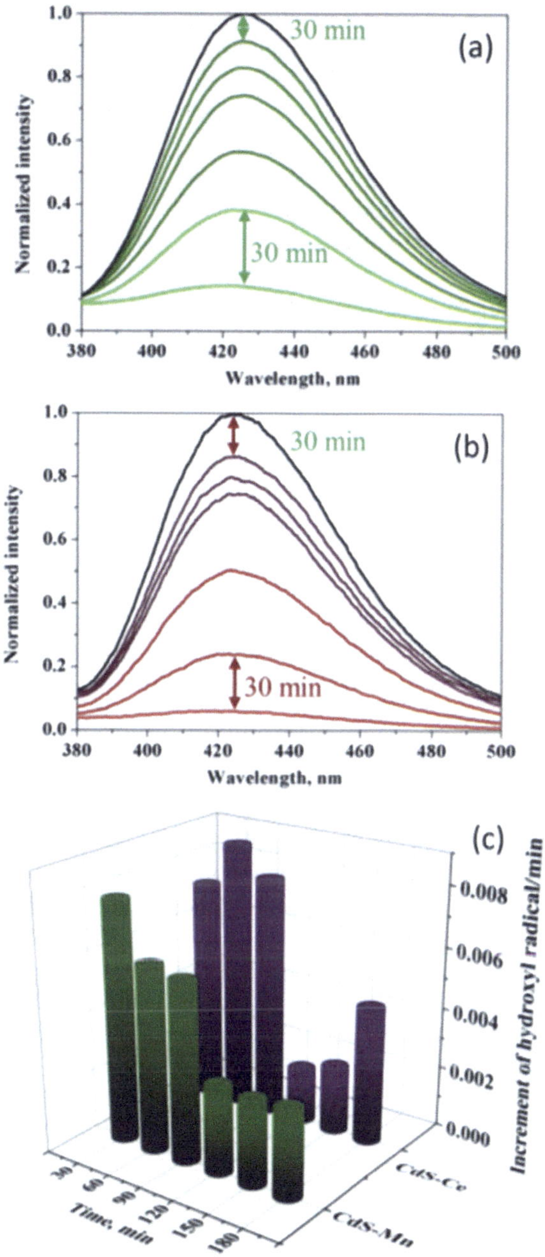

Fig. 4.18 Content change of hydroxyl radical produced on **a** ZnOeCdSeMn, **b** ZnOeCdSeCe under nonstop illumination for 180 min. **c** The variety of the augmentation of the hydroxyl extremist (the information were gathered every 30 min). For instance, the addition rate got at 180 min is the distinction in the standardized photoluminescence power somewhere in the range of 180 and 150 min separated by 30 min [Reproduced with consents from Bryan and Gamelin (2005)]

arrives at the ideal state after 30–60 min of illumination, following which a long downtrend happens that includes an unassuming decrease in photocatalytic action from 60 to 90 min of light, and a strong reduction in photocatalytic action from 90 to 120 min of illumination. Curiously, following this light stage (i.e., the most reduced pace of OH age), a slow increment happens from 120 to 180 min of illumination. For ZnOeCdSeMn, the pace of expansion in OH age step by step diminishes all through the 180 min of illumination.

Note that an unmistakable reduction in the pace of OH age happens from 90 to 120 min of light. After this point, the pace of OH age stays steady. The varieties in the pace of OH age for the two model photocatalysts likewise concur with past reports on long-term photostability tests with photocurrent thickness change (Santra and Kamat 2012; Li et al. 2014; Ertis and Boz 2017) using Ce doped CdS as model impetus showed that photodegradation movement on methyl blue encounters comparable inclination. Ranjith et al. (2018) detailed that debased Rhodamine B utilizing Ni/Ce co-doped CdS NPs as impetuses likewise affirmed a conspicuous expansion in action after 100 min illumination. Xiang et al. (2011) ever precisely and quantitatively concentrated on the OH created on different semiconductor photocatalysts, affirming that the natural properties of semiconductors (counting the band design and stage structure), pH variance impacted the creation paces of OH.

Fu et al. (2020) reported a three-dimensional (3D) photocatalyst synthesized through the self-assembly of graphitic carbon nitride (gC_3N_4), TiO_2, graphene (Gr), and carbon nanotubes (CNTs), in which both Gr and CNTs act as electron mediators, to form an all-solid-state g-C_3N_4/Gr–CNTs/TiO_2 Z-scheme photocatalytic system. In this Z-scheme photocatalyst, the highly interconnected CNTs not only can effectively promote the transfer of photogenerated charge carriers, but also built a 3D open frame network to harvest more incident light and provide more specific surface area for photoreactions. Benefiting from the 3D nanostructure and Z scheme system, phenol removal over g-C_3N_4/GrCNTs/TiO_2 reached up to 90% within 120 min using a high-pressure Xe short-arc lamp as light source, which was much higher than that achieved over pure g-C_3N_4. The 3D Z-scheme photocatalyst fabricated in this study may be an attractive material for many environmental and energy-related applications. The photocatalytic activity of g-C_3N_4/Gr–CNTs/TiO_2 was evaluated using phenol as the substrate. Before light irradiation, the mixed solution of phenol and photocatalyst was stirred for 60 min to attained absorption equilibrium. Approximately 9% of phenol was absorbed by gC_3N_4/Gr–CNTs/TiO_2. Approximately 90% of phenol was removed after light irradiation for 120 min in the presence of 3D CN–Gr–C–T_3, while 50%, 68%, and 81% of phenol was removed using 3D CN–Gr–C–T1, 3D CN–Gr–C–T2, and 3D CN–Gr–C–T4 respectively.

Apparently, there is an optimum Gr–CNTs content in g-C_3N_4/Gr–CNTs/TiO_2. When the mass ratio of g-C_3N_4 to Gr–CNTs is lower than 0.7, the photocatalytic activity of g-C_3N_4/Gr-CNTs/TiO_2 increased gradually as the Gr–CNTs content increased. However, when the Gr-CNTs content was increased further, the photocatalytic activity of g-C_3N_4/Gr–CNTs/TiO_2 decreased; this may be because excess Gr–CNTs hinder the light absorption of photocatalysts. In addition, the pure g-C_3N_4 and pure TiO_2 exhibits a lower phenol removal rate than that of 3D g-C_3N_4/GrCNTs/TiO_2

photocatalysts. In order to reveal the reactive oxygen species, the ESR spin-trapping technique has been applied using DMPO as the spin-trap reagent (Chen et al. 2018, 2019). The typical quartet peaks with a 1:2:2:1 intensity are observed for g-C_3N_4/Gr–CNTs/TiO_2 under irradiation, which could be ascribed to the DMPO-%OH adduct derived from H_2O oxidization by photogenerated holes (Huang et al. 2017). Meanwhile, it can be seen from Fig. 18b that evident six identical peaks appeared, which is assigned to DMPO-%O^{2-} adduct derived from O^2 reduction by photogenerated electrons (Li et al. 2020), whereas no ESR signals were observed in the dark. This result indicated that both hydroxyl radicals (%OH) and superoxide anion radicals (%O^{2-}) are the dominant reactive radicals in photocatalytic process.

Previous works have confirmed that both g-C_3N_4 and TiO_2 are suitable candidates to construct the Z-scheme photocatalyst because of their matched band gap positions and extremely strong anti-light corrosion (Lu et al. 2018; Wang and Liu 2019).

When the g-C_3N_4/Gr CNTs/TiO_2 is irradiated with incident light, a typical Z-scheme system is formed with g-C_3N_4, TiO_2, Gr and CNTs. In this artificial Z-scheme system, g-C_3N_4 and TiO_2 serve as the PS I and PS II, respectively. When the incident light energy is larger than the energy gap of g-C_3N_4 and TiO_2, the photogenerated electrons in the valence band (VB) would be migrated to the conduction band (CB) and left photogenerated holes in the VB. Particularly, benefiting from the super conductivity, CNTs could be served as the solid-state electron mediators and provide pathways for the photogenerated electrons in TiO_2 and the photogenerated holes in g-C_3N_4 moving to the Gr and following to recombine in the Gr. Meanwhile, the photogenerated electrons in the CB of g-C_3N_4 accumulate to form an electron rich region and the photogenerated holes in the VB of TiO_2 accumulate to form a hole-rich region. Obviously, the electrons in the CB of g-C_3N_4 could reduce O_2 to yield superoxide anion radicals (%O^{2-}), and the holes in the VB of TiO_2 could oxidize H_2O to yield hydroxyl radicals (%OH). Therefore, this artificial Z-scheme g-C_3N_4/Gr–CNTs/TiO_2 features the spatial isolation of photogenerated electrons and holes, resulting in the efficient charge separation and enhanced photocatalytic activities.

4.4 Conclusion

In modern era, researchers have developed that the semiconductors after doping of impurities could impact various catalytic, electrical, optical and magnetic properties of the materials. Deep studies have been performed all around the universe on atoms such as Mn, Cu, Co, and many other doped semiconductor materials for their novel advanced properties and their application in diverse fields. Semiconductors with magnetic impurities, also known as diluted magnetic semiconductors or semi-magnetic semiconductors, have been explored for decades.

This chapter has been an extensive review for classification of existing photocatalytic materials and the methods for development of their performance as catalysts. The potential industrial application usage of this technology is also discussed in detail.

References

Ahmad M, Ahmed E, Ahmed W, Elhissi A, Hong ZL, Khalid NR (2014) Enhancing visible light responsive photocatalytic activity by decorating Mn-doped ZnO nanoparticles on graphene. Ceram Int 40(7):10085–10097

Ali T, Ahmed A, Alam U, Uddin I, Tripathi P, Muneer M (2018) Enhanced photocatalytic and antibacterial activities of Ag-doped TiO_2 nanoparticles under visible light. Mater Chem Phys 212:325–335

Ambigadevi J, Kumar PS, Vo D-VN, Haran SH, Raghavan TS (2020) Recent developments in photocatalytic remediation of textile effluent using semiconductor based nanostructured catalyst: a review. J Environ Chem Eng, p 104881

Bao N, Shen L, Takata T, Domen K (2008) Self-templated synthesis of nanoporous CdS nanostructures for highly efficient photocatalytic hydrogen production under visible light. Chem Mater 20(1):110–117

Bi X et al (2020) Construction of g-C_3N_4/TiO_2 nanotube arrays Z-scheme heterojunction to improve visible light catalytic activity. Colloids Surf, A 603:125193

Bodke MR, Purushotham Y, Dole BN (2014) Crystallographic and optical studies on Cr doped ZnS nanocrystals. Cerâmica 60(355):425–428

Bryan JD, Gamelin DR (2005) Doped semiconductor nanocrystals: synthesis, characterization, physical properties, and applications. Prog Inorg Chem 54(47):47–126

Chen S, Pan Y (2021) Noble metal interlayer-doping enhances the catalytic activity of 2H–MoS_2 from first-principles investigations. Int J Hydrogen Energy 46(40):21040–21049

Chen F et al (2018) Thickness-dependent facet junction control of layered $BiOIO_3$ single crystals for highly efficient CO_2 photoreduction. Adv Func Mater 28(46):1804284

Chen F, Huang H, Guo L, Zhang Y, Ma T (2019) The role of polarization in photocatalysis. Angew Chem Int Ed 58(30):10061–10073

Chen C et al (2020) Synthesis of a flower-like SnO/ZnO nanostructure with high catalytic activity and stability under natural sunlight. J Alloy Compd 826:154122

Dave S, Khan AM, Purohit SD, Suthar DL (2021a) application of green synthesized metal nanoparticles in the photocatalytic degradation of dyes and its mathematical modelling using the caputo–fabrizio fractional derivative without the singular kernel. J Math. https://doi.org/10.1155/2021/9948422

Dave S, Jagtap P, Verma S, Nehra R, Dave S, Mohanty P, Das J (2021b) Mathematical modeling and surface response curves for green synthesized nanomaterials and their application in dye degradation. Photocat Degrad Dyes 1:571–591

Dave S, Dave S, Das J (2021c) Photocatalytic degradation of dyes in textile effluent: a green approach to eradicate environmental pollution. In: The future of effluent treatment plants. Elsevier, pp 199–214

Demirci S, Dikici T, Yurddaskal M, Gultekin S, Toparli M, Celik E (2016) Synthesis and characterization of Ag doped TiO_2 heterojunction films and their photocatalytic performances. Appl Surf Sci 390:591–601

Ertis IF, Boz I (2017) Synthesis and characterization of metal-doped (Ni Co, Ce, Sb) CdS catalysts and their use in methylene blue degradation under visible light irradiation. Modern Res Catal 6(01):1

Ethiraj AS, Uttam P, Varunkumar K, Chong KF, Ali GA (2020) Photocatalytic performance of a novel semiconductor nanocatalyst: copper doped nickel oxide for phenol degradation. Mater Chem Phys 242:122520

Fu Z, Wang H, Wang Y, Wang S, Li Z, Sun Q (2020) Construction of three-dimensional g-C_3N_4/Gr-CNTs/TiO_2 Z-scheme catalyst with enhanced photocatalytic activity. Appl Surf Sci 510:145494

Giahi M, Pathania D, Agarwal S, Ali GA, Chong KF, Gupta VK (2019) Preparation of Mg-doped TiO_2 nanoparticles for photocatalytic degradation of some organic pollutants. Studia UBB Chemia 64:7–18

González-Rodríguez J et al (2020) Enhanced photocatalytic activity of semiconductor nanocomposites doped with ag nanoclusters under UV and visible light. Catalysts 10(1):31

Goto H, Hanada Y, Ohno T, Matsumura M (2004) Quantitative analysis of superoxide ion and hydrogen peroxide produced from molecular oxygen on photoirradiated TiO_2 particles. J Catal 225(1):223–229

Guha S et al (2003) Temperature-dependent optical studies of $Ti_{1-x}Co_xO_2$. Appl Phys Lett 83(16):3296–3298

Guillén-Santiago A, Mayén SA, Torres-Delgado G, Castanedo-Pérez R, Maldonado A, de la L. Olvera M (2010) Photocatalytic degradation of methylene blue using undoped and Ag-doped TiO_2 thin films deposited by a sol-gel process: effect of the ageing time of the starting solution and the film thickness. Mater Sci Eng B 174(1–3):84–87

Hisatomi T, Kubota J, Domen K (2014) Recent advances in semiconductors for photocatalytic and photoelectrochemical water splitting. Chem Soc Rev 43(22):7520–7535

Houas A, Lachheb H, Ksibi M, Elaloui E, Guillard C, Herrmann J-M (2001) Photocatalytic degradation pathway of methylene blue in water. Appl Catal B 31(2):145–157

Hu J-S et al (2005) Mass production and high photocatalytic activity of ZnS nanoporous nanoparticles. Angew Chem Int Ed 44(8):1269–1273

Huang H, Tu S, Zeng C, Zhang T, Reshak AH, Zhang Y (2017) Macroscopic polarization enhancement promoting photo-and piezoelectric-induced charge separation and molecular oxygen activation. Angew Chem Int Ed 56(39):11860–11864

Ilkme ES, Soylu GSP (2021) The role of some metal ions in enhancement of photocatalytic activity of Fe_2O_3-V_2O_5 binary oxide. Turk J Chem 45(2):348–361

Khlyustova A, Sirotkin N, Kusova T, Kraev A, Titov V, Agafonov A (2020) Doped TiO_2: the effect of doping elements on photocatalytic activity. Mater Adv 1(5):1193–1201

Khurshid F, Jeyavelan M, Hudson MSL, Nagarajan S (2019) Ag-doped ZnO nanorods embedded reduced graphene oxide nanocomposite for photo-electrochemical applications. R Soc Open Sci 6(2):181764

Kumar R, Rana D, Umar A, Sharma P, Chauhan S, Chauhan MS (2015) Ag-doped ZnO nanoellipsoids: potential scaffold for photocatalytic and sensing applications. Talanta 137:204–213

Kwon YT, Song KY, Lee WI, Choi GJ, Do YR (2000) Photocatalytic behavior of WO_3-loaded TiO_2 in an oxidation reaction. J Catal 191(1):192–199

Lachheb H et al (2002) Photocatalytic degradation of various types of dyes (Alizarin S, Crocein Orange G, Methyl Red, Congo Red, Methylene Blue) in water by UV-irradiated titania. Appl Catal B 39(1):75–90

Lee H et al (2018) Comparative study of catalytic activities among transition metal-doped IrO_2 nanoparticles. Sci Rep 8(1):1–8

Li D, Haneda H, Ohashi N, Hishita S, Yoshikawa Y (2004) Synthesis of nanosized nitrogen-containing MO_x–ZnO(M=W, V, Fe) composite powders by spray pyrolysis and their visible-light-driven photocatalysis in gas-phase acetaldehyde decomposition. Catal Today 93:895–901

Li W et al (2014) Stable core/shell CdTe/Mn-CdS quantum dots sensitized three-dimensional, macroporous ZnO nanosheet photoelectrode and their photoelectrochemical properties. ACS Appl Mater Interfaces 6(15):12353–12362

Li J et al (2020) Unraveling the mechanism of binary channel reactions in photocatalytic formaldehyde decomposition for promoted mineralization. Appl Catal B 260:118130

Li K et al (2021) New insight into the mechanism of enhanced photo-Fenton reaction efficiency for Fe-doped semiconductors: A case study of Fe/g-C_3N_4. Catal Today 371:58–63
Lin H-F, Liao S-C, Hung S-W (2005) The dc thermal plasma synthesis of ZnO nanoparticles for visible-light photocatalyst. J Photochem Photobiol, A 174(1):82–87
Liu G-S et al (2018) In situ electrospinning iodine-based fibrous meshes for antibacterial wound dressing. Nanoscale Res Lett 13(1):1–7
Lu L, Wang G, Zou M, Wang J, Li J (2018) Effects of calcining temperature on formation of hierarchical TiO_2/g-C_3N_4 hybrids as an effective Z-scheme heterojunction photocatalyst. Appl Surf Sci 441:1012–1023
Masakazu A, Takahit S (1987) Photocatalytic hydrogenation of CH_3CCH with H_2O on small-particle TiO_2: size quantization effects and reaction intermediates. Chem 91:4305–4310
Moballegh A, Shahverdi HR, Aghababazadeh R, Mirhabibi AR (2007) ZnO nanoparticles obtained by mechanochemical technique and the optical properties. Surf Sci 601(13):2850–2854
Mohammed MK (2020) Sol-gel synthesis of Au-doped TiO_2 supported SWCNT nanohybrid with visible-light-driven photocatalytic for high degradation performance toward methylene blue dye. Optik 223:165607
Muñoz-Fernandez L, Sierra-Fernández A, Milošević O, Rabanal ME (2016) Solvothermal synthesis of Ag/ZnO and Pt/ZnO nanocomposites and comparison of their photocatalytic behaviors on dyes degradation. Adv Powder Technol 27(3):983–993
Nigussie GY et al (2018) Antibacterial activity of Ag-doped TiO_2 and Ag-doped ZnO nanoparticles. Int J Photoenergy 2018
Rahman MA (2014) A review on semiconductors including applications and temperature effects in semiconductors. Am Sci Res J Eng Technol Sci (ASRJETS) 7(1):50–70
Rahman R, Samanta D, Pathak A, Nath TK (2021) Tuning of structural and optical properties with enhanced catalytic activity in chemically synthesized Co-doped MoS_2 nanosheets. RSC Adv 11(3):1303–1319
Ranjith R, Krishnakumar V, Boobas S, Venkatesan J, Jayaprakash J (2018) An efficient photocatalytic and antibacterial performance of Ni/Ce–Codoped CdS nanostructure under visible light irradiation. ChemistrySelect 3(32):9259–9267
Reddy EL, Biju VM, Subrahmanyam C (2012a) Production of hydrogen from hydrogen sulfide assisted by dielectric barrier discharge. Int J Hydrogen Energy 37(3):2204–2209
Reddy EL, Biju VM, Subrahmanyam C (2012b) Production of hydrogen and sulfur from hydrogen sulfide assisted by nonthermal plasma. Appl Energy 95:87–92
Sakthivel S, Neppolian B, Shankar MV, Arabindoo B, Palanichamy M, Murugesan V (2003) Solar photocatalytic degradation of azo dye: comparison of photocatalytic efficiency of ZnO and TiO_2. Sol Energy Mater Sol Cells 77(1):65–82
Santra PK, Kamat PV (2012) Mn-doped quantum dot sensitized solar cells: a strategy to boost efficiency over 5%. J Am Chem Soc 134(5):2508–2511
Satapathy S, Acharya D, Dixit PK, Mishra G, Das J, Dave S (2021) Mechanistic aspects and rate-limiting steps in green synthesis of metal and metal oxide nanoparticles and their potential in photocatalytic degradation of textile dye. In: Photocatalytic degradation of dyes. Elsevier, pp 605–630
Senthilraja A, Subash B, Krishnakumar B, Rajamanickam D, Swaminathan M, Shanthi M (2014) Synthesis, characterization and catalytic activity of co-doped Ag–Au–ZnO for MB dye degradation under UV-A light. Mater Sci Semicond Process 22:83–91
Shah M, Dave S, Das J (2021) Photocatalytic degradation of dyes current trends and future perspectives
Sheng P et al (2020) The origin of enhanced photoelectrochemical activity in metal-ion-doped ZnO/CdS quantum dots. J Alloy Compd 822:153700
Subash B, Krishnakumar B, Pandiyan V, Swaminathan M, Shanthi M (2013) Synthesis and characterization of novel WO_3 loaded Ag–ZnO and its photocatalytic activity. Mater Res Bull 48(1):63–69

Subramanian V, Wolf EE, Kamat PV (2003) Green emission to probe photoinduced charging events in ZnO–Au nanoparticles. Charge distribution and Fermi-level equilibration. J Phys Chem B 107(30):7479–7485

Szabó-Bárdos E, Czili H, Horváth A (2003) Photocatalytic oxidation of oxalic acid enhanced by silver deposition on a TiO_2 surface. J Photochem Photobiol, A 154(2–3):195–201

Tong H, Ouyang S, Bi Y, Umezawa N, Oshikiri M, Ye J (2012) Nano-photocatalytic materials: possibilities and challenges. Adv Mater 24(2):229–251

Vallejo W, Cantillo A, Díaz-Uribe C (2020) Methylene blue photodegradation under visible irradiation on Ag-Doped ZnO thin films. Int J Photoenergy 2020

Wang X, Liu M (2019) Photocatalytic enhancement mechanism of direct Z-scheme heterojunction Og-C_3N_4@ Fe-TiO_2 under visible-light irradiation. Appl Surf Sci 485:353–360

Xiang Q, Yu J, Wong PK (2011) Quantitative characterization of hydroxyl radicals produced by various photocatalysts. J Colloid Interface Sci 357(1):163–167

Xing C, Zhang Y, Yan W, Guo L (2006) Band structure-controlled solid solution of $Cd_{1-x}Zn_xS$ photocatalyst for hydrogen production by water splitting. Int J Hydrogen Energy 31(14):2018–2024

Yang S, Lee H (2017) Determining the catalytic activity of transition metal-doped TiO_2 nanoparticles using surface spectroscopic analysis. Nanoscale Res Lett 12(1):1–8

You Y, Wan L, Zhang S, Xu D (2010) Effect of different doping methods on microstructure and photo-catalytic activity of Ag_2O–TiO_2 nanofibers. Mater Res Bull 45(12):1850–1854

Zeng X, Zhang J, Huang F (2012) Optical and magnetic properties of Cr-doped ZnS nanocrystallites. J Appl Phys 111(12):123525

Zhang F et al (2019) Recent advances and applications of semiconductor photocatalytic technology. Appl Sci 9(12):2489

Zhao G-B et al (2007) Production of hydrogen and sulfur from hydrogen sulfide in a nonthermal-plasma pulsed corona discharge reactor. Chem Eng Sci 62(8):2216–2227

Zhao L et al (2013a) Decomposition of hydrogen sulfide in non-thermal plasma aided by supported CdS and ZnS semiconductors. Green Chem 15(6):1509–1513

Zhao L, Wang Y, Li X, Wang A, Song C, Hu Y (2013b) Hydrogen production via decomposition of hydrogen sulfide by synergy of non-thermal plasma and semiconductor catalysis. Int J Hydrogen Energy 38(34):14415–14423

Zhao L, Wang Y, Wang A, Li X, Song C, Hu Y (2019) Cr–doped ZnS semiconductor catalyst with high catalytic activity for hydrogen production from hydrogen sulfide in non-thermal plasma. Catal Today 337:83–89

Zhong JB et al (2012) Improved photocatalytic performance of Pd-doped ZnO. Curr Appl Phys 12(3):998–1001

Chapter 5
Hybrid Treatment Technologies for Dye Degradation in Wastewater

Swati Singh

Abstract Industrial effluents from textile, paper, pharmaceutical and food processing industry are contaminated with high concentration of various dyes. Most of the dyes are aromatic with azo (–N=N–) and nitro (–NO_2) functional groups. The effluents containing dyes can affect aquatic life, human health and environment due to toxic, non-biodegradable and carcinogenic nature of dyes. Advanced oxidation processes (AOPs) are known to decolourize and degrade dye effluents. However, a single AOP or any other physicochemical process alone is ineffective for complete degradation of the dye. Therefore, hybrid treatment technologies comprising a combination of AOPs or physicochemical processes have gained recognition in the recent decade for the removal of recalcitrant compounds. In the present chapter, an in-depth evaluation of potential hybrid technologies is performed with a focus on previous literature on hybrid treatment methods for the degradation of dye in wastewater. In addition, limitations of the different hybrid systems are discussed for their better application during wastewater treatment.

Keywords Advanced oxidation processes · Dye wastewater · Hybrid treatment systems · Industrial wastewater treatment · Physicochemical treatment

5.1 Introduction

The rapid industrialization in the recent years has contaminated of one of the most essential natural resource—water. The increased volume of wastewater and pollution levels in natural water bodies have resulted water shortage in many parts of the world. Industries such as textile, leather dyeing, paper printing, cosmetic, pharmaceutical and food processing generate wastewater with high chemical oxygen demand (COD) and strong colour due to the presence of dyes and other chemicals (Khataee et al. 2012). Textile effluents have strong colour, high pH, suspended solids, organics,

S. Singh (✉)
Environmental Science and Engineering Department, Indian Institute of Technology Bombay, Mumbai 400076, India
e-mail: swatisingh0109@gmail.com

S. Dave and J. Das (eds.), *Trends and Contemporary Technologies for Photocatalytic Degradation of Dyes*, Environmental Science and Engineering,
https://doi.org/10.1007/978-3-031-08991-6_5

Table 5.1 Typical wastewater characteristic of textile industry effluents

Parameters	Values						
	Aydiner et al. (2019)	GilPavas et al. (2019)	Zazou et al. (2019)	Geraldino et al. (2020)	Silva et al. (2020)	Lebron et al. (2021)	Louhichi et al. (2022)
pH	6.2	9.1	8.75	5.7	7.9	7.7	6.8
Temperature (°C)	25.4	n.a	22	27.5	n.a	n.a	20.2
Colour (mg Pt-Co/L)	n.a	1366	n.a	1715	134	n.a	n.a
COD (mg/L)	2830	875	267	656	65	1771	713
BOD (mg/L)	n.a	190	n.a	158	n.a	n.a	n.a
Electric conductivity (mS/cm)	2.6	12.8	2.5	1.4	2.3	3.8	8.6

n.a. not assessed

metals and salts (Shindhal et al. 2020). The typical wastewater characteristics of a textile industry wastewater from recent literature are summarized in Table 5.1. The wastewater composition depends on the process, equipment, fabric, dye and season.

The dyes can be characterized based on their origin as natural and synthetic dyes. The synthetic dyes are further classified based on their mode of application (such as reactive, direct, disperse, basic and vat dyeing) and according to their chemical structure or functional group (i.e., azo, anthraquinone, sulphur, triarylmethane and phthalocyanine (Yaseen and Scholz 2019). Chromophore groups such as azo (–N=N–), nitro (–N=O), carbonyl (–C=O) quinoid and auxochrome groups (e.g., amine, carboxyl, sulphonate and hydroxyl) in the dye structure impart colour to the dye (Holkar et al. 2016).

Dye containing wastewater can pollute the nearby water bodies and are a serious threat to human health and aquatic life due to the toxic and carcinogenic nature of dyes. The colour of wastewater is aesthetically unpleasant and it can affect the oxygen solubility in water bodies (Gupta et al. 2015). The recalcitrant nature of dye makes it resistant to the conventional biological treatment processes. Hence, it is essential to treat dye containing effluents prior to their discharge into aquatic systems.

Literature suggests that the application of a single process may not be sufficient for the complete degradation of mixtures of dye in wastewater. Therefore, there is a growing interest in hybrid treatment technologies comprising a combination of AOPs or physicochemical processes for effective dye degradation. However, hybrid treatment methods must be evaluated for their application at large scale. The present chapter provides an overview of various existing technologies and emerging hybrid technologies for the degradation of dye in wastewater. In addition, the limitations and challenges of hybrid treatment methods are also discussed. Further, recommendations are provided for future research work in this area.

5.2 Advanced Treatment Technologies

Advanced oxidation processes (AOPs) offer a viable treatment option for treatment of refractory wastewaters having low biodegradability. They have been found to be effective for the degradation and mineralization of recalcitrant organic compounds (Oturan and Aaron 2014). AOPs are based on in situ production of highly reactive and unselective hydroxyl radicals. AOPs can be broadly classified into Fenton-type, ozone based, irradiation based, electrical, ultrasonic and thermal processes. These processes can be carried out in the presence or absence of catalysts in homogenous or heterogeneous medium. An overview of various AOPs is present in Fig. 5.1.

The selection of an AOP is dependent on the wastewater characteristics, cost effectiveness and environmental compliances which have to be met for any location. A few AOPs such as ozonation and photolysis are already well established for wastewater treatment at large scale (Miklos et al. 2018). However, many of the other AOPs must be evaluated for their performance at industrial scale. In this section, the process and reactions of the most widely used AOPs are briefly discussed. In addition, the previous literature on the degradation of dyes by the selected AOPs are also summarized.

5.2.1 Fenton-Type Processes

The traditional Fenton's oxidation process involves the formation of highly reactive hydroxyl radical ($OH^{\cdot}$) in presence of Fenton's reagent i.e., a combination of

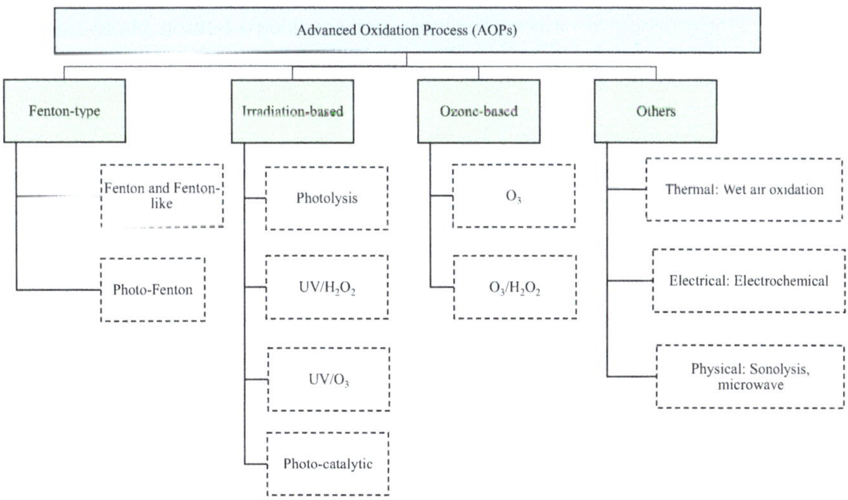

Fig. 5.1 Classification of advanced oxidation processes for wastewater treatment

hydrogen peroxide and ferrous ions in homogenous system (Babuponnusami and Muthukumar 2014). Hydroxyl and sulphate radicals can be generated from H_2O_2 and $S_2O_8^{2-}$, respectively in the presence of ferrous ions as per Eqs. 5.1 and 5.2 (Walling 1975). The redox potentials of hydroxyl ($OH^{\cdot}$) and sulphate ($SO4^{-\cdot}$) radicals are 2.8 eV and 2.5–3.1 eV, respectively (Anandan et al. 2020). The generated $OH^{\cdot}$ radical can abstract the hydrogen from organic compounds to produce organic radical ($R^{\cdot}$) as per Eq. 5.3.

$$Fe^{2+} + H_2O_2 \rightarrow Fe^{3+} + OH^- + OH^{\cdot} \tag{5.1}$$

$$S_2O_8{}^{2-} + Fe^{2+} \rightarrow SO_4{}^{\cdot-} + SO_4{}^{2-} + Fe^{3+} \tag{5.2}$$

$$OH^{\cdot} + RH \rightarrow H_2O + R^{\cdot} \tag{5.3}$$

Fenton's oxidation has been used previously for the treatment of a mixture of four reactive dyes (i.e., remazol black 5, remazol red, remazol yellow 84 and remazol brilliant blue) having total initial dye concentration of 300 mg/L by Meriç et al. (2003). The reaction pH, temperature, $FeSO_4$ concentration and H_2O_2 dose were 4, 50 °C, 500 mg/L and 1000 mg/L, respectively. Under the above mentioned conditions, a COD removal of 93% and colour removal of >99% could be achieved. In contrast, heterogeneous Fenton-like oxidation was carried out using scrap zero-valent iron for textile industry wastewater having initial COD of 875 mg/L (GilPavas et al. 2019). A colour removal of 95% and COD removal of 76% was obtained at initial pH of 3 with 2 g/L catalyst and 24.5 mM H_2O_2 dose after 1 h duration. An advantage of Fenton's process is that it can be performed at room temperature and the reagents used during the process are readily available (Silva et al. 2020). To further increase the efficiency of the process, variation of Fenton's such are electro-Fenton, photo-Fenton, sono-Fenton and Fenton-like processes have been explored.

5.2.2 *Irradiation Based Processes*

Photolysis oxidation is carried out in the presence of a light source (UV, visible or LED). The presence of UV irradiation to the oxidants such as hydrogen peroxide or persulphate can enhance their degradation to form hydroxyl ($OH^{\cdot}$) and sulphate ($SO_4{}^{\cdot-}$) radicals as per Eqs. 5.4 and 5.5 (Samsami et al. 2020).

$$H_2O_2 + hv \rightarrow 2\,OH^{\cdot} \tag{5.4}$$

$$S_2O_8{}^{2-} + hv \rightarrow SO_4{}^{\cdot-} \tag{5.5}$$

In past, photo-oxidation (UV/H_2O_2) of a mixture of dyes (CI basic red 46, malachite green and CI basic blue) showed a colour removal of 97% by Khataee et al. (2012). The optimum condition for initial concentration of dyes, H_2O_2 dose and reaction time were 4 mg/L, 48 mg/L and 30 min, respectively as determined from central composite design.

The photocatalytic degradation using semiconductors as catalysts involves the adsorption of light or photon having energy greater than bandgap of the semiconductor material forming electron–hole pairs which can result in formation of hydroxyl radicals as per Eqs. 5.6 and 5.7 (Miklos et al. 2018; Dhangar and Kumar 2020).

$$TiO_2 + h\nu \rightarrow \left(e^- + h^+\right) \tag{5.6}$$

$$h^+ + OH^-{}_{ad} \rightarrow {}^{\cdot}OH_{ad} \tag{5.7}$$

Semiconductor materials like TiO_2 and ZnO have been widely used as a catalyst due to their higher efficiency under UV irradiation in the last few decades. Juang et al. (2010) performed the photocatalytic oxidation of acid orange 7 and reactive red 2 having initial concentration of 0.086 mM each under UV/TiO_2 system. The authors have reported complete removal of dyes with 0.5 g/L TiO_2 dose after 20 min of reaction.

Photo-Fenton process is a modification of traditional Fenton's process where light energy (UV or solar) is coupled with Fenton's reagent. Complete decolourization of a mixture of dyes (remazol red RR and remazol blue RB) could be achieved in 2 h during homogeneous photo-Fenton oxidation at an initial reaction pH of 3 (Punzi et al. 2012). The initial concentration of each dye was 50 mg/L with 0.25 mM iron and 12 mM H_2O_2 dose. Photocatalytic processes have the advantages of higher efficiencies without the formation of sludge. However, the efficiency of photocatalytic process depends on the availability of photon, intensity of light, reaction pH, presence of oxidants and catalysts (Oturan and Aaron 2014).

5.2.3 *Ozone Based Processes*

Ozone is an unstable gas which can react directly as molecular ozone or through secondary oxidants i.e., hydroxyl radicals. The reaction of ozone with organic compound can be represented by Eq. 5.8 which suggests that for every three moles of ozone, two moles of $OH^{\cdot}$ radical are generated (Malik et al. 2020).

$$3O_3 + OH^- + H^+ \rightarrow 2OH^{\cdot} + 4O_2 \tag{5.8}$$

The chain termination occurs in the presence of carbonates, bicarbonates and hydroperoxyl radical (Eqs. 5.9–5.11).

$$OH^{\cdot} + CO_3^{2-} \rightarrow CO_3^{\cdot -} + OH^- \quad (5.9)$$

$$OH^{\cdot} + HCO_3^- \rightarrow HCO_3^{\cdot} + OH^- \quad (5.10)$$

$$OH^{\cdot} + OH_2^{\cdot} \rightarrow O_2 + H_2O \quad (5.11)$$

Ozonation has been used effectively for the selective degradation of organic pollutants having double bonds (e.g., C=C) and functional groups (e.g., $-CH_3$, –OH). Previously, wastewater containing sirius blue dye (initial concentration = 400 mg/L) showed 91% colour removal with ozone dose of 24.03 g/m^3 by Turhan and Turgut (2009). Ozonation can be carried out at room temperature and pressure with no sludge production (Dhangar and Kumar 2020). Therefore, it considered a cleaner process for the treatment of high strength wastewater.

5.2.4 *Other AOPs*

Electrochemical processes have been used widely for water and wastewater treatment. The effect of different electrode materials, oxidants and experimental conditions for a variety of pollutant has been reviewed by Chen (2004). These processes are not affected by the presence of salts or ions in the wastewater. Electrochemical oxidation of synthetic wastewater comprising sixteen dyes (total initial dye concentration = 361 mg/L, initial COD = 281 mg/L) using titanium-tantalum-platinum-iridium (Ti–Ta–Pt–Ir) anode and stainless steel cathode showed more than 95% decolourization within 15 min (Chatzisymeon et al. 2006). The initial reaction pH, electrolyte and current were 7.5, NaCl (0.5%) and 5 A, respectively.

Sonolytic processes are AOPs which use ultrasound wave for the degradation of organic contaminants. Ultrasound waves have frequencies ranging higher than the hearing range of humans i.e., 20 kHz to 10 MHz (Pirsaheb and Moradi 2021). Cavitation is process of formation, growth and implosion of bubble or cavities under ultrasound waves producing high pressures and temperatures in the microsystem (Gągol et al. 2018). The ultrasound waves are transmitted by acoustic cavitation where highly reactive radicals are generated as per Eq. 5.12. The hydrogen radical produced converts into hydroperoxyl radical in the presence of oxygen which later forms hydrogen peroxide as per Eqs. 5.13 and 5.14 (Anandan et al. 2020).

$$H_2O + \text{ultrasound}))))\rightarrow OH^{\cdot} + H^{\cdot} \quad (5.12)$$

$$H^{\cdot} + O_2 \rightarrow HO_2^{\cdot} \quad (5.13)$$

$$2\,HO_2^{\cdot} \rightarrow H_2O_2 + O_2 \quad (5.14)$$

Recently, sonocatalytic oxidation of methyl orange (initial concentration = 20 mg/L) was performed at 42 kHz ultrasound frequency in the presence of Au/Fe_3O_4 nanoparticles as catalysts (de Jesús Ruíz-Baltazar 2021). A removal of 92% was observed during the process with catalyst dose of 0.075 g/L. Sonolysis process can be operated at ambient conditions without the use of any chemicals. They are considered safer and cleaner processes that are not affected by the toxicity or biodegradability of the pollutant (Chakma and Moholkar 2016; Pirsaheb and Moradi 2021). Therefore, it is widely combined with other treatment techniques for enhanced removals. The advanced oxidative treatment options for dye removal are summarized in Table 5.2.

Table 5.2 Degradation of dyes using advanced oxidation processes

Process	Wastewater characteristics	Results	References
Fenton's oxidation	Total concentration of dye mixture = 300 mg/L	COD removal = 93%; colour removal = >99%	Meriç et al. (2003)
Fenton-like oxidation	Initial COD of textile industry wastewater = 875 mg/L	COD removal = 76%; colour removal = 95% in 1 h	GilPavas et al. (2019)
UV/H_2O_2	Concentration of dye mixture (CI basic red 46, malachite green and CI basic blue) = 12 mg/L	Colour removal = 97% in 30 min	Khataee et al. (2012)
Photo-Fenton process	Concentration of dye mixture (remazol red RR and remazol blue RB) = 100 mg/L	Colour removal = 100% in 2 h	Punzi et al. (2012)
UV/TiO_2	Concentration of dye (acid orange 7 and reactive red 2) = 0.086 mM	Dye removal = 100% in 20 min	Juang et al. (2010)
Ozonation	Concentration of dye (sirius blue) = 400 mg/L	Colour removal = 91%	Turhan and Turgut (2009)
Electrochemical oxidation	Total concentration of dyes = 361 mg/L, initial COD = 281 mg/L	Colour removal = 95% in 15 min	Chatzisymeon et al. (2006)
Sonocatalytic oxidation	Initial methyl orange concentration = 20 mg/L	Dye removal = 92%	de Jesús Ruíz-Baltazar (2021)

5.3 Limitations of AOPs

AOPs have shown promising results for the degradation of a variety of dyes as reviewed in the previous section. The various advantages of individual AOPs are summarized in Table 5.3. Despite the high efficiencies of AOPs, the application of these processes are limited due to various reasons. In this section, the drawbacks and limitations of different AOPs will be briefly discussed.

Fenton's oxidation processes are effective in the pH range of 2–4 and therefore, acidification of initial wastewater and neutralization of treated wastewater is required. High oxidant and iron dosage are required for higher degree of oxidation increasing the cost of the process (Babuponnusami and Muthukumar 2014). Fenton's and electrochemical process generate sludge during the process which must be disposed-off appropriately (Hai et al. 2007).

Photocatalytic processes are efficient in UV light but the efficiency is low under visible light. At higher doses of the catalyst, scattering is observed which further reduces the efficiency of the process (Zainal et al. 2007). Another drawback of this process is the difficulty in separation of catalysts from the aqueous solution. To

Table 5.3 Advantages and limitations of various advanced oxidation processes

Process	Advantages	Limitation	References
Fenton's process	• Simple and easy operation • High pressure or temperature is not required	• Effective in acidic pH • Formation of sludge • Neutralization is required	Babuponnusami and Muthukumar (2014)
Photocatalytic process	• No sludge is produced • High COD removal	• Catalyst separation and reuse is difficult • High operational cost due to lamps • Light scattering at higher catalyst doses	Zainal et al. (2007)
Ozone based processes	• No sludge is produced • No chemicals are used	• High energy and operational cost due to ozonators	Dhangar and Kumar (2020)
Electrochemical process	• Effective in COD reduction • Presence of salts do not affect the process	• Generation of sludge • High electricity cost	Hai et al. (2007)
Cavitation process	• Effective oxidation of hydrophilic and hydrophobic organics • Lower mass transfer limitation • Low operational cost as compared to other AOPs	• Difficult to achieve optimum ultrasonic frequency • Efficiency reduces for larger volumes (>10 L)	Gągol et al. (2018)

improve the efficiency in visible light doping with metal or non-metal can be done while a solid support material can be used to easily separate the catalyst.

The selective nature of ozone limits the mass transfer and wide application of the process. The sonochemical method do not produce secondary pollutants but they require long duration for complete degradation of pollutant. The removal efficiencies have been found to reduce with increase in the volume of wastewater during cavitation process (Gągol et al. 2018). The advantages and limitation of various AOPs are compiled and presented in Table 5.3.

Most of the AOPs such as photocatalytic, electro-oxidation, ozonation and sonication process are energy intensive and highly operational and maintenance costs is associated with these processes. To reduce the cost of the process without affecting its efficiency, combination of various processes can be explored.

5.4 Emerging Hybrid Treatment Technologies

The limitations of the individual AOPs suggest that a standalone AOP is not sufficient and efficient for the treatment of complex dye containing wastewater streams. It is therefore necessary to understand the potential of the combination of different techniques for the effective wastewater treatment of industrial effluents.

In this subsection, hybrid processes involving photocatalytic processes are divided into two types for easier understanding. The first combination is photochemical methods coupled with electrochemical methods while the second combination is photochemical methods coupled with sonolytic methods. Other combinations such as ozonation coupled with cavitation and sonication coupled with Fenton process have also been used for dye degradation. However, in this section, previous literature on (i) electro-photocatalytic and (ii) sono-photocatalytic hybrid process for degradation of dyes has been discussed in details.

5.4.1 Photochemical Processes Coupled with Electrochemical Processes

A combined process using electrochemistry and photochemistry has gained popularity in the last few decades. Photo-electrocatalytic degradation of turquoise blue 15 (initial concentration = 0.025 mM) dye was carried out by using Ti/TiO_2 photoanode and Pt cathode under UV light (125 W) by Osugi et al. (2005). The authors reported complete decolourization and 95% mineralization at an initial pH of 2 in 6 h duration. Similarly, electrochemical assisted photo-degradation of six dyes (i.e., methyl orange, naphthol blue black, reactive blue 2, methylene blue, rhodamine 6G and direct red 81) having total initial dye concentration of 1000 mg/L was carried out by Zainal et al. (2007). A COD removal of 73% could be obtained for the dye mixtures after 150 min of the combined treatment with TiO_2/Ti plate anode and platinum cathode under light irradiation (wavelength = 360–830 nm). In another study, photo-electro-Fenton process was combined with heterogeneous photocatalytic process using TiO_2 nanoparticles immobilized on glass plates for the treatment of CI basic red 46 dye (Khataee et al. 2011). Colour removal of 89% could be obtained in 35 min by the combined photo-electro-Fenton/UV/TiO_2 process where ferric concentration, initial dye concentration and current were 0.1 mM, 15 mg/L and 300 mA, respectively. The colour removal reported by authors for the individual AOPs i.e., UV/TiO_2, electro-Fenton and photo-electro-Fenton were 12, 32 and 84%, respectively suggesting higher efficiency of the combined treatment process.

On the other hand, Almeida et al. (2014) coupled photo-electro-catalytic process with electro-Fenton process using Pt/TiO_2 nanotubes photoanode and PTFE cathode for the degradation of acid red 29 dye (initial concentration = 85 mg/L)). A total organic carbon (TOC) removal of 98% could be obtained in the combined system (current density = 16.6 mA/cm^2, Hg lamp = 80 W) at initial pH of 3 in the presence of ferrous ions (0.5 mM) after 2 h duration. The authors suggest that the addition of TiO_2 nanotubes with Pt nanoparticles decreases the band gap energy and increases the degradation efficiency due to adsorption in visible region. Conversely, photo-electro-Fenton has also been combined with photocatalytic process for the removal of CI basic red 46 dye (initial concentration = 20 mg/L) by Zareí et al. (2010). The photocatalyst used in this study was TiO_2 while carbon nanotube-PTFE were used as cathode and Pt sheet was used as anode. A TOC removal of 99% was achieved in 6 h using the combined process (initial pH = 3, current = 100 mA and Fe^{3+} = 0.1 mM) while 92% decolourization could be observed within 1 h duration. A summary of previous literature on electrochemical processes coupled with other AOPs is presented in Table 5.4.

5.4.2 Photochemical Processes Coupled with Sonolytic Processes

The integration of sonolysis with photocatalysis (i.e., sonophotocatalysis) works on the synergistic effects of generation of radical and electron–hole pairs at high temperature and pressure. Advantages of sonophotocatalysis include improved light penetration, uniform distribution of catalysts, reduced mass transfer limitations, regeneration of catalyst surface and higher radical production (Pirsaheb and Moradi 2021).

TiO_2 has been used as a catalyst for sonophotocatalytic oxidation of methyl orange (initial concentration = 100 μM) under xenon-arc lamp (450 W) in ultrasonic bath (213 kHz) (He et al. 2011). Complete mineralization was observed at an initial pH of 2 with 1 g/L TiO_2 dose in 120 min. Sathishkumar et al. (2014) performed sonophotocatalytic oxidation of acid blue 113 (initial concentration = 0.01 mM) with Y^{3+} doped TiO_2 (catalyst dose = 1 g/L) catalyst in the presence of visible light

Table 5.4 Summary of hybrid electrochemical processes for dye degradation

Wastewater characteristics	Reaction conditions	Results	References
Initial concentration of turquoise blue 15 = 0.025 mM	Anode = Ti/TiO_2, cathode = Pt, initial pH = 2	Colour removal = 95% in 6 h	Osugi et al. (2005)
Total initial dye concentration = 1000 mg/L	Anode = TiO_2/Ti plate, Cathode = Pt	COD removal of 73% in 150 min	Zainal et al. (2007)
Initial concentration of CI basic red 46 dye = 20 mg/L	Anode = Pt sheet, cathode = carbon nanotube-PTFE, initial pH = 3, current = 100 Ma, Fe^{3+} = 0.1 mM	TOC removal = 99% in 6 h	Zarei et al. (2010)
Initial concentration of CI basic red 46 = 15 mg/L	ferric concentration = 0.1 mM, current = 300 mA	Colour removal of 89% could be obtained in 35 min	Khataee et al. (2011)
Initial concentration of acid red 29 dye = 85 mg/L	Anode = Pt/TiO_2 nanotubes, cathode = PTFE, current density = 16.6 mA/cm^2, initial pH = 3, ferrous ions = 0.5 mM	TOC removal = 98% in 2 h	Almeida et al. (2014)

and ultrasound (42 kHz). The authors have reported 65% mineralization after 5 h duration.

ZnO was used a photocatalyst for the oxidation of direct blue 71 (initial concentration = 100 mg/L) in an ultrasonic bath (20 kHz) under UV light (Ertugay and Acar 2014). The sonophotocatalytic processes resulted in complete colour removal in 20 min with 1 g/L ZnO and 75 mg/L H_2O_2 at an initial pH of 2.5. In another study, acid blue 113 degradation was carried out using zinc oxide catalysts (dose = 0.88 g/L) and persulphate as oxidant (dose = 2.43 mM) under UV light (9 W) at ultrasound frequency of 20 kHz. A removal efficiency of 98% was reported by authors at an initial pH of 6.1 in 25 min (Asgari et al. 2020).

Authors have also studied sono-photo-ferrioxalate system for the decolourization of acid red B dye (initial concentration = 20 mg/L) under UV irradiation (160 W) at ultrasound frequency of 40 kHz (Chakma and Moholkar 2016). The decolourization obtained during the process was 85% after 60 min at reaction pH of 3 with Fe^{3+}/oxalate ratio of 1:3. A summary of previous literature on electrochemical processes coupled with other AOPs is presented in Table 5.5.

Table 5.5 Summary of hybrid sono-photochemical processes for dye degradation

Wastewater characteristics	Reaction conditions	Results	References
Initial concentration of methyl orange = 100 μM	Ultrasound frequency = 213 kHz, initial pH = 2, TiO_2 dose = 1 g/L	Dye removal = 100% in 2 h	He et al. (2011)
Initial concentration of acid blue 113 = 0.01 mM	Ultrasound frequency = 42 kHz, Y^{3+} doped TiO_2 dose = 1 g/L	TOC removal = 65% in 5 h	Sathishkumar et al. (2014)
Initial concentration of direct blue 71 = 100 mg/L	Ultrasound frequency = 20 kHz, ZnO dose = 1 g/L, H_2O_2 = 75 mg/L, initial pH = 2.5	Colour removal = 100% in 20 min	Ertugay and Acar (2014)
Initial concentration of acid red B dye = 20 mg/L ()	Ultrasound frequency of 40 kHz, Fe^{3+}/oxalate = 1:3, initial pH = 3	Colour removal = 85% in 1 h	Chakma and Moholkar (2016)
Initial concentration of acid blue 113 = 200 mg/L	Ultrasound frequency = 20 kHz, ZnO dose = 0.88 g/L, persulphate = 2.43 mM, initial pH = 6.1	Dye removal = 98% in 25 min	Asgari et al. (2020)

5.5 Conclusions and Future Prospects

The discharge of coloured effluent containing dye into the natural waterbodies not only results in aesthetic pollution but it also affects aquatic life. The effective treatment of dye wastewater is a challenge due to their recalcitrant nature and carcinogenic properties. Several AOPs have been found effective for the degradation of dye wastewater and a few have been implemented at large scale. However, AOPs are bound to certain practical limitations for industrial application. Moreover, AOPs have high energy demand and operational costs making them expensive for industrial use. To overcome the shortcomings of individual AOPs, hybrid treatment processes have been found efficient for the removal of dyes in wastewaters by various researcher. The present chapter summarizes state-of-art information on hybrid treatment technologies derived from integration of AOPs. Among the various hybrid techniques, sono-photocatalysis have attracted the attention of most scholars. The advantage of sono-photocatalysis process has reduced mass transfer limitations due to increased turbulence and regeneration of catalyst surface. However, the selection of an appropriate hybrid technology will be case-specific depending on the initial wastewater characteristic and end-use of the treated effluent. Further, efforts are required for their industrial application in future. A few recommendations regarding the research on hybrid process are as follows:

- The potential of hybrid treatment technologies needs to evaluated for industrial effluents by piloting these technologies.

- A detailed discussion on the kinetics of the hybrid process must be performed for better understanding.
- Studies should be performed to find a cost effective and environmental friendly treatment option by utilizing solar energy and waste materials.

References

Almeida LC, Silva BF, Zanoni MV (2014) Combined photoelectrocatalytic/electro-Fenton process using a Pt/TiO_2NTs photoanode for enhanced degradation of an azo dye: a mechanistic study. J Electroanal Chem 734:43–52

Anandan S, Ponnusamy VK, Ashokkumar M (2020) A review on hybrid techniques for the degradation of organic pollutants in aqueous environment. Ultras Sonochem 67:105130

Asgari G, Shabanloo A, Salari M, Eslami F (2020) Sonophotocatalytic treatment of AB113 dye and real textile wastewater using ZnO/persulfate: modeling by response surface methodology and artificial neural network. Environ Res 184:109367

Aydiner C, Mert BK, Dogan EC, Yatmaz HC, Dagli S, Aksu S, Tilki YM, Goren AY, Balci E (2019) Novel hybrid treatments of textile wastewater by membrane oxidation reactor: performance investigations, optimizations and efficiency comparisons. Sci Total Environ 683:411–426

Babuponnusami A, Muthukumar K (2014) A review on Fenton and improvements to the Fenton process for wastewater treatment. J Environ Chem Eng 2(1):557–572

Chakma S, Moholkar VS (2016) Mechanistic analysis of hybrid sono-photo-ferrioxalate system for decolorization of azo dye. J Taiwan Inst Chem Eng 60:469–478

Chen G (2004) Electrochemical technologies in wastewater treatment. Sep Purif Technol 38(1):11–41

Chatzisymeon E, Xekoukoulotakis NP, Coz A, Kalogerakis N, Mantzavinos D (2006) Electrochemical treatment of textile dyes and dyehouse effluents. J Hazard Mater 137(2):998–1007

Dhangar K, Kumar M (2020) Tricks and tracks in removal of emerging contaminants from the wastewater through hybrid treatment systems: a review. Sci Total Environ 738:140320

Gągol M, Przyjazny A, Boczkaj G (2018) Wastewater treatment by means of advanced oxidation processes based on cavitation–a review. Chem Eng J 338:599–627

Hai FI, Yamamoto K, Fukushi K (2007) Hybrid treatment systems for dye wastewater. Crit Rev Environ Sci Technol 37(4):315–377

de Jesús Ruíz-Baltazar Á (2021) Sonochemical activation-assisted biosynthesis of Au/Fe_3O_4 nanoparticles and sonocatalytic degradation of methyl orange. Ultrason Sonochem 73:105521

Ertugay N, Acar FN (2014) The degradation of Direct Blue 71 by sono, photo and sonophotocatalytic oxidation in the presence of ZnO nanocatalyst. Appl Surf Sci 318:121–126

Geraldino HC, Freitas TK, Manholer DD, França F, Oliveira JH, Volnistem EA, Lima ARF, Bertotti M, Girotto EM, Garcia JC (2020) Electrochemical generation of H_2O_2 using gas diffusion electrode improved with rGO intensified with the Fe_3O_4/GO catalyst for degradation of textile wastewater. J Water Process Eng 36:101377

GilPavas E, Correa-Sanchez S, Acosta DA (2019) Using scrap zero valent iron to replace dissolved iron in the Fenton process for textile wastewater treatment: optimization and assessment of toxicity and biodegradability. Environ Pollut 252:1709–1718

Gupta VK, Khamparia S, Tyagi I, Jaspal D, Malviya A (2015) Decolorization of mixture of dyes: a critical review. Global J Environ Sci Manage 1(1):71–94

He Y, Grieser F, Ashokkumar M (2011) The mechanism of sonophotocatalytic degradation of methyl orange and its products in aqueous solutions. Ultrason Sonochem 18(5):974–980

Holkar CR, Jadhav AJ, Pinjari DV, Mahamuni NM, Pandit AB (2016) A critical review on textile wastewater treatments: possible approaches. J Environ Manage 182:351–366

Juang RS, Lin SH, Hsueh PY (2010) Removal of binary azo dyes from water by UV-irradiated degradation in TiO_2 suspensions. J Hazard Mater 182(1–3):820–826
Khataee AR, Zarei M, Ordikhani-Seyedlar R (2011) Heterogeneous photocatalysis of a dye solution using supported TiO_2 nanoparticles combined with homogeneous photoelectrochemical process: molecular degradation products. J Mol Catal a: Chem 338(1–2):84–91
Khataee AR, Naseri A, Zarei M, Safarpour M, Moradkhannejhad L (2012) Chemometrics approach for determination and optimization of simultaneous photooxidative decolourization of a mixture of three textile dyes. Environ Technol 33(20):2305–2317
Lebron YAR, Moreira VR, Maia A, Couto CF, Moravia WG, Amaral MCS (2021) Integrated photo-Fenton and membrane-based techniques for textile effluent reclamation. Sep Purif Technol 272:118932
Louhichi B, Gaied F, Mansouri K, Jeday MR (2022) Treatment of textile industry effluents by Electro-Coagulation and Electro-Fenton processes using solar energy: a comparative study. Chem Eng J 427:131735
Malik SN, Ghosh PC, Vaidya AN, Mudliar SN (2020) Hybrid ozonation process for industrial wastewater treatment: principles and applications: a review. J Water Process Eng 35:101193
Meriç S, Kaptan D, Tünay O (2003) Removal of color and COD from a mixture of four reactive azo dyes using Fenton oxidation process. J Environ Sci Health, Part A 38(10):2241–2250
Miklos DB, Remy C, Jekel M, Linden KG, Drewes JE, Hübner U (2018) Evaluation of advanced oxidation processes for water and wastewater treatment–A critical review. Water Res 139:118–131
Osugi ME, Umbuzeiro GA, Anderson MA, Zanoni MVB (2005) Degradation of metallophtalocyanine dye by combined processes of electrochemistry and photoelectrochemistry. Electrochim Acta 50(25–26):5261–5269
Oturan MA, Aaron JJ (2014) Advanced oxidation processes in water/wastewater treatment: principles and applications: a review. Crit Rev Environ Sci Technol 44(23):2577–2641
Pirsaheb M, Moradi N (2021) A systematic review of the sonophotocatalytic process for the decolorization of dyes in aqueous solution: synergistic mechanisms, degradation pathways, and process optimization. J Water Process Eng 44:102314
Punzi M, Mattiasson B, Jonstrup M (2012) Treatment of synthetic textile wastewater by homogeneous and heterogeneous photo-Fenton oxidation. J Photochem Photobiol, A 248:30–35
Samsami S, Mohamadizaniani M, Sarrafzadeh MH, Rene ER, Firoozbahr M (2020) Recent advances in the treatment of dye-containing wastewater from textile industries: overview and perspectives. Process Saf Environ Prot 143:138–163
Sathishkumar P, Mangalaraja RV, Rozas O, Mansilla HD, Gracia-Pinilla MA, Anandan S (2014) Low frequency ultrasound (42 kHz) assisted degradation of acid blue 113 in the presence of visible light driven rare earth nanoclusters loaded TiO_2 nanophotocatalysts. Ultrason Sonochem 21(5):1675–1681
Silva LG, Moreira FC, Cechinel MAP, Mazur LP, de Souza AAU, Souza SMGU, Boaventura RAR, Vilar VJ (2020) Integration of Fenton's reaction based processes and cation exchange processes in textile wastewater treatment as a strategy for water reuse. J Environ Manage 272:111082
Shindhal T, Rakholiya P, Varjani S, Pandey A, Ngo HH, Guo W, Ng HY, Taherzadeh MJ (2021) A critical review on advances in the practices and perspectives for the treatment of dye industry wastewater. Bioeng 12(1):70–87
Turhan K, Turgut Z (2009) Decolorization of direct dye in textile wastewater by ozonization in a semi-batch bubble column reactor. Desalination 242(1–3):256–263
Walling C (1975) Fenton's reagent revisited. Acc Chem Res 8(4):125–131
Yaseen DA, Scholz M (2019) Textile dye wastewater characteristics and constituents of synthetic effluents: a critical review. Int J Environ Sci Technol 16(2):1193–1226

Zainal Z, Lee CY, Hussein MZ, Kassim A, Yusof NA (2007) Electrochemical-assisted photodegradation of mixed dye and textile effluents using TiO_2 thin films. J Hazard Mater 146(1–2):73–80
Zarei M, Khataee AR, Ordikhani-Seyedlar R, Fathinia M (2010) Photoelectro-Fenton combined with photocatalytic process for degradation of an azo dye using supported TiO_2 nanoparticles and carbon nanotube cathode: neural network modeling. Electrochim Acta 55(24):7259–7265
Zazou H, Afanga H, Akhouairi S, Ouchtak H, Addi AA, Akbour RA, Assabbane A, Douch J, Elmchaouri A, Duplay J, Jada A, Hamdani M (2019) Treatment of textile industry wastewater by electrocoagulation coupled with electrochemical advanced oxidation process. J Water Process Eng 28:214–221

Chapter 6
Aerogel Nanomaterials for Dye Degradation

Sanjana Jacob, S. Kaviya, and K. Anand

Abstract Water pollution has been considered as a major environmental issue that has become a threat to human health. Therefore, an efficient solution for removing pollutants from water is needed. One such efficient solution for the removal of pollutants is by using aerogels. Due to their high absorption rate and properties, aerogels exhibit a wide range of properties and hence can be chosen for demanding applications. It has been proven that the characteristic features like exceptionally high porosity, huge specific surface area, very low density, ease of detachment from aqueous solution etc. make these materials suitable for dye removal. The current chapter deals with the study of aerogel materials, classification, synthesis (some selected aerogel materials) and their dye degradation and photocatalytic applications.

Keywords Aerogel · Nanomaterials · Synthesis · Dye degradation · Photocatalysis

6.1 Introduction

Water pollution has been considered as a major environmental issue, that has become a threat to human health. Organic pollutants and its toxic substances in the waste water not only destroy the ecosystem, but also significantly affect the aquatic life. Therefore, an efficient solution for the removal of pollutants from water is needed.

S. Jacob
Department of Nanotechnology Engineering, Srinivas Institute of Technology, Mangalore, Karnataka 574143, India

S. Kaviya
Department of Chemical Engineering, Birla Institute of Technology and Science, BITS Pilani Hyderabad Campus, Hyderabad, Telangana 500078, India

K. Anand (✉)
Department of Basic Sciences, Amal Jyothi College of Engineering, Kanjirapally, Kerala 686518, India
e-mail: anand.rrii@gmail.com

Apcotex Industries Limited, Plot No.3/1, MIDC Industrial Area, Taloja, Dist. Raigad, Maharashtra 410208, India

S. Dave and J. Das (eds.), *Trends and Contemporary Technologies for Photocatalytic Degradation of Dyes*, Environmental Science and Engineering,
https://doi.org/10.1007/978-3-031-08991-6_6

Various technologies are proposed to erase such pollutants in the water like oxidation, adsorption, photocatalysis degradation, electron degradation etc. (Wu et al. 2020). Aerogel materials due to their large surface area and absorption capacity can play a major role in dye degradation. In this chapter the usage of aerogel nanomaterials for dye degradation particularly, the photocatalytic degradation properties of different aerogel materials are discussed.

6.1.1 Aerogel—Overview

Aerogel materials have three-dimensional open network that are arranged by the structured nanoparticles/polymer molecules (Fricke 1988; Fricke and Emmerling 1998; Pierre and Pajonk 2002; Vacher et al. 1988). The aerogel was first made by Kistler in 1931. It was called "aerogel" which is air and gel because it replaced liquid element inside of the wet gel with the air not destructing solid micro structure (Du et al. 2013). Aerogels are the sol–gel traced solid material that has porousness about 80–99.8%. The advanced level porosity was gained through the supercritical drying of the alcogel or the hydrogel in autoclave (Fricke et al. 1992). In the recent developments, it states that aerogel could also be recognized as a new state of matter (Wagh et al. 1999) and also aerogel shows qualitative variation in the bulk property when compared with the other states of matter. Aerogel maintains a fixed shape and volume like that of the solids and the density of the aerogel may extend between 1000 and 1 kg/m^3. Aerogel exhibits unique properties like low density solid, reduced thermal conductivity, low sonic velocity, low refractive index, reduced dielectric constant, reduced sound speed and broad specific surface area (Fricke and Emmerling 1998; Pierre and Pajonk 2002; Schaefer and Keefer 1986; Gesser and Goswami 1989). The structure of aerogel can be defined using the electron microscope, pore size analyzer, small angle X-ray scattering etc. Properties are measured by particular instruments, for example, mechanical properties of aerogel can be tried/tested by the accurate UTM or the dynamic thermo-mechanical analyzer in compression or the three-point bend type. All kinds of ultra-light foams cannot be categorized as aerogel state (Du et al. 2013; Schaedler et al. 2011).

6.2 Classification, Properties and Applications

Aerogels can be classified using different methods: based on the appearance, they are classified into monolith, film and powder. Based on the preparation techniques, they are classified into aerogels, xerogels and cryogels. Classifications of aerogels are shown in Fig. 6.1 (Nita et al. 2020). From this figure, one understands that aerogels can be separated into two categories: inorganic and organic. Inorganic aerogels is further subdivided into five categories: Fluoride, Silica, Metal, Oxides and Chalcogens. Organic is subdivided into two: Polymer and Carbon based (Du et al. 2013).

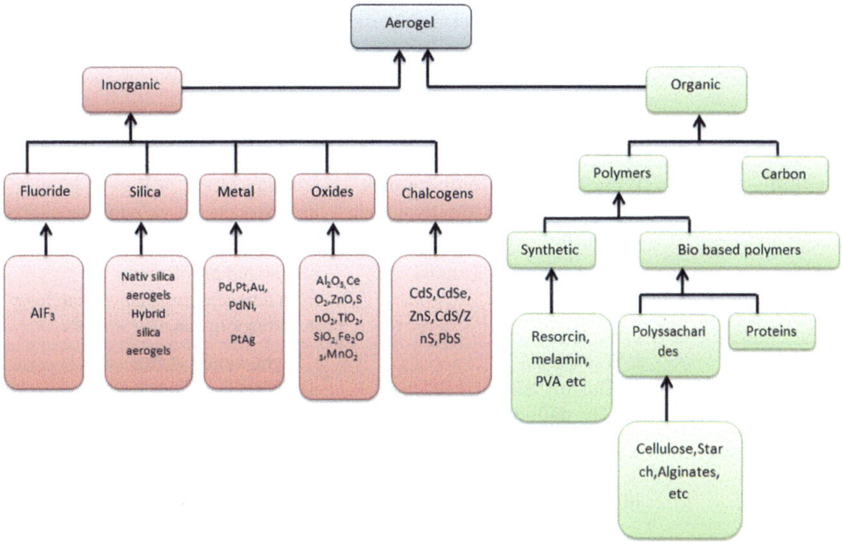

Fig. 6.1 Classification of aerogels (Nita et al. 2020). (Reproduced from reference (Nita et al. 2020) open access)

The unique properties of aerogel make them attractive in technology and science. Based on their properties, aerogels can be used for different applications (Table 6.1).

6.2.1 *Synthesis of Aerogel Materials*

Sol–gel method of synthesis

Sol–gel action of synthesis is the most popular and trusted way for the synthesis of the materials, specifically the metal oxides that have regular small particle size and with different morphologies. It consists of the transformation of the system from the liquid sol into the solid gel phase (Gurav et al. 2010).

The sol–gel way for the synthesis/preparation of aerogel can normally be separated into the following steps: as in Fig. 6.2 (Du et al. 2013).

(a) Solution—sol formation: Nanoscale sol substances are created in the precursor solution spontaneously or that are catalyzed by the catalysts through condensation and by hydrolysis reactions.
(b) Sol-gel formation (Gelation): The substances that are the sol are cross linked and arranged into a wet gel with coherent network hierarchically.
(c) Gel-aerogel formation (Drying and Densification): Solvent in the wet gel is replaced by the air without damaging the microstructure seriously, and then it is densified (Du et al. 2013).

Table 6.1 Application of aerogel with respect to properties adapted with permissions from (Gurav et al. 2010)

Property	Feature	Application
Mechanical	i. Its elastic ii. Its light weight	i. Hypervelocity particle trap ii. Energy absorber
Electrical	i. The lowest dielectric constant ii. The high dielectric strength iii. The high surface area	i. In dielectrics for ICs ii. In spacers for vacuum electrodes iii. In capacitors
Acoustic	i. The low speed of sound	i. In sound proof rooms ii. In acoustic impedance matching in ultrasonic distance sensors
Optical	i. Transparent ii. The low refractive index iii. The multiple composition	i. Light weight optics ii. Cherenkov detectors iii. Light guides
Density/porosity	i. Lightest synthetic solid ii. High surface area iii. Multiple compositions	i. For catalysis ii. In sensor iii. For fuel storage iv. In ion exchange v. For filters for pollutants gaseous vi. In targets for ICF vii. In pigment carriers viii. As template
Thermal conductivity	i. The best insulating solid ii. Transparent iii. It withstands high temperature iv. Its light weight	i. In building construction and appliance insulation ii. In storage media iii. In automobiles, space vehicles iv. In solar devices, solar ponds

The main advantage of sol–gel process is its simplicity and are economical. It is one of the effective ways of producing materials that are of high quality. The processing of sol–gel technique finds application in high-quality glass production for fibers and optical components, fine powders of oxides and thin-film coatings (Gurav et al. 2010; Matsuda et al. 2000; Torikai et al. 1994; Hamaker 1937).

Though different types of aerogel materials are available, herein, we discuss the synthesis of some selected aerogel nanomaterials as given below:

a. TiO_2@CNF aerogel
b. SiO_2 aerogel
c. 3D nitrogen-doped graphene aerogel
d. Synthesis of MoS_2 aerogel
e. PTA/ZIF-8@cellulose aerogel.

TiO_2@CNF aerogel synthesis

TiO_2@CNF aerogel is created by the combination of sol–gel method and hydrothermal method. CNF aerogels were soaked for about 30 min in the mixed solution of $Ti(OC_4H_9)_4$ and absolute ethanol, and then the aerogel was taken out. The remaining $Ti(OC_4H_9)_4$, which was not absorbed, were washed out by absolute

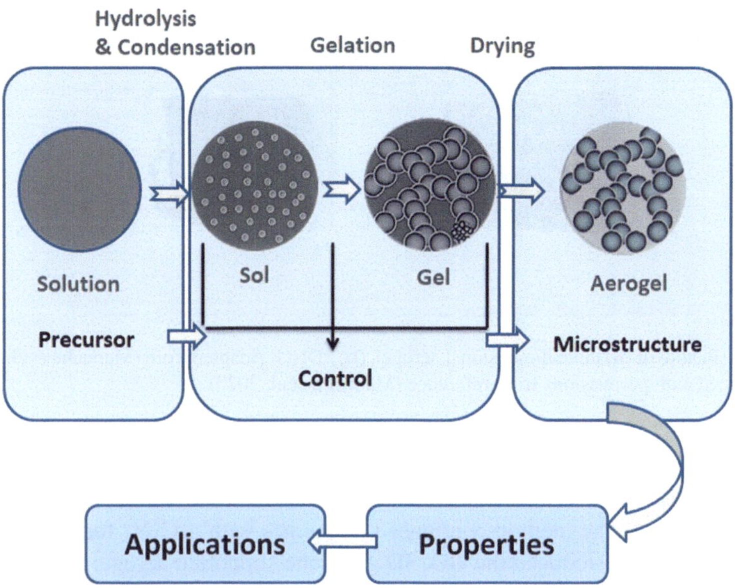

Fig. 6.2 The schematic demonstration showing the synthesis of the aerogel materials by sol–gel process (Du et al. 2013). Reproduced with permissions from reference (Du et al. 2013)

ethanol. Later the gel is immersed in a blended sol of water and absolute ethanol in a ratio of volume 4:1, and the pH adjusted to 3.0 with the dilute HCl. The gel is left undisturbed for 2 h to confirm that $Ti(OC_4H_9)_4$ gets hydrolyzed completely. The intermediate is placed onto a PTFE-lined hydrothermal reactor that is reacted at 120 °C for the variable duration to understand the optimum time of reaction. The freeze-drying process takes place to get TiO_2@CNF aerogel. The processes like absorption, hydrothermal and hydrolysis are repeated 1–5 times to get the aerogel loaded with different quantity of TiO_2 NPs (Li et al. 2021).

SiO_2 aerogel

For SiO_2 aerogel synthesis, 30 g of the rice husk ash is integrated along with 200 ml of 1 mol L^{-1} sodium hydroxide (NaOH) solution. The aqueous solution is heated to 100 °C and then is left for 3.5 h. The pH (= 5) is altered by hydrochloric acid and ammonia solution. The removal of the Na ion from the sodium silicate through the cation exchange resin is done thereafter. After 24 h, hydrophobically modified silica aerogel is obtained by replacing the solvent with absolute ethanol, n-hexane and chlorotrimethylsilane (Liu et al. 2021).

3D nitrogen-doped graphene aerogel (3DNG)

g-C_3N_4 is dispersed into the 6.21 ml of distilled water through ultrasonication for about 6 h. The solution is then added into 3.72 ml of GO earlier for further ultra-sonication for the next 2 h. After that, 37.2 mg of urea and ascorbic acid of 9.3 mg

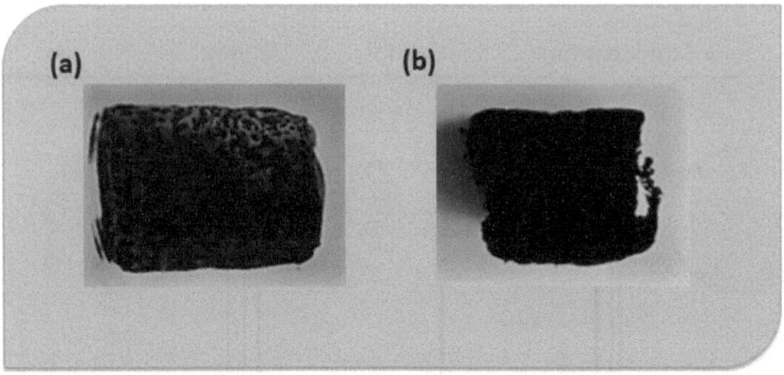

Fig. 6.3 Picture of (**a**) three dimensional aerogel, (**b**) 3DNG. Adapted from (Maouche et al. 2020). Reproduced with permissions from reference (Maouche et al. 2020)

were then added and distributed into the solution for 1 h. The solution is covered and contained at 180 °C for about 6 h in the Teflon-lined autoclave, and it is made to cool naturally. As the method continues GO merges with g-C_3N_4 forming a three-dimensional porous reduced the GO. 3D graphene-supported aerogel is obtained as shown in Fig. 6.3(a) (Maouche et al. 2020).

Three-dimensional aerogel is formed after freeze-drying the three-dimensional hydrogel. It is heated for about 1 h under the Ar atmosphere at around 1000 °C and the 3D nitrogen-doped graphene aerogel that are thermally treated is obtained as in Fig. 6.3(b) (Maouche et al. 2020).

Synthesis of MoS_2 aerogel

For the synthesis of MoS_2 aerogel, 50 ml of the exfoliated MoS_2 solution is mixed with 5 mL of the 20% nafion solution and it is kept for sonication so that it can mix together and can form a uniform gel-like form of suspension. The container is then transferred to a 20 °C refrigerator for freezing the mixture. Later the solution that is frozen is freeze-dried in a lyophilizer at 60 °C and at 10^{-3} torr chamber to make MoS_2 aerogel (Peng et al. 2021).

PTA/ZIF-8@cellulose aerogel

The PTA/ZIF-8@cellulose aerogel composite material has a multi-layer 3D network shape. It is made using a cellulose aerogel framework. MOFs are used as filler and the preparation method is shown in Fig. 6.4. Steps involve freeze-drying, composition and impregnation (Wen et al. 2020).

6.2.1.1 Different Dyes Used in Industries

Dyes have been used by various industries like textile, food, paper, cosmetic, leather, and fur with textile industry as a major consumer (Benkhaya et al. 2017). Apart from these, the dyes are used as indicators, hair coloring agents and in photography. The

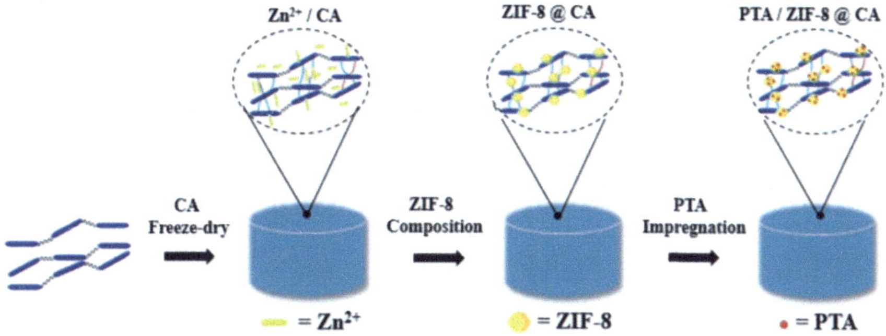

Fig. 6.4 Preparation process of PTA/ZIF-8@cellulose aerogel. Adapted from (Wen et al. 2020). Reproduced with permissions from reference (Wen et al. 2020)

dyes were categorized based on their chemical structure, usage and application. Table 6.2 shows categorization of dyes based on usage or applications (Hunger 2007).

6.2.2 Dye Degradation using Aerogel Materials

Degradation of dye involves the chemical breakdown of large dye molecules into smaller molecules. Removal of dye from industrial effluent is broadly classified into physical, chemical and biological methods, which are discussed in Table 6.3 (Siddiqui 2018). Aerogel materials are considered as one of the promising solutions for dye removal and can be synthesized in different forms using various methods. They are basically classified into inorganic aerogels, organic aerogels, and organic-inorganic hybrid aerogels materials (Hasanpour and Hatami 2020).

The inorganic aerogels are synthesized from almost all the metals or oxide semiconductors. These inorganic aerogels are traditionally prepared through hydrolysis and condensation methods. Semiconductor photocatalyst based aerogels such as TiO_2, SiO_2, ZnO and $InVO_4$ is analyzed for degradation of various pollutants. Among these, silica and titania were of greater interest because of its non-toxicity, less cost and photo-stability.

Xu et al. 2016a synthesized a binary aerogel of silicon dioxide/titanium dioxide by using sodium silicate and titanium tetrachloride as precursors. The surface modification of the material was performed to achieve the desired binary aerogel. The prepared aerogel shows notable photocatalytic activity for the degradation of methyl orange. It has achieved a photocatalytic efficiency of about 84.9% after exposure of the aerogel for 210 min under the irradiation of ultraviolet light. The degraded aerogel is reused and it shows 70% efficiency even after recycling the aerogel for about 4 times.

Zhang et al. (2018) prepared TiO_2 aerogel using a preceramic polymer as a precursor and bacterial cellulose as a bio template. The composite aerogel was

Table 6.2 Categorization of the dyes based on its use or application. Adapted from (Hunger 2007)

Class	Principal substrates	Method of application	Chemical types	Some examples
Acid	Nylon, wool, silk, paper, inks and leather	Usually from neutral to acidic dyebaths	Azo (including premetallized), anthraquinone, triphenylmethane, azine, xanthene, nitro and nitroso (Yagub et al. 2014)	Indian ink, congo red, nigrosoine
Azoic components and composition	Cotton, rayon, cellulose acetate any polyester	Fiber impregnated with coupling components and treated with solution of stabilized diazonium salt	Azo	Disperse orange 1, methyl orange
Basic	Paper, polyacrylonitrile, modified nylon, polyester and inks	Applied from acidic dyebaths	Cyanine, hemicyanine, diazahemicyanine, triarylmethane, azo, azine, xanthene, acridine, oxazine, and anthraquinone (Hunger 2007; Yagub et al. 2014; Mallakpour and Rashidimoghadam 2018; Siddiqui 2018)	Methylene blue, toluidine blue, thionine, and crystal violet
Direct	Cotton, rayon, paper, leather and nylon	Applied from neutral or slightly alkaline baths containing additional electrolyte	Azo, phthalocyanine, stilbene and oxazine (Yagub et al. 2014)	Direct light blue, direct light brilliant blue, direct copper blue 2R, direct diazo black BH
Disperse	Polyester, polyamide, acetate, acrylic and plastics	From solution dispersion or suspension in a mass	Stilbene, pyrazoles, coumarin, and naphthalimides (Aspland 1992)	Disperse orange 1, disperse red 9, disperse yellow 26

(continued)

Table 6.2 (continued)

Class	Principal substrates	Method of application	Chemical types	Some examples
Food, drug and cosmetic	The foods, the drugs and cosmetics	NA	Azo, anthraquinone, carotenoid and triarylmethane (Hunger 2007; Mallakpour and Rashidimoghadam 2018)	FD&C blue #1, D&C green #6
Mordant	Wool, leather, and anodized aluminium	Applied in conjunction with Cr salts	Azo and anthraquinone (Hunger 2007; Mallakpour and Rashidimoghadam 2018)	Mordant brown 35
Oxidation bases	Hair, fur, and cotton	Aromatic amines and phenol oxidized on the substrate	Aniline black and intermediate structures (Hunger 2007)	Basic yellow 57, basic red 76, basic blue 99
Reactive	Cotton, wool, silk and nylon	Reactive site on dye reacts with functional group on fiber to bind dye covalently under influence of heat and pH (alkaline)	Azo, anthraquinone, phthalocyanine, formazan, oxazine, and basic (Hunger 2007; Mallakpour and Rashidimoghadam 2018)	Reactive blue 74, reactive brown 2, reactive yellow 4
Solvent	Plastics, gasoline, varnishes, lacquers, strains, inks, fats, oils, waxes	Dissolves in the substrate	Azo, triphenylmethane, anthraquinone, and phthalocyanine (Hunger 2007)	Solvent red 24, solvent red 26, solvent red 164

Table 6.3 Methods for dye removal (Siddiqui 2018)

Physical methods	a. Precipitation/sedimentation b. Coagulation c. Filtration d. Adsorption e. Solvent extraction f. Photocatalytic degradation
Chemical methods	a. Catalytic degradation b. Chemical precipitation c. Reduction d. Oxidation e. Ion exchange f. Electrolysis g. Fenton reaction
Biological methods	a. Fungal degradation b. Algal treatment c. Anaerobic digestion d. Trickling filters e. Activated sludge f. Surface immobilization

prepared by using the freeze drying method and then calcinated under the argon atmosphere. The aerogel was further calcinated in the air atmosphere at 450 °C for three different times (1, 2 and 3 h) of calcination. The TiO_2 aerogel was analyzed for the degradation of methyl orange under UV light irradiation. The result showed that the aerogel synthesized with a calcination time of 2 h has maximum photocatalytic activity.

Kim et al. (2013) used sol–gel synthesis to prepare the hydrophobic TiO_2/SiO_2 binary aerogel by using sodium silicate and titanium oxychloride as sodium and titanium precursors respectively. The aerogels were prepared with different silica to titanium ratios and were calcinated at various temperatures starting from 200 to 1000 °C. The photocatalytic activity of the aerogel was evaluated for the degradation of methylene blue (MB) under UV light radiation. As demonstrated in Fig. 6.5(a), the result showed that the binary composite with the weight ratio of 2(g of Si/g of Ti) exhibited the maximum activity. The photocatalytic activity of about 85% is maintained for three cycles as shown in Fig. 6.5(b).

Another study by Chang et al. (2014) reveals the preparation of SiO_2/TiO_2 composites under ambient pressure drying using the precursors such as tetraethoxysilane and tetrabutyl trinitrate. The prepared samples were heat-treated (calcinated) at 400–800 °C and are analyzed for the photocatalytic degradation of methylene blue dye. The result shows that the samples which were calcinated at 800 °C has better photocatalytic activity.

STAB (silica titania aerogel like balls) was synthesized by using sol–gel method followed by ball doping method (Xu et al. 2007). STAB samples were heat treated at various temperatures. The result shows that the STAB which was annealed at 600 °C has a maximum photocatalytic activity for the degradation of methylene blue dye.

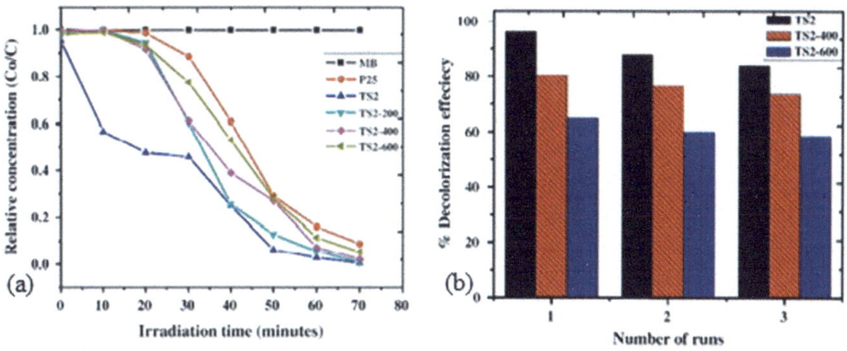

Fig. 6.5 **a** Titania–silica aerogel photocatalyst degradation concentration versus time plot in decolorization of MB solution. **b** Evaluation of the reusability of TS2, TS2-400 and TS2-600 photocatalysts (Kim et al. 2013). Reproduced with permissions from reference (Kim et al. 2013)

Chemical liquid deposition of Ti onto nanoporous Si results in Si–Ti composite aerogels. The composite was synthesized by Zu et al. (2015) with different deposition cycles, i.e., by repeating the deposition procedure 5, 10 and 15 times. The prepared sample were calcinated at various temperatures in order to examine the heat effects. The Si–Ti composite aerogel which was synthesized by using 15 deposition cycles and calcinated at 600 °C shows the highest photodegradation efficiency toward methylene blue dye degradation.

Apart from inorganic aerogel, organic aerogel has also been synthesized using organic precursors like phenol–formaldehyde, melanin-formaldehyde, resorcinol–formaldehyde etc. The main properties of organic aerogels include less brittleness, greater stability, low weight and have great covalent bonds than the inorganic aerogels (Hasanpour and Hatami 2020).

The photocatalytic degradation of acridine orange under visible light irradiation was demonstrated by Jiang et al. (2019). The impact of various nitrogen precursors in nitrogen doped graphene aerogel on photocatalytic activity of acridine orange has been investigated. The maximum degradation was obtained for tetraethylenepentamine in nitrogen doped graphene aerogel. Polyaniline enhanced cellulose aerogel was synthesized using ionic liquid via a regeneration route by Zhou et al. (2014). The prepared aerogel was evaluated for the photodegradation activity toward the methylene blue (MB) dye using sunlight irradiation. The synthesized polyaniline enhanced cellulose aerogel showed outstanding photocatalytic activity with degradation efficiency of about 95% after 2 h.

To improve the dye degradation properties of the aerogel, two or more systems were combined to produce hybrid or composite material. In recent years the synthesis of inorganic–organic hybrid aerogels has been increased. The hybrid aerogels are majorly synthesized via sol–gel method. In these hybrid aerogels, carbon-based hybrid photocatalysts considering graphene and derivate were of main interest because of their large specific surface area, the broad spectrum of adsorption, high adsorption capacity, high electrochemical stability, great electrical conductivity, outstanding mechanical strength and higher rates of adsorption–desorption (Hasanpour and Hatami 2020).

He et al. (2018) used a 2 step hydrothermal method to synthesize a novel 3D based graphene aerogel (GA)-carbon quantum dots (CQDs)/g-C_3N_4 nanosheet (CNN) composites aerogel with different mass ratios of Graphene Oxide to CNN (from 8 to 40%). The composites aerogel was evaluated for the photocatalytic degradation of the MO dye. The result showed that the GA/CQDs/CNN composite with a mass ratio of 24% has the highest MO dye removal performance of 91.1%. Factors that can contribute to the photocatalytic performance are the 3D porous structure, large specific surface area and the carbon quantum dots location on g-C_3N_4 nanosheet that considerably facilitate light collection and electron segregation.

Xiaolin et al. (2019) synthesized a composite aerogel by in situ deposition of Cu doped Cu_2O on 3D graphene oxide (RGO)/cellulose (CE) aerogels. The prepared composite was compared with Cu_2O/CE aerogels for degradation efficiency of the photocatalyst toward methyl orange under irradiation of visible light. The effect of pH, the initial concentration of methyl orange solution and the amount of catalyst were studied and demonstrated in Fig. 6.6(a–c) respectively. The result shows that

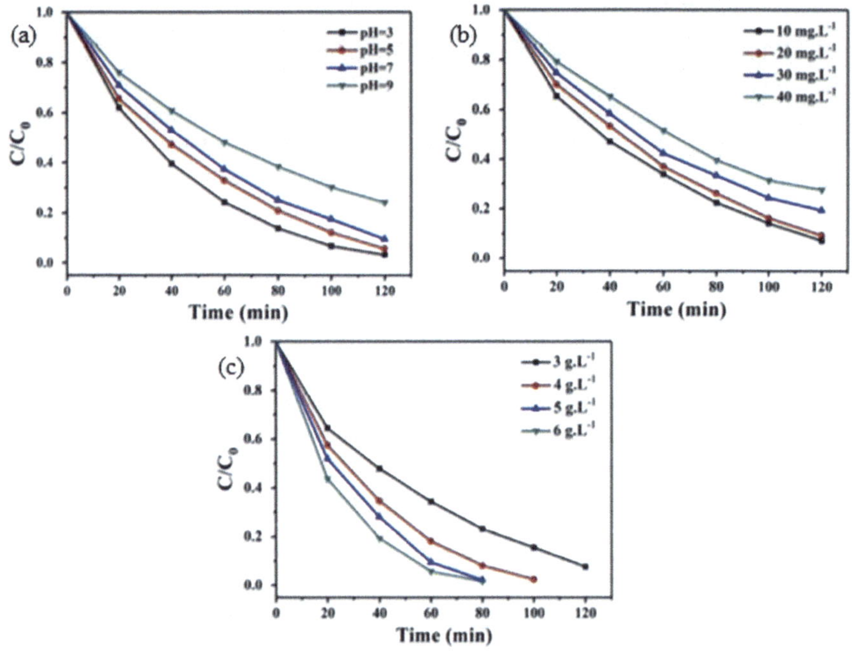

Fig. 6.6 (**a**) Cu@Cu_2O/RGO/CE—effect of pH on MO photodegradation. (**b**) Cu@Cu_2O/RGO/CE—effect of initial concentrations of MO on photodegradation. (**c**) Cu@Cu_2O/RGO/CE—effect of catalyst loading on MO photodegradation. Reproduced with permissions from reference (Xiaolin et al. 2019)

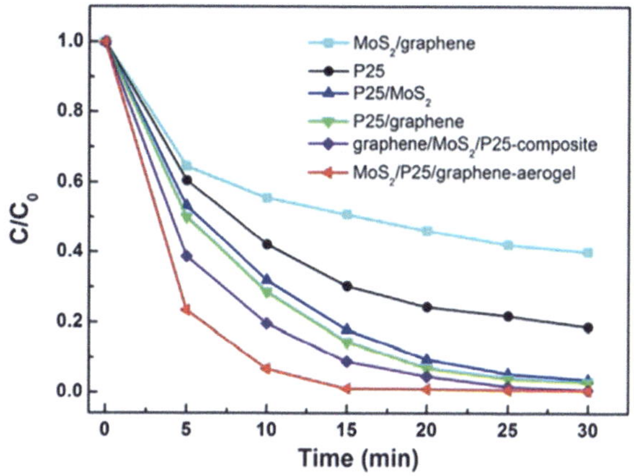

Fig. 6.7 Photo-degradation of methyl orange by $MoS_2/TiO_2(P2)_5$/graphene-aerogel(GA), graphene/TiO_2(P25)/MoS_2-composite, TiO_2(P25)/graphene, MoS_2/TiO_2(P25), MoS_2/graphene, and TiO_2(P25) with reaction time of 30 min under UV irradiation (Han et al. 2014). Reproduced with permissions from reference (Han et al. 2014)

the hybrid catalyst of Cu@Cu_2O/RGO/CE has higher photocatalytic activity when compared to Cu_2O/CE aerogels.

Fan et al. (2015) described the synthesis of novel photocatalyst material made of AgX (where X = Cl, Br) doped on the 3D graphene aerogel (GA) surface. The prepared sample was evaluated for the photodegradation efficiency toward methyl orange (MO) under visible light irradiation. The result obtained shows that the MO can almost completely be removed by AgBr/GA while the pure AgBr can only remove 65% of the MO. Similarly, the photodegradation efficiency of AgCl/GA was high when compared with bare AgCl.

Han et al. (2014) synthesized the 3D molybdenum disulfide (MoS_2)/TiO_2(P25)/graphene aerogel (GA) by hydrothermal method. The synthesized composite was compared with TiO_2/GA, MoS_2/GA, MoS_2/TiO_2 by evaluating the degradation efficiency against the removal of methyl orange under ultraviolet light radiation. As shown in Fig. 6.7, the degradation efficiency of almost 100% was achieved by MoS_2/TiO_2(P25)/graphene aerogel (GA) within 15 min.

Wei et al. (2019a) created a new 3D based cadmium sulfide (CdS)/graphene aerogel (GA) by a one-pot hydrothermal method. The CdS-GA's photocatalytic activity was analyzed for the elimination of organic pollutants in water. The removal of organic pollutant efficiency was high for the CdS/GA composite hybrid aerogel in comparison with pure CdS and GA. In the case of methyl orange removal, the CdS/GA hybrid shows 98.8% efficiency within 60 min, according to their finding shown in Fig. 6.8.

Bin et al. (2015) synthesized the novel 3D graphene-Cobalt(II,III) oxide (Co_3O_4) composite by using freeze-drying and calcination techniques. Synthesized graphene oxide by hummer's method was added to the Cobalt nitrate solution and then freeze-dried followed by calcination in order to produce 3D GA/Co_3O_4 hybrid aerogels. The hybrid aerogels were prepared by varying the GO to Cobalt nitrate weight ratios (0, 1, 2, 3%). The hybrid aerogels were analyzed for the photocatalytic activity of methyl orange under a xenon arc lamp as the light source. The highest degradation

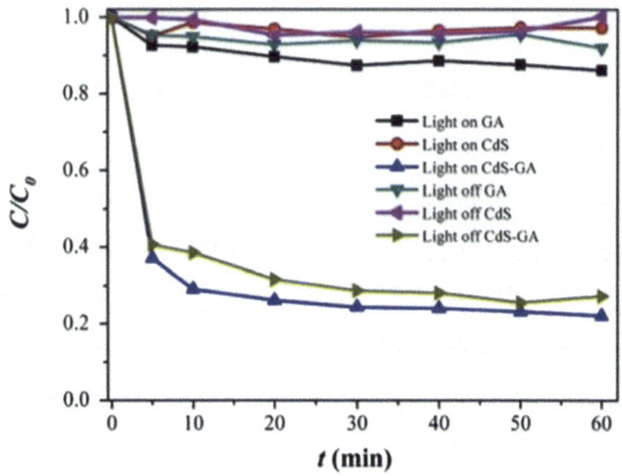

Fig. 6.8 The degradation performance of CdS, GA and the CdS-GA hybrids for methyl orange dye in water (Wei et al. 2019a). Reproduced with permissions from reference (Wei et al. 2019a)

efficiency of 95.3% was shown by3D GA/Co_3O_4 hybrid aerogels synthesized with a 2% weight ratio of GO to Cobalt nitrate. They also demonstrated that magnetic decantation could efficiently separate samples from the reaction media.

Yang et al. (2017) synthesized a novel 3D-3D Bismuth tungstate (Bi_2WO_6)-Graphene hydrogel (GH) by using the hydrothermal and freeze-drying method as demonstrated in Fig. 6.9(a). The hydrogels were prepared by changing the molar quantities of precursors used for making Bi_2WO_6 and keeping the graphene oxide solution volume constant. The Bi_2WO_6/GH samples were analyzed for photocatalytic activity against the degradation of methylene blue under the region of UV and visible light. The hydrogel shows refer to Fig. 6.9(b) better photocatalytic activity than pure Bismuth tungstate because the presence of graphene in the composite improves the photocatalytic efficiency.

Melone et al. (2013) prepared hybrid aerogels by mixing cellulose nanofibers (CNF) with either TiO_2(T) sol or TiO_2/SiO_2(TS) sol. The prepared aerogels were

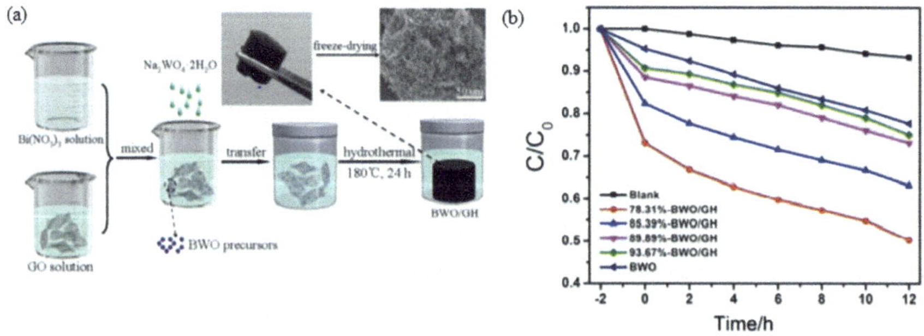

Fig. 6.9 (**a**) Schematic demonstration of the preparation process of 3D-Bi_2WO_6/GH composite. (**b**) The removal of Methylene Blue by Bi_2WO_6 and Bi_2WO_6/GH composites under visible light source irradiation (Yang et al. 2017). Reproduced with permissions from reference (Yang et al. 2017)

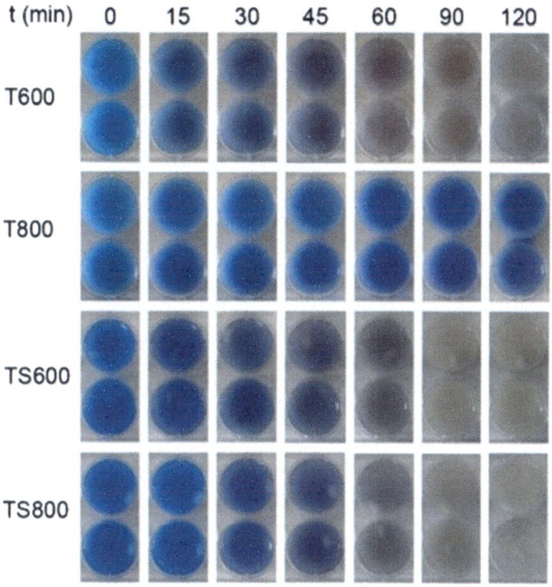

Fig. 6.10 MB degradation analysis under ultra violet irradiation on TX and TSX aerogels where X is the calcination temperature (Melone et al. 2013). Reproduced with permissions from reference (Melone et al. 2013)

calcinated at various temperatures. The absorption experiment with methylene blue (MB) and Rhodamine B (RhB) dyes was conducted in order to check degradation effects. In the case of MB degradation, the CNF-TiO_2 aerogel which is calcinated 600 °C (T600) shows higher efficiency than the sample calcinated at 800 °C (T800) and the result are shown in Fig. 6.10. The CNF-TiO_2/SiO_2 shows better efficiency when compared to CNF-TiO_2 at both calcination temperatures for the degradation of MB. In the case of RhB degradation, the degradation by CNF-TiO_2 was negligible at both the calcination temperatures and the CNF-TiO_2/SiO_2 aerogel shows that by increasing the annealing temperature, the efficiency increases.

Wei et al. (2019b) effectively synthesized unique 2D based nanorods of bismuth phosphate (BP)/3D biomass-based carbonaceous aerogels (BCA) composite by hydrothermal process. The aerogels were prepared at different BP/BCA weight percentages of 5, 10, 20, 30 and 40. The composite aerogels were evaluated for the degradation of methylene blue dye under UV light irradiation for about 180 min. The BP/BCA composite aerogel with 20 weight percent shows the highest photocatalytic activity of about 71.9%, followed by 10 weight percent composite which shows about 52.8% activity.

Zhang et al. (2019) fabricated a Graphene Oxide (GO)-Titanium dioxide (TiO_2)-Carbon Quantum Dots (CQD) foam by a simple and low-cost solvothermal method. The prepared form was analyzed in order to check the degradation efficiency of the methyl orange (MO), methylene blue (MB) and Rhodamine B (RhB) dyes under the light source of a xenon lamp. The result shows that GO-TiO_2-CQD foam has high degradation efficiency of 92.48% toward MO, 95.5% toward MB (shown in Fig. 6.11) and 92.84% toward are RhB when compared to Graphene Oxide foam,

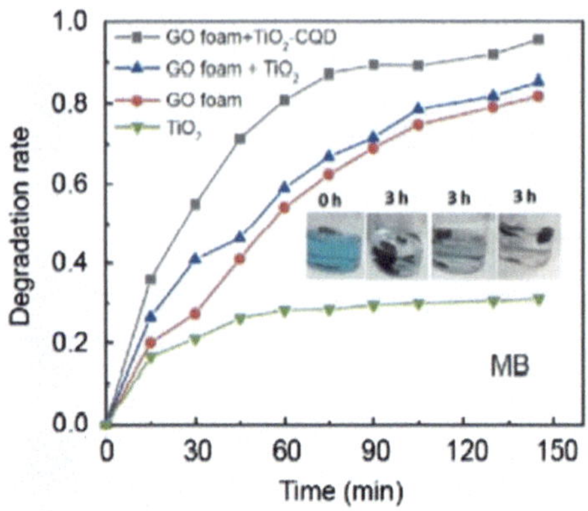

Fig. 6.11 Photocatalytic degradation kinetics of MB (Zhang et al. 2019). Reproduced with permissions from reference (Zhang et al. 2019)

Graphene Oxide-TiO_2 foam and TiO_2 powder. The composite foam maintains a high degradation rate and stability even after many cycles of usage.

6.2.2.1 Photocatalysis Using Aerogel Materials

Photocatalysis is a process that can be used to resolve related environmental issues originated by organic pollutants, posing a big threat to human life and ecosystem (Qi and Yu 2020). Photocatalysis carried out with powder samples normally have the attitude to cluster and for recycling the intricate procedure, it will limit the practical applications.

The 3D porous solid network aerogels are the ones that have very low density, huge specific surface area and high porosity. It consists of nano-scale characteristics with an overall monolith size of various cm. Therefore, aerogels are considered as a bridge connecting the nano and the macro within that building block; it not only holds the pristine properties, but it does form a fresh function through the three-dimensional action of the building block (Aegerter et al. 2011). Aerogel is very special for photocatalytic applications. After the improvement in synthesis technologies, the forms of aerogel photocatalysts have increased from the conventional oxide and the chalcogenide aerogels to the recent composite aerogel. And also, their application extended from the primary physical adsorption to the current photochemical reaction that includes the environmental correction and clean energy manufacture (Serhan et al. 2019).

In order to understand the photocatalytic activity of aerogel materials, let us consider the example of graphene aerogel and CdS aerogel. Table 6.4 shows the photocatalytic degradation of the RhB organic dyes in the water with different graphene aerogel photocatalysts (Long et al. 2020).

Graphene aerogel under photocatalysis can also be used for the separation of uranium from the water below visible light and in the air atmosphere. For this process, a novel graphene aerogel is prepared, GA-200. Uranium has photochemical property, due to which it plays a major role to create H_2O_2 with O_2 in the air, and it forms

Table 6.4 Photocatalytic degradation of the RhB organic dyes in the water with the different graphene aerogel supported photocatalysts (Long et al. 2020). Reproduced with permissions from reference (Long et al. 2020)

Photocatalyst (mg)	Pollutant (mg/L)	Light (W)	Degradation (%)	Time (min)	Cycle	Cyclic effect	Refs.
Ag_2O/ALG-rGO (30)	RhB (Vacher et al. 1988)	500 W Xe lamp	96	150	5	89%	Ma et al. (2018)
Bi_2WO_6/ GA (20)	RhB (Gesser and Goswami 1989)	300 W Xe lamp	99.6	45	ND	ND	Xu et al. (2016b)
BiOBr/RGO aerogel	RhB	300 W Xe lamp	Over 68	ND	5	68%	Liu et al. (2015)
CeO_2/RGO	RhB	150 W Xe lamp	85	120	3	No significant changes	Choi et al. (2016)
BiOBr/RGO	RhB	300 W Xe lamp	50	60	ND	ND	Yu et al. (2017a)
g-C_3N_4-TiO_2-GA (5)	RhB (Peng et al. 2021)	500 W Xe lamp	98.40	60	4	75.60%	Zhang et al. (2018a)
Fe_2O_3-TiO_2-GA	RhB (Peng et al. 2021)	500 W Xe lamp	97.70	60	4	81.80%	Zhang et al. (2018b)
$W_{18}O_{49}$-RGO	RhB	500 W Xe lamp	100	25	30	No significant changes	Li et al. (2014)
TiO_2-GA (5)	RhB (Peng et al. 2021)	300 W Xe lamp	98.7	180	5	70%	Zhang et al. (2016)
3D Ag/Ag@ Ag_3PO_4/GA (7.5)	RhB (Torikai et al. 1994)	400 W Xe lamp	100.0%	15	6	No significant changes	Chen et al. (2018)
3D g-C_3N_4/BiOBr/RGO	RhB	300 W Xe lamp	66	60	3	No significant changes	Yu et al. (2017b)

RGO reduced graphene oxide, *GA* graphene aerogel, *RhB* rhodamine B

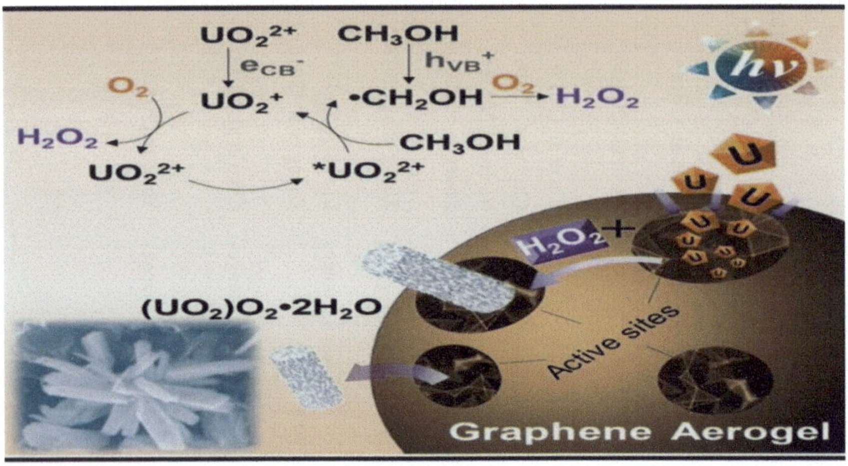

Fig. 6.12 Schematic representation of uranium removal using graphene aerogel (Wang et al. 2020). Reproduced with permissions from reference (Wang et al. 2020)

$(UO_2)O_2.2H_2O$. The products that are neutral can empty from the surface of the GA-200, and thus, it regenerates the active sites in situ. This results on the separation capacity of 1050 mg/g (Wang et al. 2020) (Fig. 6.12).

CdS aerogels synthesized by the normal sol–gel assembly of the individual nanocrystals into a porous channel, which is accompanied by a supercritical drying method, can give high surface area for photocatalytic degradation. The evaluation to understand the quality of CdS aerogel for the dye degradation of organic dyes was done using methylene blue and methyl orange as the test molecule. After analysis, it was found that the CdS aerogel material showed impressive photocatalytic activity for degradation of dyes when compared with the typical ligand capped CdS nanocrystals. The catalytic efficiency of the CdS aerogel can be increased by decreasing the length of the chain and extending the surface organics that lead to more and higher hydrophilic reachable surface area (Korala et al. 2017).

6.3 Conclusion

Aerogels, due to their high absorption rate and properties, can be explored for a wide range of applications, and it has been proven that they can be used as a promising solution for dye removal. Dyes are used in different industries like Textile, Food, Paper, Cosmetic, Leather, Fur with the textile industry as a major consumer. Researchers believe photocatalysts with the 3D porous network structures (aerogels) to be the most likely materials for water purification in recent periods. Characteristic features like exceptionally high porosity, huge specific surface area, very low density, and ease of detachment from aqueous solution etc. make these materials suitable for dye removal. Photocatalytic characteristics of the aerogels can be changed through combining them with diverse materials such as organic, inorganic and hybrid. Because of the large specific surface area, the high sorption capacity, great electrical conductivity, advanced electrochemical stability, superior mechanical strength, and advanced adsorption–desorption rates, new studies have displayed that the carbon-based hybrid aerogels considering graphene and derivatives as photocatalysts can effectively remove pollutants, especially the organic dyes from wastewater.

References

Aegerter MA, Leventis N, Koebel MM (2011) Aerogels handbook. Springer

Aspland JR (1992) A series on dyeing. Chapter 8: disperse dyes and their application to polyester. Text Chem Color 24:8–13

Bin Z, Hui L (2015) Three-dimensional porous graphene-Co_3O_4 nanocomposites for high performance photocatalysts. Appl Surf Sci 357:439–444

Chen F et al (2018) 3D graphene aerogels-supported Ag and Ag@Ag_3PO_4 heterostructure for the efficient adsorption-photocatalysis capture of different dye pollutants in water. Mater Res Bull 105:334–341

Cheng S, Liu X, Yun S, Luo H, Gao Y (2014) SiO_2/TiO_2 composite aerogels: preparation via ambient pressure drying and photocatalytic performance. Ceram Int 40:13781–13786

Choi J, Reddy DA, Islam MJ, Ma R, Kim TK (2016) Self-assembly of CeO_2 nanostructures/reduced graphene oxide composite aerogels for efficient photocatalytic degradation of organic pollutants in water. J Alloys Compd 688:527–536

Du A, Zhou B, Zhang Z, Shen J (2013) A special material or a new state of matter: a review and reconsideration of the aerogel. Materials (Basel) 6:941–968

Fan Y et al (2015) Convenient recycling of 3D AgX/graphene aerogels (X = Br, Cl) for efficient photocatalytic degradation of water pollutants. Adv Mater 27:3767–3773

Fricke J (1988) Aerogels—highly tenuous solids with fascinating properties. J Non Cryst Solids 100:169 173

Fricke J, Emmerling A (1998) Aerogels—recent progress in production techniques and novel applications. J Sol-Gel Sci Technol 13:299–303

Fricke J, Emmerling A (1992) Aerogels—preparation, properties, applications:37–87. https://doi.org/10.1007/bfb0036965

Gesser HD, Goswami PC (1989) Aerogels and related porous materials. Chem Rev 89:765–788

Gurav JL, Jung IK, Park HH, Kang ES, Nadargi DY (2010) Silica aerogel: synthesis and applications. J Nanomater 2010

Hamaker HC (1937) A general theory of lyophobic colloids. II. Recl des Trav Chim des Pays-Bas 56:3–25

Han W et al (2014) Enhanced photocatalytic activities of three-dimensional graphene-based aerogel embedding TiO_2 nanoparticles and loading MoS_2 nanosheets as co-catalyst. Int J Hydrogen Energy 39:19502–19512

Hasanpour M, Hatami M (2020) Photocatalytic performance of aerogels for organic dyes removal from wastewaters: review study. J Mol Liq 309:113094

He H, Huang L, Zhong Z, Tan S (2018) Constructing three-dimensional porous graphene-carbon quantum dots/g-C_3N_4 nanosheet aerogel metal-free photocatalyst with enhanced photocatalytic activity. Appl Surf Sci 441:285–294

Hunger K (2007) Industrial dyes: chemistry, properties, applications. Wiley

Jiang Y, Chowdhury S, Balasubramanian R (2019) New insights into the role of nitrogen-bonding configurations in enhancing the photocatalytic activity of nitrogen-doped graphene aerogels. J Colloid Interface Sci 534:574–585

Kim YN et al (2013) Sol–gel synthesis of sodium silicate and titanium oxychloride based TiO_2–SiO_2 aerogels and their photocatalytic property under UV irradiation. Chem Eng J 231:502–511

Korala L et al (2017) CdS aerogels as efficient photocatalysts for degradation of organic dyes under visible light irradiation. Inorg Chem Front 4:1451–1457

Li K, Zhang X, Qin Y, Li Y (2021) Construction of the cellulose nanofibers (Cnfs) aerogel loading tio2 nps and its application in disposal of organic pollutants. Polymers (Basel) 13

Li X et al (2014) Tungsten oxide nanowire-reduced graphene oxide aerogel for high-efficiency visible light photocatalysis. Carbon 78:38–48

Liu W, Cai J, Li Z (2015) Self-assembly of semiconductor nanoparticles/reduced graphene oxide (RGO) composite aerogels for enhanced photocatalytic performance and facile recycling in aqueous photocatalysis. ACS Sustain Chem Eng 3:277–282

Liu Z et al (2021) Synthesis of mesoporous carbon nitride by molten salt-assisted silica aerogel for rhodamine B adsorption and photocatalytic degradation. J Mater Sci 56:11248–11265

Long S et al (2020) 3D graphene aerogel based photocatalysts: synthesized, properties, and applications. In: Colloids and surfaces A: physicochemical and engineering aspects, vol 594. Elsevier B.V.

Ma Y, Wang J, Xu S, Feng S, Wang J (2018) Ag_2O/sodium alginate-reduced graphene oxide aerogel beads for efficient visible light driven photocatalysis. Appl Surf Sci 430:155–164

Mallakpour S, Rashidimoghadam S (2018) Carbon nanotubes for dyes removal. In: Composite nanoadsorbents. Elsevier Inc. https://doi.org/10.1016/B978-0-12-814132-8.00010-1

Maouche C et al (2020) A 3D nitrogen-doped graphene aerogel for enhanced visible-light photocatalytic pollutant degradation and hydrogen evolution. RSC Adv 10:12423–12431

Matsuda H, Kobayashi N, Kobayashi T, Miyazawa K, Kuwabara M (2000) Room-temperature synthesis of crystalline barium titanate thin films by high-concentration sol-gel method. J Non Cryst Solids 271:162–166

Melone L et al (2013) Ceramic aerogels from TEMPO-oxidized cellulose nanofibre templates: synthesis, characterization, and photocatalytic properties. J Photochem Photobiol A Chem 261:53–60

Nita LE, Ghilan A, Rusu AG, Neamtu I, Chiriac AP (2020) New trends in bio-based aerogels. Pharmaceutics 12

Peng YH, Kashale AA, Lai Y, Hsu FC, Chen IWP (2021) Exfoliation of 2D materials by saponin in water: aerogel adsorption/photodegradation organic dye. Chemosphere 274:129795

Pierre AC, Pajonk GM (2002) Chemistry of aerogels and their applications. Chem Rev 102:4243–4265

Qi K, Yu J (2020) Modification of ZnO-based photocatalysts for enhanced photocatalytic activity. Interface Sci Technol 31:265–284

Said B, El Harfi S, El Harfi A (2017) Classifications, properties and applications of textile dyes: a review. Appl J Environ Eng Sci 3:311–320

Schaedler TA et al (2011) Ultralight metallic microlattices. Science 334:962–965

Schaefer DW, Keefer KD (1986) Structure of random porous materials: silica aerogel. Phys Rev Lett 56:2199–2202

Serhan M et al (2019) Total iron measurement in human serum with a smartphone. In: AIChE annual meeting of conference on proceeding, 2019 Nov

Siddiqui SI et al (2018) Decolorization of textile wastewater using composite materials. Nanomater Wet Process Text:187–218. https://doi.org/10.1002/9781119459804.ch6

Torikai D et al (1994) Comparison of high-purity H_2/O_2 and LPG/O_2 flame-fused silica glasses from sol-gel silica powder. J Non Cryst Solids 179:328–334

Vacher R, Woignier T, Pelous J, Courtens E (1988) Structure and self-similarity of silica aerogels. Phys Rev B 37:6500–6503

Wagh PB, Begag R, Pajonk GM, Venkateswara Rao A, Haranath D (1999) Comparison of some physical properties of silica aerogel monoliths synthesized by different precursors. Mater Chem Phys 57:214–218

Wang Z et al (2020) Graphene aerogel for photocatalysis-assist uranium elimination under visible light and air atmosphere. Chem Eng J 402:126256

Wei XN et al (2019a) One-pot self-assembly of 3D CdS-graphene aerogels with superior adsorption capacity and photocatalytic activity for water purification. Powder Technol 345:213–222

Wei W et al (2019b) $BiPO_4$ nanorods anchored in biomass-based carbonaceous aerogel skeleton: a 2D–3D heterojunction composite as an energy-efficient photocatalyst. J Supercrit Fluids 147:33–41

Wen J, Liu H, Zheng Y, Wu Y, Gao J (2020) A novel of PTA/ZIF-8@cellulose aerogel composite materials for efficient photocatalytic degradation of organic dyes in water. Zeitschrift Fur Anorg Und Allg Chemie 646:444–450

Wu Y et al (2020) Zeolitic imidazolate framework-67@cellulose aerogel for rapid and efficient degradation of organic pollutants. J Solid State Chem 291:121621

Xiaolin D et al (2019) High photocatalytic activity of Cu@Cu_2O/RGO/cellulose hybrid aerogels as reusable catalysts with enhanced mass and electron transfer. React Funct Polym 138:79–87

Xu Z et al (2007) Preparation and characterization of silica-titania aerogel-like balls by ambient pressure drying. J Sol-Gel Sci Technol 41:203–207

Xu H, Zhu P, Wang L, Jiang Z, Zhao S (2016a) Structural characteristics and photocatalytic activity of ambient pressure dried SiO_2/TiO_2 aerogel composites by one-step solvent exchange/surface modification. J Wuhan Univ Technol Mater Sci Ed 31:80–86

Xu X, Ming F, Hong J, Xie Y, Wang Z (2016b) Three-dimensional porous aerogel constructed by Bi_2WO_6 nanosheets and graphene with excellent visible-light photocatalytic performance. Mater Lett 179:52–56

Yagub MT, Sen TK, Afroze S, Ang HM (2014) Dye and its removal from aqueous solution by adsorption: a review. Adv Colloid Interface Sci 209:172–184

Yang J, Chen D, Zhu Y, Zhang Y, Zhu Y (2017) 3D–3D porous Bi_2WO_6/graphene hydrogel composite with excellent synergistic effect of adsorption-enrichment and photocatalytic degradation. Appl Catal B Environ 205:228–237

Yu X, Shi J, Feng L, Li C, Wang L (2017a) A three-dimensional BiOBr/RGO heterostructural aerogel with enhanced and selective photocatalytic properties under visible light. Appl Surf Sci 396:1775–1782

Yu X et al (2017b) Ternary-component reduced graphene oxide aerogel constructed by g-C_3N_4/BiOBr heterojunction and graphene oxide with enhanced photocatalytic performance. J Alloys Compd 729:162–170

Zhang JJ et al (2016) Synergetic adsorption and photocatalytic degradation of pollutants over 3D TiO_2-graphene aerogel composites synthesized: via a facile one-pot route. Photochem Photobiol Sci 15:1012–1019

Zhang JJ et al (2018a) High-efficiency removal of rhodamine B dye in water using g-C_3N_4 and TiO_2 co-hybridized 3D graphene aerogel composites. Sep Purif Technol 194:96–103

Zhang JJ et al (2018b) Three-dimensional Fe_2O3–TiO_2–graphene aerogel nanocomposites with enhanced adsorption and visible light-driven photocatalytic performance in the removal of RhB dyes. J Ind Eng Chem 61:407–415

Zhang X, Wei W, Zhang S, Wen B, Su Z (2019) Advanced 3D nanohybrid foam based on graphene oxide: facile fabrication strategy, interfacial synergetic mechanism, and excellent photocatalytic performance. Sci China Mater 62:1888–1897

Zhang B et al (2018) Bacterial cellulose derived monolithic titania aerogel consisting of 3D reticulate titania nanofibers. Cellulose 25:7189–7196

Zhou Z, Zhang X, Lu C, Lan L, Yuan G (2014) Polyaniline-decorated cellulose aerogel nanocomposite with strong interfacial adhesion and enhanced photocatalytic activity. RSC Adv 4:8966–8972

Zu G et al (2015) Silica-titania composite aerogel photocatalysts by chemical liquid deposition of titania onto nanoporous silica scaffolds. ACS Appl Mater Interfaces 7:5400–5409

Chapter 7
Effective Materials in the Photocatalytic Treatment of Dyestuffs and Stained Wastewater

Rahul Bhattacharjee, Tamoghni Mitra, Priya Mitra, Soumya Biswas, Saikat Ghosh, Soham Chattopadhyay, Sumira Malik, and Abhijit Dey

Abstract Organic dyes utilised in textile industry possess poisonous toxins and thus are subjected to treatment to avoid being released into the environment. Several methodologies, including physical, chemical and biological methods, have been used to solve this challenge. Photocatalysis is now one of the most advanced oxidation processes, which is carried out using light and proper photocatalytic materials. Heterogeneous photocatalysis offers multiple advantages, including being environmentally acceptable, cost-effective and effective in degrading different colours and organic pollutants found in waste effluent. For the treatment of textile effluent including dye components, a variety of photocatalysts, as well as modified or composite photocatalysis, are utilised. The most promising component appears to be titanium dioxide (TiO_2), and various modified variants of TiO_2 exhibit increased photocatalytic activity on dye degradation. This chapter discusses the many traditional procedures for removing dyestuffs from textile effluents, as well as the photocatalysis process in detail. Photocatalysis was used to remove many dye components, which was also covered in this chapter.

Rahul Bhattacharjee and Tamoghni Mitra—Contributed equally.

R. Bhattacharjee · T. Mitra · P. Mitra · S. Biswas · S. Ghosh
KIIT School of Biotechnology, Kalinga Institute of Industrial Technology (KIIT-DU), Bhubaneswar, Odisha, India

S. Chattopadhyay
Maulana Azad College, Kolkata, West Bengal, India

A. Dey (✉)
Department of Life Sciences, Presidency University, 86/1 College Street, Kolkata, West Bengal 700073, India
e-mail: abhijit.dbs@presiuniv.ac.in

S. Malik
Amity Institute of Biotechnology, Amity University Jharkhand, Ranchi, Jharkhand 834001, India

S. Dave and J. Das (eds.), *Trends and Contemporary Technologies for Photocatalytic Degradation of Dyes*, Environmental Science and Engineering,
https://doi.org/10.1007/978-3-031-08991-6_7

7.1 Introduction

The growth of the global population is the indicator of the development of numerous establishments leading to industrialisation for nurturing the population demands. Handicrafts and textile industries, tanneries, pulp and paper industries have all grown along with the development of industries fulfilling the requirements of essential supplies like medicine, food, fuel, etc.; produces a massive volume of coloured effluents, which is responsible for causing global environmental problems if dumped into neighbouring waterbodies without ample treatment for decolorizing the dyes. Natural colourants prepared from crushing fruits and plant products have been used extensively for fabrics colouration before globalisation, but the revolution necessitates the use of more advanced and synthetic colouring agents of many variations to fulfil the demands according to the requirement during the dyeing practices of not only in the tannery and textile industries but in the food industries and pharmaceutical industries also. Different synthetic dyes used for fabrics colouration are methylene blue, malachite green, methyl orange, etc.; causes deprivation of the aquatic environment because of their stable configurations, when dumped without appropriate precautions (Khan and Mondal 2021). There are over a million distinct types of dyes commercially available, which are frequently used in the textile industry, producing 50,100 L wastewater effluents per kilogram of finished material (Khan and Mondal 2021; Manu and Chaudhari 2002; Lai 2014). Since dyes are non-biodegradable and remain in the environment for prolonged durations, they instigate adverse consequences, such as the blocking of sunlight by the higher colour intensity in freshwater bodies, causing the death of aquatic life forms. Apart from that, it was also accountable for lessening the safety of drinking water. Considering the persistent nature of these organic molecules which are light-resistant and withstand aerobic degradation, the management of wastewater effluents containing dyes and pigments, often known as 'new and emerging contaminants,' is extremely challenging. Due to incomplete dye degradation during wastewater treatment, the formation of highly poisonous, carcinogenic properties bearing naphthalene, benzidine and other aromatic intermediate products are escalating (Suteu et al. 2011; Borker and Salker 2006; Dave et al. 2021; Satapathy et al. 2021; Dave 2021), requiring a method for destroying dyes and pigments contained in wastewater into simple, non-toxic molecules.

Wastewater containing dyes and colouring pigments is treated using a variety of physical, chemical, biological, electrochemical and photochemical techniques, while photocatalysis is presumed to be a reasonable solution among them for removing dissolved organic pollutants. In contrast to physical procedures (adsorption process and aslocoagulation process), When used in conjunction with photocatalysis, waste effluents may be readily transferred to other media and generally create impurities, but organic dye molecules can be mineralised into water, mineral acids and CO_2. The employment of biological processes is restricted due to the long retention period and the death of living bacteria in very hazardous conditions. The chemical reactions mediated by light in the presence of a photocatalyst, a substance capable of light absorption and speeding up the reaction rate, are known as photocatalysis. A photocatalyst upon light absorption produces an electron–hole pair that triggers chemical conversions and regenerates its chemical configuration (Fox and Dulay 1993). This photocatalytic approach has captured the curiosity of experts in recent years, having the potential to address a variety of environmental concerns. This approach is widely regarded as a fantastic model for wastewater treatment and the green movement, allowing for a green economic climate. There are no oxidants required and the process takes place under normal atmospheric conditions (Khan and Mondal 2021).

7.2 Classification of Dyes

Prior to the development of synthetic dyes, there were just a few natural dyes available. As a result of the rise in yearly worldwide dye output, it has become necessary to classify dyes. According to estimates, they might be in the millions of tonnes (Gürses et al. 2016). There are numerous structural variations in dyes, making it impossible to classify them all based on just one criterion. This is also useless from a practical standpoint.

In the colour index, synthetic dyes and textile dyes may be classed based on their colour, chromophore structure and manner of application (Allen 2013; Rauf and Hisaindee 2013; Zollinger 2003). Dyes are categorised into numerous classes based on their chromophores (for example, xanthene, azo, triphenylmethane, oxazine, diphenylmethane and so on) (Rauf et al. 2011). Each colour in textile dyes has a distinct chemical structure (Christie a2014; Raman and Kanmani 2016). Their chromophoric group's chemical structure may be used to give them a name. The chromogenic chromophore is a collection of atoms found in dye compounds that are responsible for the dye's colour. Auxochrome is a chromophore that attaches to a group of atoms, enhancing the chromogen's colour (Zinatloo-Ajabshir and Salavati-Niasari 2015, 2016). The most often used chromophores are ($-NO_2$), ($-C=N-$), ($-N=N-$), ($-C=O-$), ($-C=C-$) and quinoid rings (Antoniotti and Duñach 2002). The auxochromes accessible include ($-OH$), ($-CO_2H$), ($-SO_3H$) and ($-NH_3$) (Christie 2014; Raman and Kanmani 2016; Welham 2000). Additionally, some dyes are soluble in a variety of solvents, including basic, direct, mordant, reactive dyes and acid. Vat, azo, disperse and sulphur dyes are all examples of insoluble dyes. Nonionic and ionic dyes are classified according to their charge upon dissolution. The term 'catalytic ionic dyes' refers to any basic dyes, while the term 'anionic dyes' refers to dyes that are acidic, reactive as well as direct dyes. To put it simply, nonionic dyes are just dispersed dye in their most fundamental form. Due to the fact that dyes have a variety of applications, they may be categorised as azo dyes for dyeing polymers, sulphur dyes for dyeing cotton and drug dyes for use in the creation of pharmaceuticals and so on.

No categorisation of dyes can satisfy everyone, since many substances satisfy chemical requirements for inclusion in many groups. The following diagram (Fig. 7.1) highlights the general classification of each dye family.

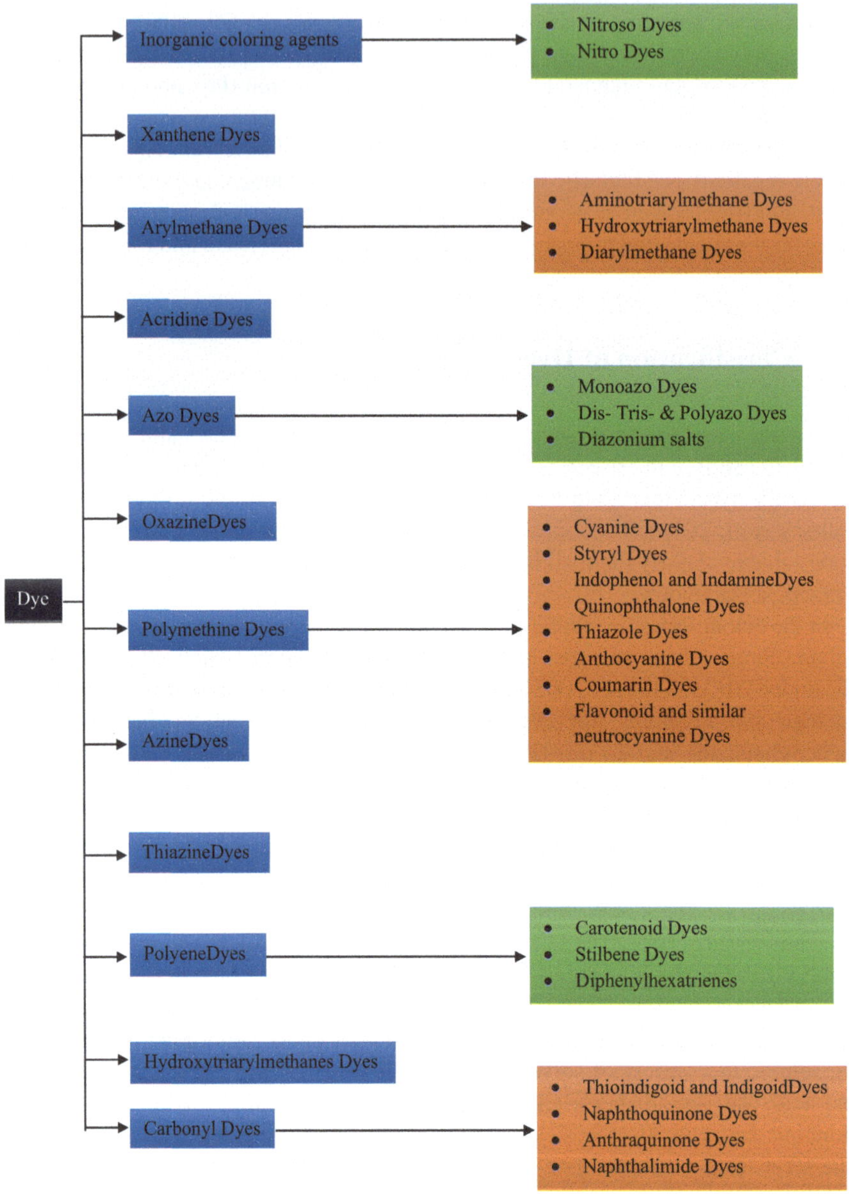

Fig. 7.1 Schematic representation of the classification of dyes

7.3 Photocatalysis

Photocatalysis is a combination of two words, i.e. photo which refers to photon and catalyst which is a substance that can change the rate of reaction. It is phenomenon in which a semiconductor utilises light energy to generate electron–hole pairs to drive chemical reactions. The catalyst that utilises the light energy to carry the reaction is known as a photocatalyst (Yang and Wang 2018).

Based on physical state of reactants, there are two sorts of photocatalysis, that is homogenous and heterogenous photocatalysis. It is possible to have homogenous photocatalysis when the reactant and semiconductor are in same state of the reaction. In heterogeneous photocatalysis, the semiconductor and reactant are in distinct phases, like, liquid, gas or solid (Ameta et al. 2018). The photocatalysis method has recently piqued the interest of scientists, who believe it has the potential to help solve a number of environmental problems. The elimination of dyestuffs from textile wastewater discharges using photocatalysis is often considered as the most successful approach. In the green movement, this strategy has been considered a paradigm and a cleaner alternative for garbage and environmental remediation. On the basis of how the reactants look in their physical state, two kinds of photocatalytic reactions may be distinguished.

- Homogeneous photocatalysis: Homogeneous photocatalysis is the name given to photocatalytic reactions in which the reactant and the semiconductor are both in the same state of matter (solid, liquid or gas) (Manu and Chaudhari 2002).
- Heterogeneous photocatalysis: Heterogeneous reactions are those in which the reactant and semiconductor are in various phases of the reaction cycle (Lai 2014).

Photocatalysis is defined as the acceleration of chemical transformations in the presence of a material called a photocatalyst that absorbs light and participates in the process. It forms an electron–hole pair when photons are absorbed by a component, which causes chemical reactions in reaction contributors and rejuvenates its biochemical arrangement after every series (Suteu et al. 2011).

7.3.1 The Process of Photocatalysis

One must employ a semiconductor light source or light irradiation in order to activate a solid material through photocatalysis, which is explained by 'redox mechanism' chemistry. All organic dyestuffs contained in a wastewater sample may be degraded using photocatalytic materials. However, the bandgap of these materials is critical. Rather than the band gap, the position of the photocatalyst's band determines the photocatalyst's oxidation or reduction strength, which in turn increases photocatalytic degradation. Several semiconductors have bandgaps that fall between 0.7 and 5 eV. Semiconductor bands are separated by a bandgap (also known as an energy gap), which is a difference in energy between a solid's conduction and valence bands.

Particularly important is the use of photocatalytic processes to remove pollutants from waste, particularly those with a bandgap of 2.0–3.3 eV (Borker and Salker 2006). The photocatalyst's surface is often where the oxidation and reduction reactions take place. Azo dye molecules are a common kind of dye used in staining applications.

The degradation of dyes has been attributed to radical species created during the photoexcitation of the semiconductor. In a broad sense, the stages involved may be shown as follows:

$$\text{photons}(hv) + \text{Semiconductor} \rightarrow h^{+}(\text{VB}) + e^{-}(\text{CB}) \tag{7.1}$$

$$\text{H}_2\text{O} + h^{+} \rightarrow \text{OH}^{-} + \text{H}^{+} \tag{7.2}$$

$$\text{OH}^{-} + h^{+} \rightarrow \text{OH}\cdot \tag{7.3}$$

$$\text{O}_2 + e^{-} \rightarrow \text{O}_2. \tag{7.4}$$

$$\text{H}_2\text{O} + \text{O}_2 + 2e^{-} \rightarrow \text{H}_2\text{O}_2 \tag{7.5}$$

$$\text{H}_2\text{O}_2 + e^{-} \rightarrow \text{OH}^{-} + \text{OH}\cdot \tag{7.6}$$

$$\begin{aligned} &\text{Organic contaminant} + \text{O}_2 + \text{OH}\cdot \rightarrow \text{H}_2\text{O} + \text{CO}_2 \\ &\quad + \text{additional decomposition products.} \end{aligned} \tag{7.7}$$

The deterioration of the dye is caused by an excited electron and a hole in the semiconductor. Nanotechnology has made use of many different semiconductors, with the enhanced surface area and favourable quantum size impact of most of them.

It is necessary for the photocatalytic mediated degradation of waste-produced azo dye molecules to be carried out in the presence of UV irradiation or in the presence of visible light. The steps are as follows:

1. ·OH radicals and electrons on the semiconductor's CB largely oxidise or decrease the dye element (Dave et al. 2021).
2. The darkening of the wastewater indicates the breakage of the azo link (Satapathy et al. 2021).
3. ·OH radicals trigger oxidation processes that open aromatic rings.
4. Compounds including amines, short-chain organic acids and phenols are formed when aromatic rings are broken down into their constituents.
5. By-products like CO_2 and H_2O are signs that the dye molecule has been completely degraded by photocatalysis (Dave 2021).

7.3.2 Direct Mechanism for Dye Degradation

Additional photocatalytic dye degradation processes may take place in the presence of visible light due to their tendency to absorb a part of visible light. This method includes dye excitation from the ground state (Dye) to the triplet excited state (Dye*) using visible light photons ($\lambda > 400$ nm) (Borker and Salker 2006). In this case, TiO_2 is used to put an electron in the conduction band. These excitons dye varieties are then turned into a (Dye^+)semi-oxidised radical cation. These trapped electrons interact with the system's dissolved oxygen, resulting in the formation of superoxide radical anions (O_2^-) and hydroxyl radicals (OH^-). OH^- radicals have a substantial impact on oxidation of organic molecules as shown in the equations.

$$h\nu + \text{Dye} \rightarrow \text{Dye}* \tag{7.8}$$

$$\text{TiO}_2 + \text{Dye}* \rightarrow \text{TiO}_2 + \text{dye}^+ \tag{7.9}$$

It's widely accepted that the dye degradation process is more heavily influenced by the indirect mechanism than by the visible light-initiated mechanism. Indirect mechanisms, on the other hand, are thought to have a much slower response time.

7.3.3 The Mechanism of Indirect Degradation of Dye

The following summarises the indirect heterogeneous photocatalytic oxidation mechanism employing semiconducting materials.

a. **Photoexcitation**: To put it simply, a photocatalytic reaction begins when light passes through a semiconductor and moves a photoelectron from its fully-charged valence band into its empty conduction band over. In order for the photon to be absorbed, it must have an energy level (*hn*) equal to or higher than the semiconductor photocatalyst's band gap. The valence band (*h*VB+) is left empty after the excitation process (Fox and Dulay 1993). As a consequence, as shown in the equation, a pair of electrons and holes (*e*/*h*+) is formed as a consequence.

$$h\nu(\text{UV}) + \text{TiO}_2 \rightarrow \text{TiO}_2\left(h^+(\text{VB}) + e^-(\text{CB})\right) \tag{7.10}$$

b. **Ionisation of water**: ·OH is produced as a result of a reaction between water molecules and the holes generated at the VB, as shown by

$$h^+(\text{VB}) + \text{H}_2\text{O(ads)} \rightarrow \text{H}^+(\text{ads}) + \text{OH}^\cdot(\text{ads}) \tag{7.11}$$

Irradiated semiconductors produce hydroxyl radicals (OH) that are very powerful oxidising agents. Depending on the structure and stability of the organic compounds, the closer the organic compounds are to the catalyst surface, the more likely they are to mineralize (Gürses et al. 2016).

c. **Oxygen ionosorption**: To create an anionic superoxide radical (O_2), the electron in the conduction (eCB) is picked up by oxygen and interacts with OH or surface-bound H_2O to form the photogenerated hole (hVB+).

$$e^{-}(\mathrm{CB}) + \mathrm{O_2} \rightarrow \mathrm{O_2^{-}}(\mathrm{ads}) \tag{7.12}$$

This superoxide ion may prevent the electron-hole recombination of TiO_2, which can participate in the subsequent oxidation process as well.

d. **Protonation of superoxide**: Upon protonation of the superoxide formed, a hydroperoxyl radical ($HO_2\cdot$) is formed, followed by hydrogen peroxide (H_2O_2), which is reduced to very reactive OH radicals (Allen 2013). The chain of events is shown as follows:

$$\cdot\mathrm{O_2^{-}} + \mathrm{H^{+}} \rightarrow \mathrm{HO_2^{\cdot}} \tag{7.13}$$

$$2\,\mathrm{HO_2} \rightarrow \mathrm{O_2} + \mathrm{H_2O_2} \tag{7.14}$$

$$\mathrm{H_2O_2} \rightarrow 2\mathrm{OH} \tag{7.15}$$

$$\mathrm{OH} + \mathrm{Dye} \rightarrow \mathrm{H_2O} + \mathrm{CO_2}(\text{dye intermediate}) \tag{7.16}$$

$$h^{+}(\mathrm{VB}) + \mathrm{Dye} \rightarrow \text{oxidation product} \tag{7.17}$$

$$e^{-}(\mathrm{CB}) + \mathrm{Dye} \rightarrow \text{reduction product} \tag{7.18}$$

The surface of the photoexcited semiconductor photocatalyst is often the site of oxidation and reduction reactions (Fig. 7.2).

7.4 Factors Affecting Photocatalytic Performances

A number of variables influence the photocatalytic degradation process, which may be summarised as follows: The rate of reaction is affected in numerous dimensions by pH, which is an incredibly essential component. Since generated water is so complicated, it's impossible to quantify variables like pH (Rauf and Hisaindee 2013) of the solution, catalyst dose, temperature and pollutant concentrations. Since the catalyst surface charge fluctuates with solution pH, adsorption on semiconductor material

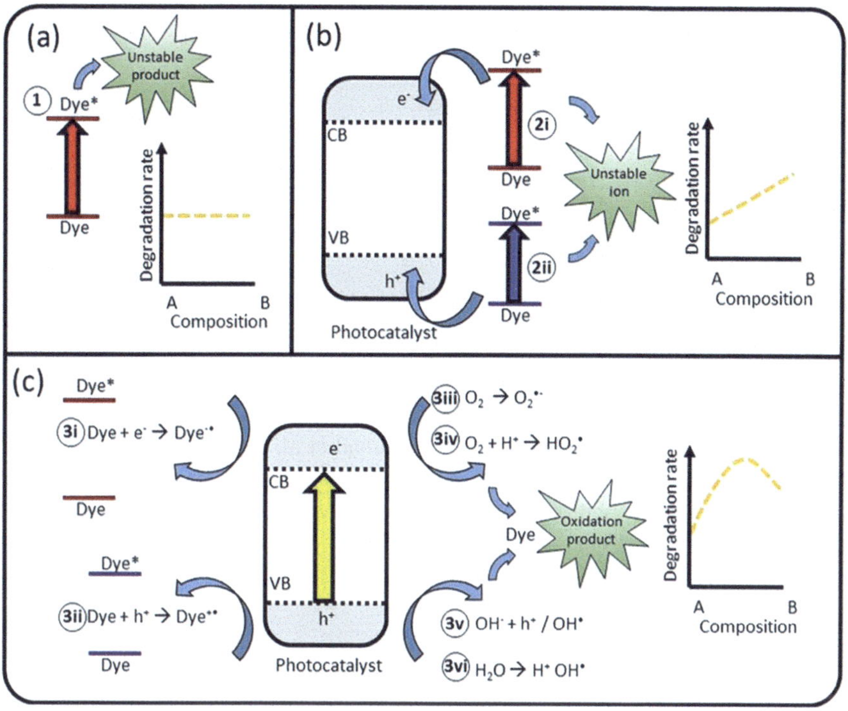

Fig. 7.2 Photodegradation mechanisms. **a** Direct degradation by (1) direct photolysis. **b** Dye-sensitisation leading to degradation through either electron (2i) or hole injection (2ii). **c** Indirect degradation of dyes by various photocatalytic reactions. (3i) and (3ii) involve direct oxidation/reduction of the dye by the photoexcited catalyst, while (3ii–3vi) proceed via radical intermediates. Reproduced with open access permission from ref. Pingmuang et al. (2017)

surfaces is greatly impacted. The pH of the solution has a significant impact on the pace of degradation and the overall photocatalytic dye degradation mechanism. During heterogeneous photocatalysis, the adsorption of pollutants (elements that need to be destroyed) and water molecules plays a crucial role in the degradation process. To effectively remove dyestuffs from polluted water, it is critical to figure out the optimal dose or photocatalyst loading (Zollinger 2003). It is important to consider light intensity, the type of dye component being removed, the topology of the photocatalyst, as well as the reactor layout when estimating the catalyst required for a given removal application. According to several studies, there are several reports of better dye adsorption by increasing catalyst loading. Hydrogen peroxide (H_2O_2) and potassium peroxydisulfate ($K_2S_2O_8$), both powerful oxidising chemicals, may be added to the catalyst solution to increase the photooxidation rate (Rauf et al. 2011). Light intensity and wavelength are linked to catalytic light absorption ranges and reduce the influence of the light source on photocatalysis, in addition to aromatic chemicals in generated water having strong UV absorption. The efficiency of the photocatalytic reaction is highly reliant on basic parameters such as amount of photocatalyst and the

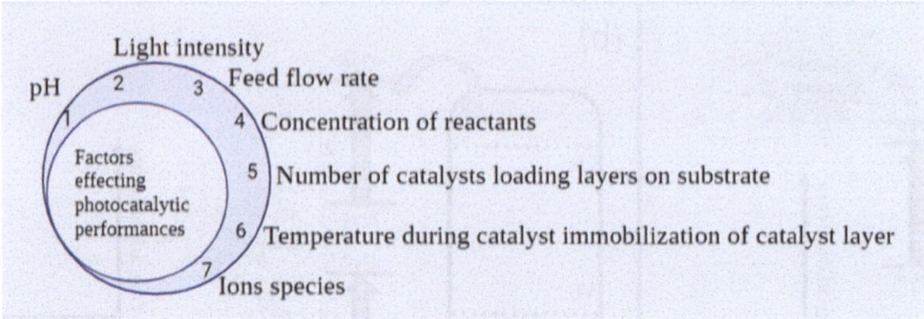

Fig. 7.3 Factors affecting photocatalytic performances

concentration of the substrate as well as operational parameters such as temperature and pH of the reaction medium, feed flow rates, intensity and wavelength of the light, concentration of salt and ion species, etc. (Ahmad et al. 2016; Kumar and Pandey 2017). Also, the nature and structure of the photocatalyst, as well as the substrate, play a significant role in the photocatalytic process (Lin et al. 2020). Several operational parameters have been discussed further (Fig. 7.3).

7.4.1 pH

According to a technique of synthesis for nano TiO_2, the surface charge characteristics of nano TiO_2 and the hydroxyl radical are regulated by pH (Jyothi et al. 2014). Nano TiO_2 photocatalytic activity is greatly influenced by the pH of the solution, which affects the interaction between photo-induced holes and hydroxides on the TiO_2 surface. pH levels that are either neutral or high are thought to be the most favourable conditions for hydroxyl radicals (Jallouli et al. 2017). Positive holes are thought to be a major cause of oxidation processes at lower pH levels (Shifu and Gengyu 2005). The photo redox process, as well as the phenomenon of interfacial electron transport, are influenced by pH changes induced by differences in molecular charge distribution, which in turn may modify the ease of site attack and bond breaking during photocatalysis (Jyothi et al. 2014).

7.4.2 Light Intensity

The ideal light intensity must be selected in photocatalytic reactions if high efficiency is to be achieved. Increasing light intensity may boost the rate of nano TiO_2 photocatalytic reaction, as the propensity of electron and hole interaction is insignificant at low-intensity levels (0–20 mW/cm^2) (Mozia 2010). At low intensities of light (about 25 mW/cm^2), the dissociation of charge carriers (electron-hole pairs) operates in a highly competitive with the hybridisation of charge carriers. This means that the rate of photocatalysis seems to be proportional to the square root of the intensity of light (Zangeneh et al. 2015). At extremely high intensities, the rate of reaction becomes independent of light intensity because the reaction rate is solely determined by mass transfer, which is limited by the existence of fully saturated materials at the catalyst surface for both adsorption and desorption (Lee et al. 2016).

7.4.3 Feed Flow Rate

Taking into account the flow rate of feed that is related to the time it spends in the system, it could have an effect on how well the system works. There is less exterior transfer of mass and less boundary layer rigidity in the liquid phase when flow speeds are increased over a photocatalyst that is immobilised. This could slow down pollutant degradation if flow speeds are enhanced above the ideal level (Lin and Valsaraj 2005). A high flow rate may limit penetration of light and reduce surface response efficiency because of the sheer growth in liquid viscosity and the reduction in holding time (Ananpattarachai and Kajitvichyanukul 2014).

7.5 Concentration of Reactants

According to the Arrhenius equation, increase in the concentration of contaminants in water enhances photocatalytic efficiency, which follows first-order kinetics. However, high concentrations outside the ideal range have a harmful impact on photocatalytic efficiency (Zainal et al. 2005). Because the number of active sites on the nanoparticle surface stays constant for a certain irradiation time, light intensity, catalyst dosage and at high concentrations owing to insufficient generation of reactive species, photocatalytic degradation is reduced. UV scattering increases with solution concentration and light penetration decreases as a result (Nam et al. 2002). In certain studies, it was discovered that increasing the concentration of reactants such as phosphamidon, phenol, reactive orange 14, lindane and MeO reduced the pace of decomposition (Rabindranathan et al. 2003).

7.5.1 Number of Catalysts Loading Layers on Substrate

The quantity of nano TiO_2 accessible for photocatalytic reactions may be increased by coating the substrate with several layers, which can improve photocatalytic activity. However, heavy loading can reduce the process' efficiency. Photocatalytic activity drops because the nano TiO_2's external surface blocks the internal layers, resulting in an increase in hole and electron recombination, which lowers light quantisation efficiency (Henderson 2011). The effectiveness can also be hampered by the dispersion of radical species via the layers. According to (Yu and Zhao 2000), the apparent pace was further increased by layering on up to five coats of film. When the number of coatings crossed 5, there was a slower growth.

7.6 Temperature During Catalyst Immobilisation

The temperature of the photocatalyst during immobilisation has a significant impact on the physiochemical characteristics of the immobilised film. However, lowering the calcination temperature reduces the critical loading, which has a negative impact on the coating film's mechanical stability. However, elevating the calcination temperature significantly increases the distribution of support molecules into the nano film (Chen and Dionysiou 2006). Temperature is also important for adherence to the support.

7.6.1 Ions Species

Ionic species in water can have an impact on photocatalytic degradation (Qamar et al. 2005). Ionic chemicals have an influence on the amount of electrons photogenerated, the prevention of electron-hole recombination and the scavenging of hydroxyl radicals. It has been shown that the presence of inorganic anions (such as HCO_3, PO_4^{3-}, NO_2, NO_3, SO_4^{2-} and Cl) in water reduces photocatalytic activity since the interaction of hydroxyl radicals and holes with the inorganic ions forms inorganic radicals, which impair photocatalytic properties (Hassan et al. 2016) upon adsorption of nano TiO_2 surface. The influence of inorganic anions on photocatalytic activity has been studied extensively (Ahmed et al. 2011). Bicarbonate (HCO_3), carbonate (CO_3^{2-}), and chloride (Cl) ions may interfere with photocatalytic activity by interacting with the hydroxyl radical. It is possible that chloride (Cl) ions, along with other inorganic anions (e.g. NO_2, HCO_3, NO_3, and PO_4^{3-}), might have an adverse effect on photocatalytic remediation by taking away the hydroxyl radicals and holes and producing reduced dichloride radicals ($Cl_{2\cdot}$) (Eq. 7.21) and reactive chloride radicals (Cl·) (Eq. 7.19) (Sirtori et al. 2010).

$$\mathrm{Cl} - + \mathrm{hvb} + \leftrightarrows \mathrm{Cl.} \tag{7.19}$$

$$\mathrm{Cl} - + \mathrm{HO.} \leftrightarrows \mathrm{ClOH.}^{-} \tag{7.20}$$

$$\mathrm{Cl} - + \mathrm{Cl.} \rightarrow \mathrm{Cl2\ .}^{-} \tag{7.21}$$

In the presence of chloride ions, the photocatalytic activity of a TiO_2/graphene oxide composite immobilised on a polymeric membrane decreased from 61 to 27% (Pastrana-Martinez et al. 2015). In another investigation, the presence of chloride ions had a deleterious influence on photocatalytic degradation at acidic pH (Yap and Lim 2011).

7.7 Photocatalysis of Wastewater

Photocatalysis, a much-admired advanced oxidation process (AOP) being used for wastewater treatment for dye removal purposes, produces water and carbon dioxide, in the overall process of decomposition of organic contaminants during cleaning up industrial pollutants (Lin et al. 2020). The mechanism of photocatalysis is grounded on redox reactions and the behaviour of semiconductors (Khan and Mondal 2021). An important role is played by the semiconductor bandgap, which may be described as the amount of energy that is required to destroy all organic dyes contained in industrial effluents via photocatalysis. However, oxidation-reduction strength is strongly dependent on the photocatalyst's band location instead of its bandgap and this has a significant effect on the degradation potential (Khan and Mondal 2021). Absorption of a wavelength containing higher energy than the bandgap creates electron-hole pairs just on the catalyst by releasing an electron in the CB and producing a positive hole in the VB, which allows photogenerated electron-hole pairs to move into more energetically advantageous locations (Lin et al. 2020). Upon exposure to UV/Vis light, semiconductor produces hydroxyl and superoxide anionic radicals, which are the principal oxidising species, able to cause the oxidation as well as mineralisation of a diverse array of organic molecules (Lin et al. 2019a, b). Diffusion and drift currents, as well as the rates of charge carrier, generated and recombined, all are having an impact on the charge separation equilibrium. The oxidation-reduction reactions taking place on the surface of the photocatalyst, causes the hole to be captured by surface water molecules, producing a reactive hydroxyl radical along with hydrogen ions, which delocalises on neighbouring water molecules, influencing the production of hydrogen peroxide (Rueda-Marquez et al. 2020). Furthermore, the generated hydroxide radicals potentially oxidise volatile elements in wastewater, such as methane (László et al. 2016).

Organic pollutant + O_2 semiconductor, $hv \rightarrow CO_2 + H_2O +$ inorganic ions/acids

$$
\begin{aligned}
&TiO_2 hv \rightarrow TiO_2(h^+, e^-)\\
&TiO_2(h^+, e^-) \leftrightarrow TiO_2(h^+) + TiO_2(e^-)\\
&(h^+) + H_2O \rightarrow OH\cdot + H^+\\
&O_2 + (e^-) + H^+ \rightarrow HOO\cdot\\
&HOO\cdot + HOO\cdot \rightarrow H_2O_2 + O_2
\end{aligned}
$$

Organic (particularly azo dyes) molecules are one of the most frequently used dyes in industries. The photocatalytic breakdown of azo dyes present in industrial effluent includes many crucial processes when exposed to UV light or visible light irradiation. The steps are:

(1) When the dye component is in the CB, electrons move through it and make it oxidise or reduce itself.
(2) The decoloration of effluent signifies the disruption of the azo bond.
(3) Oxidation processes fuelled by ·OH radicals disintegrate the aromatic structure of dye molecules.
(4) Amines, short-chain organic acids and phenolic substances are formed due to configuration changes and structural disintegration.
(5) Carbon dioxide and water production as a by-product is the indicator of the completion of photocatalytic degradation of the dye molecule.

The output of investigations on photocatalytic remediation of wastewater is inadequate, while the majority of them are employed for synthetically produced water (Jiménez et al. 2019). The primary classes of organic pollutants include glutaraldehyde, xylene, toluene, phenol, acetic acid and naphthalene are found in produced water, while some of them are readily degraded despite having a sluggish mineralisation rate for most of the elements, which necessitates a superior remediation method (Lin et al. 2020; Hong et al. 2018). The effects of several AOPs (photocatalysis, ozonation) on the treatment of synthetic wastewater imply that photocatalysis is successful of all these AOPs strategies (Jiménez et al. 2019), with less than 20% of the organic matters like carbon (TOC) being eliminated after only four hours of treatment. Variations in the TOC removal rate were identified as minor at higher P25 TiO_2 concentrations. No significant decline in TOC was seen for relatively high molecular organic materials such as grease, oil and natural organic matter in both photocatalysis and UV photolysis treatments, suggesting that the remediation of volatile, tiny organic pollutants like as toluene, naphthalene and xylene, along with the declination of phenol concentration, was verified using gas chromatograph-mass spectroscopy (Liang et al. 2011). Graphene-like TiO_2 nanocomposites (rGO-TiO_2) show better photocatalytic efficiency than TiO_2 for treating synthetic wastewater with high salt concentrations and various compositions of intractable dissolved organic substances, with a catalytic reaction rate for only 22% TOC elimination after 5 h of treatment increasing in the order of Phenols, Acetic acid, Toluenes, Xylenes and Naphthalene (Andreozzi et al. 2018). A novel approach was used to investigate the use of vertically stacked ZnO nanorods by using substrates that are specifically designed for optimal solar photocatalytic degradation by trapping visible light of HPAM (partially hydrolysed polyacrylamide), a well-known polymeric compound with oil recovery potential. After 6 h of treatment, the elimination of 150, 100, 50 and 25 mg/L of HPAM was found to be 45%, 56%, 62% and 68%, respectively, followed a decrease of 37% and 20% of TOC during 14 h and 7 h of reactivity permitting mineralisation (Dave 2021; Neppolian et al. 2008).

Precisely, photocatalysis is a well-planned efficient method for the degradation of refractory organic substances except for the proper mineralisation of wastewater. And for further reduction of organic pollutants and photocatalytic by-products, various technologies need to be utilised in succession for wastewater treatment. Toxicity, as well as biodegradability, should be given more consideration because of their substantial impact on the designing of treatment procedures (Lin et al. 2020). For toxicity assessment (both chronic as well as acute) of wastewater contaminants that are targeted to undergo the treatment of photocatalysis, a variety of approaches have been used, including bioassays with bacteria, plants (phytotoxicity), mammalian cells (genotoxicity), microalgae and seawater as well as freshwater invertebrates (He et al. 2016; Hasegawa et al. 2014; Di 2015; Tsoumachidou et al. 2017; Saverini et al. 2012; Calleja et al. 1986). Despite the development of several assessments for the investigation of photocatalytic analysis, only a few studies have included toxicity analysis for wastewater treatment, following successful remediation in minimising wastewater toxicity (Rueda-Marquez et al. 2020). For the purposes of toxicological evaluation, the standard effective concentration (EC50) is the level of a substance present in an environmental medium that causes unfavourable effects in at least half of the participants in an experiment (Lin et al. 2020). *Vibrio fischeri* was utilised as an experimental organism for the assessment of acute aquatic toxicity, which is leading to the implementation of four classes of toxicity according to their effectivity: class I indicates extremely toxic level with an EC50 value of 25%; class II indicates as toxic with an EC50 range of 25–75%, class III is known as somewhat toxic having an EC50 of 75%; and class IV is non-toxic with an EC50 of more than 75%. Photocatalysis is able to reduce the toxicity of the discharge up to 13–16%, indicating the efficiency of this method in comparison to ozone and H_2O_2 treatment for toxicity reduction of industrial effluent (Jiménez et al. 2019).

Ozone/TiO_2/UV photocatalytic oxidation was conducted in a batch mode in a laboratory-scale reaction chamber or reactor to determine the efficacy of these processes in removing pollutants and reducing the ecotoxicity of petrochemical industry-generated fluids (Corrêa et al. 2010). *Vibrio fischeri* and *Poecilia vivipara* were utilised to conduct screening for bacterial luminescence attenuation and fish toxicity, respectively, for acute toxicity analysis of the raw effluents, indicating a high toxicity level having EC50 value less than 1.55%, with lesser toxicity toward bacteria (EC50 = 30.9%), but higher toxicity induction in fish (EC50 = 1.9%), after 60 min of exposure. High tolerance of bacterial specimens to photocatalysed wastewater samples, but toxic effects in fish due to the presence of ammonia and other metal in photocatalysis treated wastewater, supports these experimental observations, insisting a post-treatment (biological treatment, sorption) for their removal (Corrêa et al. 2010). Even though distinguishing the effects of photocatalytic degradation from ozone treatment is problematic, combining photocatalysis with ozone reduces the toxicity of wastewater. Additional investigation concerning photocatalysis-induced toxicity as well as wastewater toxicity is intended to enhance the implementation of these methodologies (Lin et al. 2020; Corrêa et al. 2010).

7.8 Dye Removal From Wastewater May Be Accomplished Using a Variety of Methods

As the world's population grows, so does the growth of diverse industries to meet their demands. The quantity and volume of these operations are rapidly increasing across the world, resulting in massive volumes of coloured effluents. When these effluents are spilled into adjacent lakes and rivers without being properly treated or dyestuffs decoloured, they pose major environmental issues (Manu and Chaudhari 2002). In freshwater rivers and lakes, dyes with high colour intensity block out sunlight, killing aquatic life and plants. Water quality is also harmed, making it dangerous to drink. Because dyes are refractory organic molecules that resist aerobic digestion and are light stable, treating waste effluents containing dyestuff is particularly challenging (Suteu et al. 2011). Dyestuff-containing wastewater is treated using a variety of physical, chemical, electrochemical, biological and photochemical techniques. Treatment using photocatalytic means is one of these ways that is thought to be a viable option for removing soluble organic composites. Adsorption, advanced oxidation process (AOP), bioremediation and biodegradation, electrochemical approaches, ion exchange and membrane filtering technology are some of the dye removal technologies available. In the next section, the approaches are succinctly outlined.

7.8.1 Adsorption Technique

A common method for eliminating contaminants from waste effluent is adsorption. Numerous different metals or constituents are deposited all across the interface of a liquid or solid molecule in this surface phenomenon. Pre-treatment is not required when using liquid-phase adsorption. A well-known separation method for wastewater purification and water reuse is based on its simplicity and controllability, lower investment costs and ability to manage highly concentrated contaminants (Dąbrowski 2001). The extensive aspect of pollutant removal via the adsorption method is that no harmful components are likely to be generated during the reaction phase. When utilising adsorption methods, the decolorization operation is controlled by a variety of elements, such as the adsorbent's effective surface area, particle size, dye element-adsorbent interactions, process temperature and pH and contact length (Kumar et al. 1998). To remove dyes from wastewater, a variety of activated carbons have been utilised, including pelletised carbon, granular activated carbon (GAC), powdered activated carbon (PAC), and fibre-type activated carbon. Adsorption also involves the phase transition of pollutants into some secondary wastes, necessitating extra treatment for the proper discharge route.

7.8.2 Advanced Oxidation Process (AOP)

Textile effluent may be decolored using this method. Hydroxyl radicals (·OH) are one of the most potent oxidising agents and are used in this process. It is described as the generation of ·OH and highly reactive oxidations that particularly harm the target component. As a result of their ability to remove organic dyestuffs without creating additional elemental deposition, AOPs outperform standard physical and chemical approaches. It has been shown that ozone (O_3) treatment and UV irradiation, H_2O_2/UV combination method, Fenton and Photo-Fenton techniques have shown increased removal efficacy and/or broken up to a less complex substance that may be quickly biodegradable (Szpyrkowicz et al. 2001). Thus, there are several combinations of AOPs, like as (Fe^{2+}/H_2O_2, O_3, O_3/H_2O_2/UV, O_3/UV and H_2O_2/UV,), that might produce superior outcomes for textile industry effluent treatment.

7.8.3 Bioremediation and Biodegradation

Biodegradation is recognised as one of the most cost-effective methods for reducing a variety of organic dye components. Various microorganisms, including algae, fungus, yeast and bacteria, have been used for the biodegradation of dyestuff. They use a variety of ways to clean wastewater, including fungal decolorization, microbial degradation, adsorption through microbial biomass and bioremediation, all of which are often used in industrial wastewater treatment (Szpyrkowicz et al. 2001; Fu and Viraraghavan 2001). Aerobic and anaerobic methodologies might be successfully applied to the treatment of textile effluents using biological methods. There are several limitations to this procedure, such as the presence of excessive harmful substances, which significantly diminish the number of active microorganisms present in the biological technique (Wallace 2001). Accordingly, despite its many advantages, the biodegradation process was not able to efficiently remove the colour from certain types of organic dyestuffs and break down their complex chemical structure.

7.8.4 Electrochemical Method

It has been shown that the electrochemical method is a viable alternative to chemical treatment for textile industrial wastewaters. Electrochemical approaches are categorised based on the electrode material utilised, the chemicals employed and the contaminant removal procedure (Riyanto et al. 2013).

Electro-oxidation is a well-established electrochemical process in which an electron serves as the primary reagent, removing organic components from textile effluents without forming any secondary harmful pollutants. Electrochemical oxidation techniques are unique in that they are simple to use, do not require additional chemicals and are a cost-effective technology that might be a superior alternative for waste-effluent treatment. Organic and hazardous pollutants in effluent water, such as dye components, are frequently torn apart by direct oxidation or indirect oxidation routes in this treatment process. Contaminants are adsorbed on the anode surface and then reduced by an electron transport process in the direct anodic oxidation technique. Strong oxidants such as hypochlorite, ozone and hydrogen peroxide, on the other hand, might be regenerated by electrochemical methods during electrolysis in the indirect oxidation mode (Khan and Mondal 2021).

Electrocoagulation is a straightforward approach that produces less sludge than other procedures (Nguyen et al. 2013). The process is explained as the formation of metal ions from anodes that have been hydrolysed into metal hydroxides, as well as the development of hydrogen (H_2) bubbles at the cathode that allows the coagulated materials to float (Mollah et al. 2004). The electrocoagulation method involves three steps in order to remove contaminants from wastewater: (a) coagulants are produced in an electrolytic solution by oxidising electrodes, (b) by compressing and neutralising the diffuse double layer, pollutants are disturbed and interrupted and (c) coagulants collect contaminates that have been disturbed (Singh et al. 2014). Electrochemical methods work at a lower temperature, need no extra chemicals and produce no hazardous by-products.

7.8.5 Ion-Exchange Method

Ion exchange is a versatile process that may be used to treat polluted wastewater from a variety of industrial processes, as well as textile effluents containing a variety of complicated dyestuffs. Pollutants that are hazardous or harmful may be reduced in impact by ion exchange, which transforms them into a state that can be recycled. Resins that can exchange anions and cations with the surrounding material are known as ion exchangers. As their name implies, ion-exchange adsorbents may bind to contaminants with an opposing charge. The capacity to scatter as well as concentrate pollutants is another key feature of the ion-exchange process. An ion-exchange process can deal with reactive dyes, which are considered one of the most difficult types of dyes to work within the textile business. By employing an ion-exchange approach, it is possible to remove contaminants from waste management systems up to PPB levels (Khan and Mondal 2021).

7.8.6 Membrane Filtration Technique

While there are a variety of conventional techniques available, the membrane-based filtration process is considered to be one of the most complex technological approaches available and it is used extensively in the wastewater treatment and textile discharge treatment. On the basis of pore size and applied pressure, pressure-driven membrane filtration methods are divided into four categories: microfiltration (MF), ultrafiltration (UF), nanofiltration (NF) and reverse osmosis (RO) membranes. The semipermeable membrane has an impact on the membrane-separation concept, which is straightforward. While water molecules are allowed to pass through the semipermeable membrane, suspended particles and other pollutants are prevented from entering. Variables such as composition, material (including polymers and ceramics), surface (including homogeneous or heterogeneous, porous or dense) and shape all have an effect on the membrane filtering performance (porosity and thickness). Tertiary stage of wastewater treatment utilises membrane filtration to its fullest potential. MF and UF membranes with pore diameters ranging from 0.1 to 10 μm and 0.1–0.01 μm are utilised to extract biologically degraded products and moderately large elements from the water. To separate dye waste effluent from dissolved salts including monovalent anions and cations, NF membranes with nanopores with pore diameters typically ranging from 1 to 10 nm are utilised. Using a RO membrane with a pore capacity of 1 nm, all inorganic and organic contaminants are successfully removed from the water (Khan and Mondal 2021).

Although membrane-based operations have a number of drawbacks, including greater membrane preparation costs, incompatibility with low effluent flow rates, the need for a comparatively higher working pressure, the cost of intermittent membrane replacement and the inability to reduce dissolved solids. The membrane-separation approach, on the other end, is one of the possible wastewater treatment ways since it requires frequent cleaning and membrane-module replacement in order to maintain the efficacy of dye removal from effluent water.

7.9 Modulation of Photocatalysis of Wastewater by Dyes

7.9.1 Methyl Orange

Sodium 4-[4-(dimethylamino) phenyl] diazenyl benzene-1-sulfonate or methyl orange is also known as the universal indicator. Acidic media gives it a vivid red hue, whereas alkaline media gives it a yellow tint. Additionally, it is widely used in the chemical industry. Because of its mutagenic qualities, studies have been conducted on the treatment of effluents containing it. Photocatalytic decomposition of methyl orange in an aqueous medium was studied by He et al. (2010) using a WO_3/TiO_2 combination film. They studied the photocatalytic activity in relation to pH and other inorganic anions. According to the results, the clearance ratio decreased as pH increased and vice versa. Inorganic anions have two types: facilitators and inhibitors. The facilitators helped remove methyl orange, while the inhibitors slowed the process (reduced the elimination of methyl orange).

While anions such as HCO_3^+, NO_3^+ and PO_{43}^+ interacted with water to form ·OH, anions such as Cl and SO_{42}^+ trapped the ·OH and thereby hindered the elimination of methyl orange under UV irradiation. Nonmetallic components like F and S may be added to a TiO_2 nanotube array, Liao et al. (2012) examined the photocatalytic degradation of methyl orange. However, calcination for two hours at 500 °C greatly enhanced the photocatalytic activity of doped TiO_2 with a large aspect ratio. To treat methyl orange, *N*-doped TiO_2 was also used to make nanotube array films. Similarly, doped TiO_2 was shown to have better photocatalytic activity than undoped TiO_2 (Murugan et al. 2019). To get the highest level of photocatalytic activity, *N*-doped TiO_2 nanotubes were produced in an electrolyte solution containing 1.0 M NH_4NO_3 and HF. When methyl orange was exposed to visible light, nearly 80% of it was destroyed.

7.9.2 Indigo Carmine

Acid Blue 74, also known as indigo carmine (3, 3′-dioxo-2, 2′-bis-indolyden-5, 5′-disulfonic acid disodium salt), is a dye for textiles, a drug additive (tablets and capsules) and a diagnostic aid in medicine. Barrett's oesophagus is diagnosed with the use of acetic acid and indigo carmine. Although it's a natural colourant, the dye is very harmful to human health and may cause cancer (Barka et al. 2008). The dye has been linked to serious skin reactions and even death in certain cases of accidental ingestion. The removal of indigo carmine from contaminated effluents has been investigated. The degradation of dye was evident by the drop in COD levels and the resulting discolouration of the treated dye solution.

Depending on how much photocatalyst was employed and the concentration of dye at the start, degradation of the dye was affected. The author also concluded that TiO_2 commercially available was more effective than TiO_2 infused on activated carbon surfaces. According to Barka et al. (2008), Indigo carmine dye was effectively removed from non-oven fibres by using TiO_2 coatings. The photocatalyst surface and dye structure were both affected by pH, which was the primary governing factor in the reaction. At a pH of less than 4, the highest dye adsorption occurred. Also investigated under outdoor light was the photocatalytic degradation of indigo carmine and the efficacy of ZnO coated with Ag_2O_4 (Chowdhury et al. 2020). Because of ZnO and TiO_2's slower photogenerated electron-hole pair separation rates under outdoor light, the composite was shown to have a much greater photocatalytic activity. ZnO covered with Ag_2O_4 destroyed indigo carmine most effectively in natural settings. Indigo carmine photocatalytic degradation was also examined by Lubis et al. utilising multiple factors of pH, concentration of photocatalyst, starting dye concentration and irradiation time (Lubis and Sitompul 2019). When the dye and photocatalyst concentrations were both 10 mg L^{-1} and 250 mg, the composite showed the greatest photocatalytic activity at pH 1.

7.9.3 *Malachite Green*

Triphenylmethane dye, sometimes known as malachite green, is a cationic dye. The chemical name for this compound is *N,N*-dimethylcyclohexa-2,5-dien-1-iminium chloride. It is used extensively in the leather, textile, food and paper industries as a colourant. Even in its diluted state, this dye is particularly harmful to the environment because of its carcinogenicity and long-term persistence (Chen 2011). As a result, it is critical to eliminate these chemicals before releasing them into the environment. A Bi_2WO_6 nanoplate was used for the photocatalytic degradation of malachite green. According to the scientists, malachite green and Bi_2WO_6 were used in this investigation for the first time (Tahmasebi et al. 2019). Dye and Photocatalyst concentrations of 0.1 and 1.0 g L^{-1} at pH2 were found to be optimal. Meena et al. used ZnO nanoparticles to successfully remove malachite green from effluent water. The cap of ZnO with EDTA was investigated to see what effect it had on photocatalysts (Meena et al. 2016). According to the results, malachite green deteriorated by 94.14%, when there is a visible source of light. According to one theory, the ZnO photocatalyst may have been more stable and more active because of the capping molecules. ZnS nanoparticles with a ZnS core/shell configuration have also been studied as an alternative to EDTA in the dissolution of malachite green under visible light. The results of UV-Vis absorption spectroscopy revealed that dye may be effectively decolored. C and Fe (III) doped TiO_2 photocatalysts were studied by Ismael et al. for their synergistic effect on TiO_2 photocatalysts (Ismael 2020). Adding co-doping to TiO_2 made it more able to absorb visible light and make electrons and holes separate, which led to more malachite green being broken down. For the photocatalytic breakdown of malachite green, Ca-doped ceria nanoparticles have recently been used as well. At 35 °C with a dye concentration of 0.6 g L^{-1} and a photocatalyst concentration of 0.1 g, 93% dye degradation was observed after 90 min of UV irradiation (Amar et al. 2020).

7.9.4 *Rhodamine B*

Rhodamine B ([9-(2carboxyphenyl)-6-diethylamino-3xanthenelidene] diethylammonium chloride) is a brilliant red xanthene dye that is often used in food colouring. It is most often used as a water-based tracer dye. Fluorescence microscopy, flow cytometry and labelling acid-fast bacteria are just a few of the biotechnological uses for Rhodamine B as an analytical reagent. Despite this, rhodamine B has been banned as a food colourant because of its long-term toxicity and carcinogenic properties. The removal of rhodamine B from polluted water has been studied. TiO_2 photocatalyst immobilised on silicone was used to decolorize rhodamine B. For best results, the TiO_2 concentration should be 210.7 g/L^{-1} (Kim and Park 2006). A round disc TiO_2/Ti photoelectrode with a wedge shape was developed by scientists in a photoelectrocatalytic reactor. Using a wedge construction may have enhanced light utilisation while also carrying more pollutants to the electrode for treatment. In 60 min, the wedge surface electrode removed 100% of the dye, whereas in 150 min, it removed 93% of the COD (Li et al. 2011). Some indigenously manufactured photocatalysts, including TiO_2NWs, TiO_2 coated multi-walled carbon nanotubes (MWCNTs) and ZnO, were used in the photocatalytic degradation of rhodamine B. Compared with the other two photocatalysts, TiO_2NWs had the greatest degrading efficiency of 96.44% (Khan and Mondal 2021). Rhodamine B may be efficiently removed from water samples using simple photocatalysis at room temp using indigenously generated $BaWO_4$. Barium nitrate and sodium tungstate were used to manufacture the photocatalyst (Khan and Mondal 2021). The authors' proposed process for dye degradation is as follows:

$$\text{Step 1}: h\nu + \text{RB} \rightarrow \text{RB}^1 (\text{Singlet RB is in a high-intensity condition}) \tag{7.22}$$

$$\text{Step 2}: \text{RB}^1 \rightarrow \text{RB}^3 (\text{RB in 3rd excited state})(\text{Inter system crossing}) \tag{7.23}$$

$$\text{Step 3}: h\nu + \text{BaWO}^4 \rightarrow e^-_{\text{BaWO4}} + h^+_{(\text{BaWO4})} \tag{7.24}$$

$$\text{Step 4}: h^+_{(\text{BaWO4})} + \text{OH}^- \rightarrow h + \text{OH}^* \tag{7.25}$$

$$\text{Step 5}: \text{OH}^* + \text{RB}^3 \rightarrow \text{Degraded products like } \text{NO}_2,\ \text{CO}_2, \text{H}_2\text{O, etc.} \tag{7.26}$$

7.9.5 Methylene Blue

3,7-bis(dimethylamine)-phenothiazin-5-ium-chloride or methylene blue, also known as methylthionium chloride. This colourant is also commonly utilised in the pharmaceutical industry. It's injected intravenously to stain surgical tissues, monitor CSF leakage and help treat vasodilatory shock, among other things (Diaz-Uribe et al. 2018). However, when dye-laden water from industry and medical institutes is dumped into water bodies without treatment, it is a significant cause of environmental contamination, resulting in harmful results. It is also used in a variety of industries such as plastics, food, rubber, paper, textile and cosmetics.

As a result, it is imperative that these dye elements be removed from the wastewater before it is disposed of in the environment. Adeleke et al. investigated the photocatalytic degradation of methylene blue dye by $ZnO/NiFe_2O_4$ nanocomposites under UV light (Adeleke et al. 2018). To fabricate the solid-state nanocomposites, the co-precipitated $NiFe_2O_4$ and ZnOwere oxidised for 10 h at 800 °C. According to the researchers, the dye was eliminated as a result of the synthesis of *h*+ and ·OH ions during the photocatalytic activity. The following equations were used by the authors to explain how the dye degrades:

$$\text{NiFe}_2\text{O}_4/\text{ZnO} + \text{Dye (in aqueous medium)} \rightarrow \text{Dye} - \text{ZnO}/\text{NiFe}_2\text{O}_4 \tag{7.27}$$

$$\text{Dye} - \text{NiFe}_2\text{O}_4/\text{ZnO} + h\nu \rightarrow \text{Dye} - \text{ZnO}\left(h^+ + \text{e}^-\right)/\text{NiFe}_2\text{O}_4\left(h^+ + e^-\right) \tag{7.28}$$

$$\begin{aligned}&\text{Dye} - \text{ZnO}/\text{ZnO}\left(h^+ + e^-\right)/\text{NiFe}_2\text{O}_4\left(h^+ + e^-\right)\\&\rightarrow \text{Dye} - \text{ZnO}\left(e^-\right)/\text{NiFe}_2\text{O}_4\left(h^+\right)\end{aligned} \tag{7.29}$$

$$\text{O}_2 + \text{e}^- \rightarrow \text{O}_2^{\cdot-} \tag{7.30}$$

$$\text{H}^+ + \text{O}_2^{\cdot-} + e^- \rightarrow \text{OH}^- + \cdot\text{OH} \tag{7.31}$$

$$\text{H}_2\text{O} + h^+ + \text{H}^+ + \cdot\text{OH} \tag{7.32}$$

$$\begin{aligned}&\text{Dye} - \text{ZnO}/\text{NiFe}_2\text{O}_4 + \cdot\text{OH}/h^+/\text{O}_2^{\cdot-} \rightarrow \text{ZnO}/\text{NiFe}_2\text{O}_4\\&+ \text{H}_2\text{O} + \text{CO}_2 + \text{by - products}\end{aligned} \tag{7.33}$$

Photocatalytic removal of methylene blue dye by nano TiO_2 has been studied by Bastruck et al. Catalyst dose, pH impact, starting dye concentration and irradiation intensity were all aspects that were optimised in this research as well (Basturk et al. 2019). According to the findings, 66% dye degradation was achieved at pH 6.5 and 9, a catalyst dosage of 0.25 g L^{-1} and an irradiation of 50 Wm^{-2}. The photocatalytic impact of TiO_2 and graphene oxide was also studied by several researchers. It was found that the GO/TiO_2 complexes may be employed to degrade methylene blue in synthetic wastewater. When 200 mg of GO/TiO_2 composites (containing GO by 15% weight percent) were added to 200 mL of synthetic wastewater with an initial pH of 8, degradation rates of about 95% were found after 2.5 h (Anwer et al. 2019; Nemiwal et al. 2021) (Fig. 7.4).

7.10 Conclusion

Organic as well as inorganic colouring agents, whether ionic or nonionic, are extensively used in several industries today; and a lack of effective treatment for these dyes is producing serious environmental concerns owing to their negative impact on the ecosystem and society. New environmental laws governing the discharge of dyestuffs into the environment are implemented to resolve this issue, while experts are working in the development of new processes for budget-friendly and environment-friendly wastewater treatment, one of which is photocatalysis (homogenous photocatalysis and heterogeneous photocatalysis) participating in both direct and indirect dye degradation mechanism(photoexcitation, followed by ionisation and oxygen ionosorption

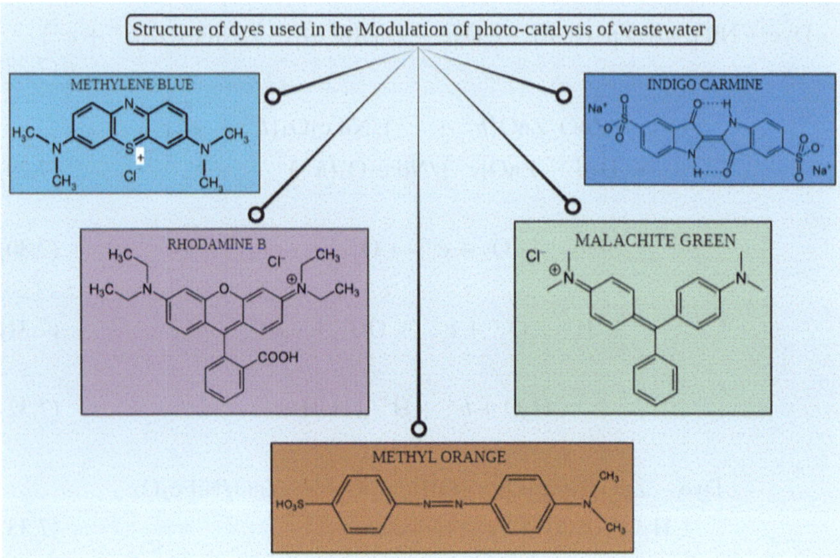

Fig. 7.4 Schematic diagrams of the structure of dyes used for phytoremidiation

and lastly protonation of superoxide). pH, light intensity, feed flow rate, the concentration of reactants, number of catalysts loading layers on the substrate, temperature during catalyst immobilisation and ionic species have the potency to influence photocatalytic degradation of dyestuffs. The development of innovative composite materials and enhanced photocatalysts should be the focus of future research to increase the efficiency of dyestuff removal and degradation. Even though different dye removal techniques exist, such as (i) adsorption, (ii) advanced oxidation process (AOP), (iii) bioremediation and biodegradation, (iv) electrochemical method, (v) ion exchange, (vi) membrane filtration technique; photocatalysis is modulated for successful wastewater treatment following degradation of industrial dyes, namely methyl orange, indigo carmine, malachite green, rhodamine B and methylene blue.

Conflict of Interest The authors declare no conflict no interest among themselves.

References

Adeleke J et al (2018) Photocatalytic degradation of methylene blue by $ZnO/NiFe_2O_4$ nanoparticles. Appl Surf Sci 455:195–200

Ahmad R et al (2016) Photocatalytic systems as an advanced environmental remediation: recent developments, limitations and new avenues for applications. J Environ Chem Eng 4(4):4143–4164

Ahmed S et al (2011) Influence of parameters on the heterogeneous photocatalytic degradation of pesticides and phenolic contaminants in wastewater: a short review. J Environ Manage 92(3):311–330

Allen RL (2013) Colour chemistry. Springer Science & Business Media

Amar IA et al (2020) Photocatalytic degradation of malachite green dye under UV light irradiation using calcium-doped ceria nanoparticles. Asian J Nanosci Mater 3(1):1–14

Ameta R et al (2018) Photocatalysis, in advanced oxidation processes for waste water treatment. Elsevier, pp 135–175

Ananpattarachai J, Kajitvichyanukul P (2014) Kinetics and mass transfer of fixed bed photoreactor using *N*-doped TiO_2 thin film for tannery wastewater under visible light. Chem Eng 42

Andreozzi M et al (2018) Treatment of saline produced water through photocatalysis using rGO-TiO_2 nanocomposites. Catal Today 315:194–204

Antoniotti S, Duñach E (2002) Direct and catalytic synthesis of quinoxaline derivatives from epoxides and ene-1, 2-diamines. Tetrahedron Lett 43(22):3971–3973

Anwer H et al (2019) Photocatalysts for degradation of dyes in industrial effluents: opportunities and challenges. Nano Res 12(5):955–972

Barka N et al (2008) Photocatalytic degradation of indigo carmine in aqueous solution by TiO_2-coated non-woven fibres. J Hazard Mater 152(3):1054–1059

Basturk E, Işık M, Karatas M (2019) Removal of aniline (methylene blue) and azo (reactive red 198) dyes by photocatalysis via nano TiO_2. Desalin Water Treat 143:306–313

Borker P, Salker A (2006) Photocatalytic degradation of textile azo dye over $Ce1-xSnxO_2$ series. Mater Sci Eng B 133(1–3):55–60

Calleja A, Baldasano J, Mulet A (1986) Toxicity analysis of leachates from hazardous wastes via microtox and *Daphnia magna*. Toxicity Assessment 1(1):73–83

Chen Y et al (2012) Photodegradation of malachite green by nanostructured Bi_2WO_6 visible light-induced photocatalyst. Int J Photoenergy 2011

Chen Y, Dionysiou DD (2006) Effect of calcination temperature on the photocatalytic activity and adhesion of TiO_2 films prepared by the P-25 powder-modified sol–gel method. J Mol Catal a: Chem 244(1–2):73–82

Chowdhury MF et al (2020) Current treatment technologies and mechanisms for removal of indigo carmine dyes from wastewater: a review. J Mole Liquids:114061

Christie R (2014) Colour chemistry. Royal Society of Chemistry

Corrêa AX et al (2010) Use of ozone-photocatalytic oxidation ($O_3/UV/TiO_2$) and biological remediation for treatment of produced water from petroleum refineries. J Environ Eng 136(1):40–45

Dąbrowski A (2001) Adsorption—from theory to practice. Adv Coll Interface Sci 93(1–3):135–224

Dave S et al (2021) Mathematical modeling and surface response curves for green synthesized nanomaterials and their application in dye degradation. Photocatalytic Degradation Dyes:571–591

Dave S, Dave S, Das J (2021) Photocatalytic degradation of dyes in textile effluent: a green approach to eradicate environmental pollution. In: The future of effluent treatment plants. Elsevier, pp 199–214

Di Ç (2015) Optimization of suspended photocatalytic treatment of two biologically treated textile effluents using TiO_2 and ZnO catalysts

Diaz-Uribe CE et al (2018) Photocatalytic degradation of methylene blue by the Anderson-type polyoxomolybdates/TiO_2 thin films. Polyhedron 149:163–170

Fox MA, Dulay MT (1993) Heterogeneous photocatalysis. Chem Rev 93(1):341–357

Fu Y, Viraraghavan T (2001) Fungal decolorization of dye wastewaters: a review. Biores Technol 79(3):251–262

Gürses A et al (2016) Classification of dye and pigments. In: Dyes and pigments. Springer, pp 31–45

Hasegawa MC et al (2014) COD removal and toxicity decrease from tannery wastewater by zinc oxide-assisted photocatalysis: a case study. Environ Technol 35(13):1589–1595

Hassan M, Zhao Y, Xie B (2016) Employing TiO_2 photocatalysis to deal with landfill leachate: current status and development. Chem Eng J 285:264–275

He J et al (2010) Photocatalytic removal of methyl orange in an aqueous solution by a WO_3/TiO_2 composite film. Korean J Chem Eng 27(2):435–438

He Y et al (2016) Construction of the elements based on lifted multiwavelet and its applications. Integr Ferroelectr 172(1):132–141

Henderson MA (2011) A surface science perspective on TiO_2 photocatalysis. Surf Sci Rep 66(6–7):185–297

Hong S et al (2018) Photolysis of glutaraldehyde in brine: a showcase study for removal of a common biocide in oil and gas produced water. J Hazard Mater 353:254–260

Ismael M (2020) Enhanced photocatalytic hydrogen production and degradation of organic pollutants from Fe(III) doped TiO_2 nanoparticles. J Environ Chem Eng 8(2):103676

Jallouli N et al (2017) Photocatalytic degradation of paracetamol on TiO_2 nanoparticles and TiO_2/cellulosic fiber under UV and sunlight irradiation. Arab J Chem 10:S3640–S3645

Jiménez S et al (2019) Produced water treatment by advanced oxidation processes. Sci Total Environ 666:12–21

Jyothi K, Yesodharan S, Yesodharan E (2014) Ultrasound (US), ultraviolet light (UV) and combination (US + UV) assisted semiconductor catalysed degradation of organic pollutants in water: oscillation in the concentration of hydrogen peroxide formed in situ. Ultrason Sonochem 21(5):1787–1796

Khan AA, Mondal M (2021) Effective materials in the photocatalytic treatment of dyestuffs and stained wastewater. In: Photocatalytic degradation of dyes. Elsevier, pp 91–111

Kim DS, Park YS (2006) Photocatalytic decolorization of rhodamine B by immobilized TiO_2 onto silicone sealant. Chem Eng J 116(2):133–137

Kumar A, Pandey G (2017) A review on the factors affecting the photocatalytic degradation of hazardous materials. Mater Sci Eng Int J 1(3):1–10

Kumar V et al (1998) Decolorization and biodegradation of anaerobically digested sugarcane molasses spent wash effluent from biomethanation plants by white-rot fungi. Process Biochem 33(1):83–88

Lai CW et al (2014) An overview: recent development of titanium oxide nanotubes as photocatalyst for dye degradation. Int J Photoenergy 2014

László B et al (2016) Photo-induced reactions in the CO_2-methane system on titanate nanotubes modified with Au and Rh nanoparticles. Appl Catal B 199:473–484

Lee KM et al (2016) Recent developments of zinc oxide based photocatalyst in water treatment technology: a review. Water Res 88:428–448

Li K et al (2011) Degradation of rhodamine B using an unconventional graded photoelectrode with wedge structure. Environ Sci Technol 45(17):7401–7407

Liang X, Zhu X, Butler EC (2011) Comparison of four advanced oxidation processes for the removal of naphthenic acids from model oil sands process water. J Hazard Mater 190(1–3):168–176

Liao HD et al (2012) Synthesis and photoelectrocatalytic property of two-nonmetal-codoped TiO_2 nanotube arrays with high aspect ratio. Adv Mater Res

Lin H, Valsaraj KT (2005) Development of an optical fiber monolith reactor for photocatalytic wastewater treatment. J Appl Electrochem 35(7):699–708

Lin L et al (2019a) Enhanced visible light photocatalysis by TiO_2-BN enabled electrospinning of nanofibers for pharmaceutical degradation and wastewater treatment. Photochem Photobiol Sci 18(12):2921–2930

Lin L et al (2019b) Adsorption and photocatalytic oxidation of ibuprofen using nanocomposites of TiO_2 nanofibers combined with BN nanosheets: degradation products and mechanisms. Chemosphere 220:921–929

Lin L et al (2020) Treatment of produced water with photocatalysis: recent advances, affecting factors and future research prospects. Catalysts 10(8):924

Lubis S, Sitompul DW (2019) Photocatalytic degradation of indigo carmine dye using α-Fe_2O_3/bentonite nanocomposite prepared by mechanochemical synthesis. In: IOP conference series: materials science and engineering. IOP Publishing

Manu B, Chaudhari S (2002) Anaerobic decolorisation of simulated textile wastewater containing azo dyes. Biores Technol 82(3):225–231

Meena S, Dipti V, Das B (2016) Photocatalytic degradation of malachite green dye by modified ZnO nanomaterial. Bull Mater Sci 39(7):1735–1743

Mollah MY et al (2004) Fundamentals, present and future perspectives of electrocoagulation. J Hazard Mater 114(1–3):199–210

Mozia S (2010) Photocatalytic membrane reactors (PMRs) in water and wastewater treatment. A review. Separation Purification Technol 73(2):71–91

Murugan C et al (2019) Improving hole mobility with the heterojunction of graphitic carbon nitride and titanium dioxide via soft template process in photoelectrocatalytic water splitting. Int J Hydrogen Energy 44(59):30885–30898

Nam W, Kim J, Han G (2002) Photocatalytic oxidation of methyl orange in a three-phase fluidized bed reactor. Chemosphere 47(9):1019–1024

Nemiwal M, Zhang TC, Kumar D (2021) Recent progress in g-C_3N_4, TiO_2 and ZnO based photocatalysts for dye degradation: strategies to improve photocatalytic activity. Sci Total Environ:144896

Neppolian B et al (2008) Degradation of textile dye by solar light using TiO_2 and ZnO photocatalysts. J Environ Sci Health Part A Toxic/hazard Subst Environ Eng 34:1829–1838

Nguyen DD, Ngo H, Yoon Y (2013) A new hybrid treatment system of bioreactors and electrocoagulation for superior removal of organic and nutrient pollutants from municipal wastewater. Biores Technol 153C:116–125

Pastrana-Martinez LM et al (2015) Graphene oxide based ultrafiltration membranes for photocatalytic degradation of organic pollutants in salty water. Water Res 77:179–190

Pingmuang K et al (2017) Composite photocatalysts containing BiVO4 for degradation of cationic dyes. Sci Rep 7(1):1–11

Qamar M, Saquib M, Muneer M (2005) Photocatalytic degradation of two selected dye derivatives, chromotrope 2B and amido black 10B, in aqueous suspensions of titanium dioxide. Dyes Pigm 65(1):1–9

Rabindranathan S, Devipriya S, Yesodharan S (2003) Photocatalytic degradation of phosphamidon on semiconductor oxides. J Hazard Mater 102(2–3):217–229

Raman CD, Kanmani S (2016) Textile dye degradation using nano zero valent iron: a review. J Environ Manage 177:341–355

Rauf M, Hisaindee S (2013) Studies on solvatochromic behavior of dyes using spectral techniques. J Mol Struct 1042:45–56

Rauf M, Meetani M, Hisaindee S (2011) An overview on the photocatalytic degradation of azo dyes in the presence of TiO_2 doped with selective transition metals. Desalination 276(1–3):13–27

Riyanto E, Norazizi N, Mohamed R (2013) Textiles industries wastewater treatment by electrochemical oxidation technique using metal plate. Int J Electrochem Sci 8:11403–11415

Rueda-Marquez JJ et al (2020) A critical review on application of photocatalysis for toxicity reduction of real wastewaters. J Clean Prod 258:120694

Satapathy S et al (2021) Mechanistic aspects and rate-limiting steps in green synthesis of metal and metal oxide nanoparticles and their potential in photocatalytic degradation of textile dye. In: Photocatalytic degradation of dyes. Elsevier, pp 605–630

Saverini M et al (2012) Genotoxicity of citrus wastewater in prokaryotic and eukaryotic cells and efficiency of heterogeneous photocatalysis by TiO_2. J Photochem Photobiol B 108:8–15

Shifu C, Gengyu C (2005) Photocatalytic degradation of organophosphorus pesticides using floating photocatalyst $TiO_2 \cdot SiO_2$/beads by sunlight. Sol Energy 79(1):1–9

Singh S, Srivastava VC, Mall ID (2014) Electrochemical treatment of dye bearing effluent with different anode–cathode combinations: mechanistic study and sludge analysis. Ind Eng Chem Res 53(26):10743–10752

Sirtori C et al (2010) Effect of water-matrix composition on trimethoprim solar photodegradation kinetics and pathways. Water Res 44(9):2735–2744

Suteu D, Zaharia C, Malutan T (2011) Removal of orange 16 reactive dye from aqueous solutions by waste sunflower seed shells. J Serb Chem Soc 76(4):607–624

Szpyrkowicz L, Juzzolino C, Kaul SN (2001) A comparative study on oxidation of disperse dyes by electrochemical process, ozone, hypochlorite and Fenton reagent. Water Res 35(9):2129–2136

Tahmasebi N, Maleki Z, Farahnak P (2019) Enhanced photocatalytic activities of Bi_2WO_6/BiOCl composite synthesized by one-step hydrothermal method with the assistance of HCl. Mater Sci Semicond Process 89:32–40

Tsoumachidou S et al (2017) Greywater as a sustainable water source: a photocatalytic treatment technology under artificial and solar illumination. J Environ Manage 195:232–241

Wallace TH (2001) Biological treatment of a synthetic dye water and an industrial textile wastewater containing azo dye compounds. Virginia Tech

Welham A (2000) The theory of dyeing (and the secret of life)

Yang X, Wang D (2018) Photocatalysis: from fundamental principles to materials and applications. ACS Appl Energy Mater 1(12):6657–6693

Yap P-S, Lim T-T (2011) Effect of aqueous matrix species on synergistic removal of bisphenol-A under solar irradiation using nitrogen-doped TiO_2/AC composite. Appl Catal B 101(3–4):709–717

Yu J, Zhao X (2000) Effect of substrates on the photocatalytic activity of nanometer TiO_2 thin films. Mater Res Bull 35(8):1293–1301

Zainal Z et al (2005) Removal of dyes using immobilized titanium dioxide illuminated by fluorescent lamps. J Hazard Mater 125(1–3):113–120

Zangeneh H et al (2015) Photocatalytic oxidation of organic dyes and pollutants in wastewater using different modified titanium dioxides: a comparative review. J Ind Eng Chem 26:1–36

Zinatloo-Ajabshir S, Salavati-Niasari M (2015) Nanocrystalline Pr6O11: synthesis, characterization, optical and photocatalytic properties. New J Chem 39(5):3948–3955

Zinatloo-Ajabshir S, Salavati-Niasari M (2016) Facile route to synthesize zirconium dioxide (ZrO_2) nanostructures: structural, optical and photocatalytic studies. J Mol Liq 216:545–551

Zollinger H (2003) Color chemistry: syntheses, properties, and applications of organic dyes and pigments. Wiley

Chapter 8
Optimizing Nanocatalyst's and Technological Factors Influencing on Photocatalytic Degradation of Organic and Inorganic Pollutants

Sushma Dave and Pratik Jagtap

Abstract In recent decades, water pollution with organic and inorganic contaminants is a chief environmental concern. Enormously increasing industrial sectors such as textile dyes, pigments, ink, polymers, plastics, medicine and cosmetics effluent discharges are the main contributors toward the depleting water reservoir qualities. Which potentially have toxic effect on the organism depending on it that also includes humans too. Many of them are destruction resistant by conventional degradation methods. Thus, optimization of photocatalytic degradation methods has proven a new era in global water pollution remediation and cleaning related fields. Photocatalysts are considered as a great potential, economical, eco-friendly, sustainable and show promising role in water remediation of wastewater without generating secondary waste. However, the development of such an advanced system in large scale is still in the optimization phase. The design and development of efficiency, photocatalyst with optimum operational parameters, configuration and integration of photocatalysis is still need to be provoked generously. This chapter will be emphasized on the mechanism and diversifying factors influencing on photocatalysis degradation that should be taken into consideration during optimization and development of the photocatalysis system depending upon the load of the organic and inorganic contaminants. Various electromagnetic spectrum-based excitable nanoparticles or nanomaterials have been studied as photocatalyst that potentially degrade the contaminants from water. TiO_2, Fe_2O_3, CuO, ZnO, CdS, SnO_2, ZnS, etc. were extensively studied for the efficiency controlled optimizing factors in the degradation of pollutants. Optimization of Influencing factors and their effect on photocatalytic degradation activity of the photocatalyst on organic and inorganic water pollutants will be the point of convergence of this chapter. Within this frame of reference surface area, morphology of photocatalyst, higher light intensity, presence of oxidant and doping agents were

S. Dave (✉)
Department of Chemistry, JIET, Jodhpur, Rajasthan, India
e-mail: drsushmadave@gmail.com

P. Jagtap
Institute of Science, Mumbai, India

S. Dave and J. Das (eds.), *Trends and Contemporary Technologies for Photocatalytic Degradation of Dyes*, Environmental Science and Engineering,
https://doi.org/10.1007/978-3-031-08991-6_8

perceived to be effective toward photocatalytic activity. Moreover, the general mechanism of photocatalysis and recent development in nanoparticles/nanomaterials and composite material will be discussed here.

Keywords Photocatalysis · Photocatalyst · Nanoparticles · Nanomaterials · Pollutants · Factors influencing · Water treatment

8.1 Introduction

Population explosion and the rapidly increasing industrial development sectors nearly hit the highly complex and difficult to manageable water pollution that ultimately threatened the massive effect on the climate (Lai et al. 2014) changes in many regions of the world. Also draining the increase in water consumption throughout the globe. Thus, it becomes more challenging to figure out an eco-friendly and more sustainable approaches to remediate the water pollution. Treated waste water becomes a general necessity of the human needs these days. Although the sewer systems, collection of waste water systems and treatment were not much accommodated as per the generated ratio. In addition to this, the present infrastructure facing the increasing demands to produce safer quality of water utilizing less amount of energy inputs (Soutsas et al. 2010). According to the existing planning it is practically difficult to design and develop such a huge system in the rapidly developing regions of the world. These challenges acquired attention of the more advanced but effective system toward the sustainable applications.

In the current scenario of the safe water available for the consumption is continuously deteriorated due to the direct or indirect discharge of the effluents from the industries such as pharmaceutical, food processing, textile, dyes, ink, rubber, plastic, polymers, chemical batteries and domesticated sewers as well as agricultural fertilizers or pesticide. All these industrial sectors contribute the contaminants are containing organic (industrial or textile dyes, herbicides or pesticides, halogenated hydrocarbons, pharmaceuticals by-products, some aromatic compounds such as polycyclic aromatic hydrocarbons) and inorganic (heavy metal ions, free radicals, carcinogens) pollutant molecules which are usually difficult to degrade by traditional methods and can cause the nuisance to the natural water reservoirs (Cacho et al. 1999; Gupta et al. 2006). These pollutants can cause the severe health damages to vital organs like kidneys, reproductive system, nervous system, lungs, lever even at very minute quantities (at ppm and ppb levels).

Several methods (in Fig. 8.1) already been applied to the waste water treatment before discharging out of the mainstreams. Each of processes from the method manifests with advantages and disadvantages.

Meanwhile, photocatalytic degradation treatment of pollutants is one of the best sustainable methods and proven a promising approach to clean the water sustainably using renewable sources and nanohybrid semiconductor materials. For efficient

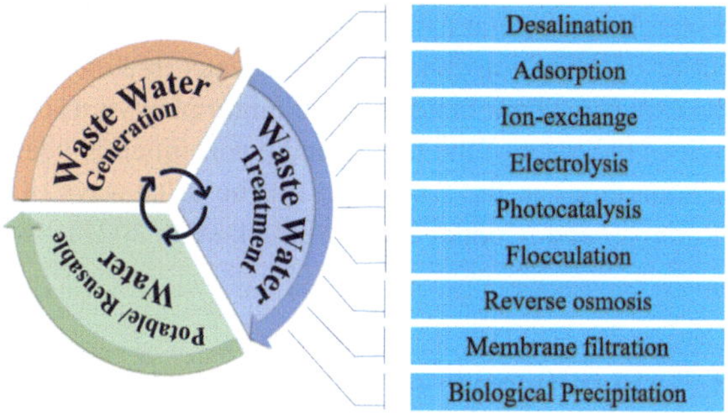

Fig. 8.1 Current scenario of applied processes involved in waste water treatment

optimum photocatalytic degradation of waste water pollutants, influencing operational factors of the reaction should be taken into consideration. This chapter reviews the most important properties of the nanoparticles as photocatalysts and also focuses on the rigorous introduction of photocatalysis mechanism and factors that influencing the rate of photocatalytic degradation reaction of pollutants present in wastewater (Dahiya and Patel 2021) (Fig. 8.2).

In the context, Nanotechnology offers better opportunities that can develop hyphened techniques relay for the next generation water supply systems. This chapter reviews the promising role of the nanoparticles/nanomaterials and nanocomposites in the water treatment processes to transform or degrade the possible organic and inorganic pollutants from the wastewater (Dave et al. 2021a). The tremendous versatility of the nanoparticles harnesses the great abilities such as high surface area, photosensitivity, catalytic activity, electromagnetic properties, anti-microbial effects and the regulatory pore size capacities etc. are the aspects in the many applications. This application may include quality monitoring sensors, catalytic sites, disinfection agents and distinctive performing matrix/membrane (Gupta and Mondal 2021) s. The development of such a nanotechnological-based application in the waste water treatment processes must be conjointly soothe with the environmental health and safety

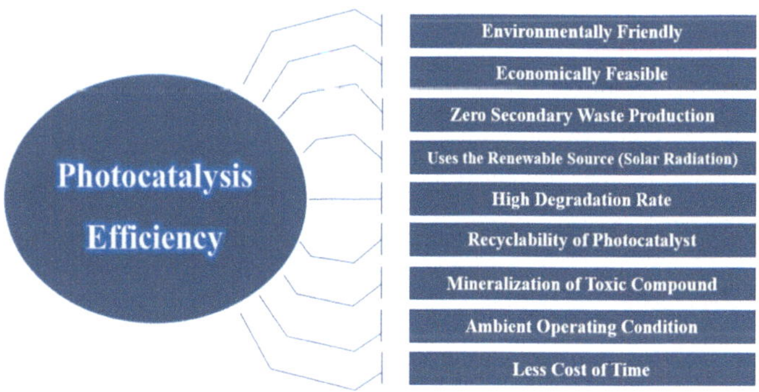

Fig. 8.2 Efficiency of photocatalysis reaction supporting in the sustainable development waste water treatment

research values that hand out the sustainable development. Highly expected prosecution of the nanotechnology in the water treatment may require the consideration of the expensive nanomaterials that might play vital role in obtaining reusable water to relieve the risks of the public and environmental health by potential minimizing the nanoparticles in the water and lift the safer ways of purification/decontaminations (Prajapati and Mondal 2021; Mallick et al. 2021).

8.2 Significance of Photocatalysis in Water Treatment

Photocatalysis is mediated through the light irradiations on photocatalyst particles have vital significance that leads to degradation of toxic pollutants from waste water. Photons are absorbed to obtain the charge which takes place in the redox reaction of toxic pollutant oxidation. Surface area plays a vital role in the mechanism. Succession of the reaction ultimately gives rise to form a hydroxyl free radical and acts as a potent oxidant to degrade the toxicants (Sarangapany and Mohanty 2021; Díez et al. 2021; Merouani and Hamdaoui 2021) (Fig. 8.3).

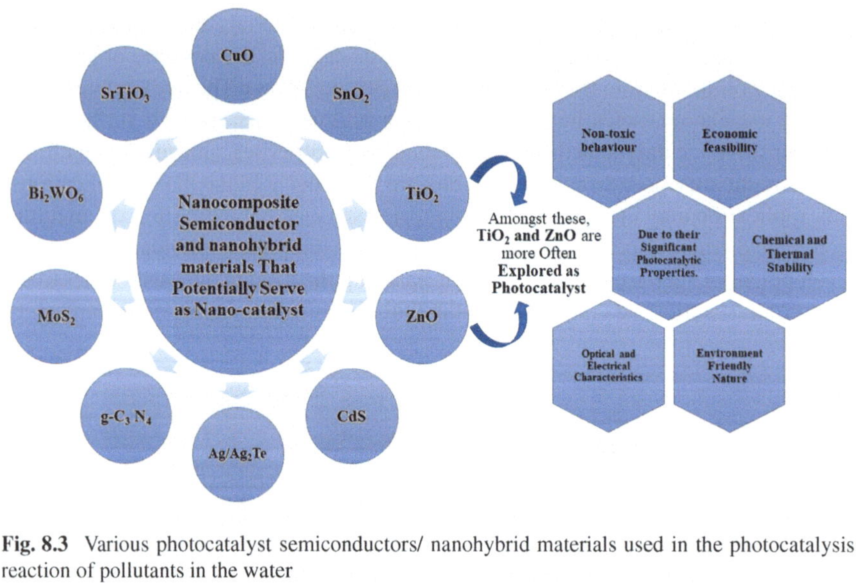

Fig. 8.3 Various photocatalyst semiconductors/ nanohybrid materials used in the photocatalysis reaction of pollutants in the water

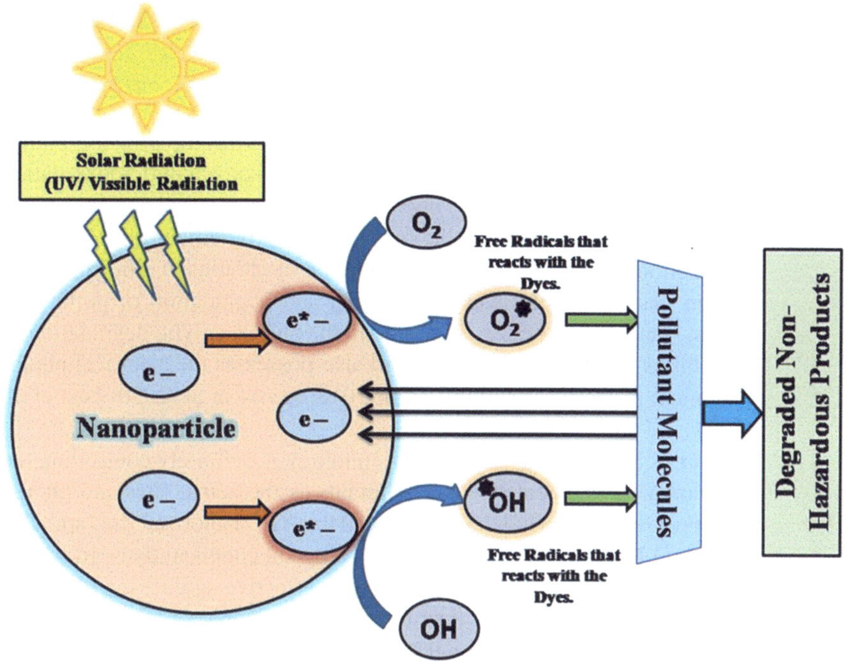

Fig. 8.4 Photochemical mechanism of nanoparticle for the formation and redox reaction causing waste water pollutant degradation

8.3 Mechanism of Nanoparticles (Photocatalyst) Involved in Photocatalysis of Wastewater Pollutants

A general mechanism of photocatalysis is driven by the photochemical transformation organic and inorganic pollutants from the waste water by absorbing the light radiation through the surface chemistry of the photocatalyst that leads to redox reaction. Nanocatalyst semiconductor material captures the energy released by the charge carriers which transfer the charge to the redox reaction of pollutants (Kumar et al. 2021). Apparently, photocatalyst can absorb the light radiation only on an appropriate illumination causing the excitation of charge carriers. Photochemical reaction mechanism produced light induced by the charge carriers and separated as Holes (H^+) and Electrons (e^-) which are responsible for oxidation of pollutants and degradation of pollutants from the waste water respectively (as shown in Fig. 8.4).

8.4 Factors Influencing on the Photocatalysis Mechanism

(A) **Ultrastructure of Photocatalyst**:

a. **Effect of Size and Surface area**: Size and surface area of the photocatalyst are most crucial factor to affect the photocatalytic degradation as the adsorption is directly involved. As the smaller size corresponds to higher surface area available for direct toward number of active sites, adsorption of contaminants and absorption of photons in efficient mechanism that ultimately lead to oxidation and degradation of pollutants. Thus, Nanoparticles play a vital role in the photocatalytic degradation as it exhibits better optical activity and also possesses the electrical photochemical properties (Zulkifili et al. 2018; Leroy et al. 2020; Isac et al. 2019).

b. **Effect of Morphology(shape)**: Morphology is definitely controlling the efficiency of photocatalysis of pollutants as the active sites and photon capture are based on it. It has been tested that ZnO nanorods, ZnO spindles and ZnO Nanoflowers can degrade the several chemical dyes. In which, the ZnO Nanorods found to be the most effective against the photocatalytic degradation. This is due to the higher number of reactive species and active site interaction on the ZnO Nanorods (Haruna et al. 2020).

(B) **Photocatalyst Doping Element**

Some Impurities are artificially added to the photocatalyst during synthesis that can potentially increase the photochemical activities. Such impurities were called as Doping agent or dopants. It increases the photochemical activity by means of elevated energy levels of dopants, better trapping of electrons, efficiently generating the oxygen deficient sites, producing more active sites for the adsorption of pollutant molecules and most importantly altering the light capturing bandgaps for more reactivity. Doping photocatalyst can be broadly classified as follows (Song et al. 2021; Ojha et al. 2020; Wang et al. 2019; Nada et al. 2020; Li et al. 2020; Teh et al. 2020; Zhang et al. 2019):

a. **Noble Metal Doping**
b. **Metal Doping**
c. **Rare Earth Metal Doping**
d. **Non-Metallic Doping**
e. **Co-Doping**
f. **Self-Doping**

(C) **Reactant Accessibility**

a. **Amount of Catalyst**: Photocatalysis rate of reaction is directly proportional to the hydroxyl ion free radical and the positive holes under the radiation. These two primarily take a part in the photocatalytic reaction of the pollutants present in the water. Photocatalysis may increase with the amount of catalyst but exceeding to certain limit it inhibits the rate of reaction. With the higher concentration of photocatalyst turbidity of the solution increases and prohibited the light entry in the water, also causes the scattering of UV–Vis radiation. Agglomeration is also a main cause of high concentration of nanocatalyst which ultimately blocks the active site and leads to lowering down the photocatalysis (Laurenti et al. 2020; Anju Chanu et al. 2019; Deveci and Mercimek 2019; Tichapondwa et al. 2020).

b. **Concentration of Pollutants**: Surface area is associated with the amount of pollutant where the photocatalysis depends upon pollutants adsorbs. This adsorption is also depending upon the initial concentration pollutants and gradually increases in the water. However, higher than a certain limit, here is significant decrease in the photocatalysis observed due to the higher initial concentration of pollutants adsorbed (Hu et al. 2016). As the photocatalyst's surface is totally covered with the adoption of pollutants, thus light photons were absorbed by the pollutant molecules instead of photocatalyst. Ultimately this reduces the production of hydroxyl ion free radicals and the positive hole. The occupied active sites of photocatalyst by pollutants show reduction in photochemical process (Zelekew et al. 2021).

(D) **Heterogeneity of Oxidants**

Effect of H_2O_2, $KBrO_3$, $(NH_4)_2S_2O_8/K_2S_2O_8$, HNO_3 was seen while photocatlytic dye degradation (Sadik et al. 2004; Sadik et al. 2004; Ovhal et al. 2021). The study on the effect of oxidants such as $(NH_4)_2S_2O_8$, $KBrO_3$ and H_2O_2 on the photooxidation of AR18 reveals that the addition of $(NH_4)_2S_2O_8$ and $KBrO_3$ increases the dye removal whereas the addition of H_2O_2 decreases the photocatalytic degradation. The unusual decrease by the addition of H_2O_2 is due to its low adsorption on the ZnO surface (Sobana and Swaminathan 2007)

(E) **Miscellany Factors**

a. **Effect of pH**: One of the preliminary factors in the photocatalysis is pH. At higher pH (alkaline) free radicals are the predominantly active, while at lower pH (acidic) oxidation of pollutants and positives holes with high oxidation potential are the key factors for photocatalysis of contaminants in the wastewater (Yeganeh et al. 2020; Vijay et al. 2019; Deshmukh et al. 2020; Laishram et al. 2018). As the pH value increases in the solution hydroxyl ion was generated in between the reaction of positive holes and hydroxyl free radicals which cause increase in the degradation rate. If the pH reaches too high ($pH > 11/12$) the excessively formed hydroxyl ions it starts competing with the pollutants and get absorbed on the photocatalyst, as a result the active site for acidic pollutants gets blocked. Vice-versa, when pH lowers, surface of the photocatalyst gets protonated causing the reduction in the cationic pollutants adsorption which end up with significant decrease in the photodegradation (Suthar et al. 2021; Dave et al. 2021b; Purohit et al. 2021).

b. **Effect of Temperature**: Photocatalysis can even occur at room temperature. However, degradation and recombination of electrons and positive hole pairs, releases energy, thus the temperature of the reaction medium gradually increases. Artificially increase in the temperature can gradually increase the rate of photocatalytic degradation reaction. But higher than 80 °C temperature can reduce the lifespan of the charge carriers through recombination with each other. Also, in temperature lower than 20 °C the reaction medium causes increase in the apparent activation of energy. Therefore, 20–80 °C is considered optimum temperature range for the photocatalysis which shows minimum apparent activation energy and low dependency of the rate of degradation (Dahiya and Patel 2021; Gupta and Mondal 2021; Prajapati and Mondal 2021; Mallick et al. 2021; Dave et al. 2021c; Satapathy et al. 2021).

c. **Effect of Light Intensity available**: The photocatalytic dye degradation in presence of sunlight when conducted in between 10 a.m. to 4 p.m. has resulted in the inference that when the maximum quantity of sunlight is available with high intensity and energy then the catalytic activity is more. The reports demonstrate that sunlight has capability to degrade the dye in presence of these photo catalyst which clearly indicates its energy economy approach thus making it more economic and green method for the industries (Mehta et al. 2016; Ali and Ameta 2013; Borhade et al. 2020; Adhikari et al. 2020).

d. **Effect of Inorganic Ions**: Waste water significantly consists of various inorganic, anionic and cationic pollutants. Presence of such dissolved inorganic ions are large enough to affect the photocatalysis efficiency. These inorganic ions may compete with the pollutant molecules and adsorbed on the surface of the photocatalyst ultimately reduced the active site for photochemical reaction. Inorganic ions affect the photocatalysis particularly by precipitation on the surface, blocking active sites, reacting with catalyst and scavenging radical and positive holes. Also, the Cl^-, SO_4^{2-}, HCO_3^- and PO_4^{3-} inorganic anions were revealed as holes and radical scavengers and reduce the rate of photocatalysis, along with inorganic cations such as Mg^{2+}, Zn^{2+}, Fe^{3+}, Cu^{2+} etc., in wastewater can also significantly affect the photodegradation activity.

8.5 Conclusion

As a future perspective there will be emphasis on the mechanism and diversify factors influencing on photocatalytic degradation that should be taken into consideration during optimization and development of the photocatalysis system depends upon the load of the organic and inorganic contaminants. Various electromagnetic spectrum-based excitable nanoparticles or nanomaterials are been studied as photocatalyst that potentially degrade the contaminants from water.

References

Adhikari C, Kaur M, Ravichandran (2020) Sunlight assisted degradation of methylene blue as a model dye using bismuth oxychloride nanoparticles: ecofriendly and industry efficient photocatalysis for waste chemical treatment. Asian J Chem 32(1)

Ali Y, Ameta A (2013) Degradation and decolouration of amaranth dye by photo-fenton and fenton reagents: a comparative study. Int J Chem Sci 11(3

Anju Chanu L, Joychandra Singh W, Jugeshwar Singh K, Nomita Devi K (2019) Effect of operational parameters on the photocatalytic degradation of Methylene blue dye solution using manganese doped ZnO nanoparticles. Results Phys 12

Borhade A, Tope D, Kushare S (2020) Mercenaria shell powder as a cost-effective and eco-friendly photocatalyst for the degradation of Eriochrome Black T Dye. Iran J Sci Technol Trans A Sci 44(1)

Cacho J, Fierro I, Deban L, Vega M, Pardo R (1999) Monitoring of the photochemical degradation of metamitron and imidacloprid by micellar electrokinetic chromatography and differential-pulse polarography. Pestic Sci

Dahiya A, Patel BK (2021) Photocatalytic degradation of organic dyes using heterogeneous catalysts. In: Photocatalytic degradation of dyes

Dave S, Jagtap P, Verma S, Nehra R, Dave S, Mohanty P et al (2021a) Mathematical modeling and surface response curves for green synthesized nanomaterials and their application in dye degradation. In: Photocatalytic degradation of dyes

Dave S, Khan AM, Purohit SD, Suthar DL (2021b) Application of green synthesized metal nanoparticles in the photocatalytic degradation of dyes and its mathematical modelling using the caputo-fabrizio fractional derivative without the singular Kernel. J Math 2021b

Dave S, Dave S, Das J (2021c) Photocatalytic degradation of dyes in textile effluent: a green approach to eradicate environmental pollution. In: The future of effluent treatment plants

Deshmukh SP, Kale DP, Kar S, Shirsath SR, Bhanvase BA, Saharan VK et al (2020) Ultrasound assisted preparation of rGO/TiO_2 nanocomposite for effective photocatalytic degradation of methylene blue under sunlight. Nano-Struct Nano-Objects 21

Deveci İ, Mercimek B (2019) Performance of SiO_2/Ag Core/Shell particles in sonocatalalytic degradation of Rhodamine B. Ultrason Sonochem 51

Díez AM, Pazos M, Sanromán MA (2021) Hybrid systems to improve photo-based processes and their importance in the dye degradation. In: Photocatalytic degradation of dyes

Gupta GK, Mondal MK (2021) Fundamentals and mechanistic pathways of dye degradation using photocatalysts. In: Photocatalytic degradation of dyes

Gupta AK, Pal A, Sahoo C (2006) Photocatalytic degradation of a mixture of crystal violet (basic violet 3) and methyl red dye in aqueous suspensions using Ag+ doped TiO_2. Dye Pigment

Haruna A, Abdulkadir I, Idris SO (2020) Photocatalytic activity and doping effects of $BiFeO_3$ nanoparticles in model organic dyes. Heliyon 6

Hu E, Wu X, Shang S, Tao XM, Jiang SX, Gan L (2016) Catalytic ozonation of simulated textile dyeing wastewater using mesoporous carbon aerogel supported copper oxide catalyst. J Clean Prod 112

Isac L, Cazan C, Enesca A, Andronic L (2019) Copper sulfide based heterojunctions as photocatalysts for dyes photodegradation. Front Chem 7

Kumar M, Swain G, Sonwani RK, Singh RS, Verma A, Rai B (2021) Effect of operating parameters on photocatalytic degradation of dyes by using graphitic carbon nitride. In: Photocatalytic degradation of dyes

Lai CW, Juan JC, Ko WB, Bee Abd Hamid S (2014) An overview: recent development of titanium oxide nanotubes as photocatalyst for dye degradation. Int J Photoenergy

Laishram D, Shejale KP, Gupta R, Sharma RK (2018) Heterostructured HfO_2/TiO_2 spherical nanoparticles for visible photocatalytic water remediation. Mater Lett 231

Laurenti M, Garino N, Garino N, Canavese G, Hernandéz S, Cauda V (2020) Piezo- and photocatalytic activity of ferroelectric ZnO:Sb thin films for the efficient degradation of rhodamine-β dye pollutant. ACS Appl Mater Interfaces 12(23)

Leroy S, Blach JF, Huvé M, Léger B, Kania N, Henninot JF et al (2020) Photocatalytic and sonophotocatalytic degradation of rhodamine B by nano-sized $La_2Ti_2O_7$ oxides synthesized with sol-gel method. J Photochem Photobiol A Chem 401

Li Z, Chen Q, Lin Q, Chen Y, Liao X, Yu H et al (2020) Three-dimensional P-doped porous g-C3N4 nanosheets as an efficient metal-free photocatalyst for visible-light photocatalytic degradation of Rhodamine B model pollutant. J Taiwan Inst Chem Eng 114

Mallick A, Patil PD, Tiwari MS, Kane P, Khonde D (2021) Green and sustainable methods for dye degradation employing photocatalytic materials. In: Photocatalytic degradation of dyes

Mehta A, Mishra A, Sharma M, Singh S, Basu S (2016) Effect of silica/titania ratio on enhanced photooxidation of industrial hazardous materials by microwave treated mesoporous SBA-15/TiO_2 nanocomposites. J Nanoparticle Res 18(7)

Merouani S, Hamdaoui O (2021) Sonophotocatalytic degradation of refractory textile dyes. In: Photocatalytic degradation of dyes

Nada AA, El Rouby WMA, Bekheet MF, Antuch M, Weber M, Miele P et al (2020) Highly textured boron/nitrogen co-doped TiO_2 with honeycomb structure showing enhanced visible-light photoelectrocatalytic activity. Appl Surf Sci 505

Ojha N, Bajpai A, Kumar S (2020) Enhanced and selective photocatalytic reduction of CO_2 by H_2O over strategically doped Fe and Cr into porous boron carbon nitride. Cataly Sci Technol 10(8)

Ovhal SD, Rodrigues CSD, Madeira LM (2021) Photocatalytic wet peroxide assisted degradation of Orange II dye by reduced graphene oxide and zeolites. J Chem Technol Biotechnol 96(2)

Prajapati AK, Mondal MK. (2021) Emerging nanocomposites as highly efficient materials for photocatalysis of dyes: synthesis routes, characterization, and reaction mechanism. In: Photocatalytic degradation of dyes

Purohit SD, Khan AM, Suthar DL, Dave S (2021) The impact on raise of environmental pollution and occurrence in biological populations pertaining to incomplete H-function. Natl Acad Sci Lett 44(3)

Sadik WA, Sadek OM, El-Demerdash AM (2004) The use of heterogenous advanced oxidation processes to degrade neutral red dye in aqueous solution. Polym Plast Technol Eng 43(6)

Sadik WA, El-Demerdash AM, Nashed AW (2004) UV-induced decolorization of indophenol by heterogenous advanced oxidation processes. Polym Plast Technol Eng 43(6)

Sarangapany S, Mohanty K (2021) A facile biogenic-mediated synthesis of Ag nanoparticles over anchored ZnO for enhanced photocatalytic degradation of organic dyes. In: Photocatalytic degradation of dyes

Satapathy S, Acharya D, Dixit PK, Mishra G, Das J, Dave S (2021) Mechanistic aspects and rate-limiting steps in green synthesis of metal and metal oxide nanoparticles and their potential in photocatalytic degradation of textile dye. In: Photocatalytic degradation of dyes

Sobana N, Swaminathan M (2007) The effect of operational parameters on the photocatalytic degradation of acid red 18 by ZnO. Sep Purif Technol 56(1)

Song H, Liu L, Wang H, Feng B, Xiao M, Tang Y et al (2021) Adjustment of the band gap of co-doped $KCl/NH_4Cl/g\text{-}C_3N_4$ for enhanced photocatalytic performance under visible light. Mater Sci Semicond Process 128

Soutsas K, Karayannis V, Poulios I, Riga A, Ntampegliotis K, Spiliotis X et al (2010) Decolorization and degradation of reactive azo dyes via heterogeneous photocatalytic processes. Desalination

Suthar DL, Purohit SD, Khan AM, Dave S (2021) Impacts of environmental pollution on the growth and conception of biological populations involving incomplete I-function. In: Lecture notes on data engineering and communications technologies

Teh YW, Chee MKT, Kong XY, Yong ST, Chai SP (2020) An insight into perovskite-based photocatalysts for artificial photosynthesis. Sustain Energy Fuels 4(3)

Tichapondwa SM, Newman JP, Kubheka O (2020) Effect of TiO_2 phase on the photocatalytic degradation of methylene blue dye. Phys Chem Earth 118–119

Vijay S, Balakrishnan RM, Rene ER, Priyanka U (2019) Photocatalytic degradation of Irgalite violet dye using nickel ferrite nanoparticles. J Water Supply Res Technol AQUA 68(8)
Wang K, Fu J, Zheng Y (2019) Insights into photocatalytic CO_2 reduction on C_3N_4: Strategy of simultaneous B, K co-doping and enhancement by N vacancies. Appl Catal B Environ 254
Yeganeh FE, Yousefi M, Hekmati M, Bikhof M (2020) Photocatalytic degradation of coomassie blue G-250 by magnetic $NiFe_2O_4$/ZnO nanocomposite. Comptes Rendus Chim 23(6–7)
Zelekew OA, Fufa PA, Sabir FK, Duma AD (2021) Water hyacinth plant extract mediated green synthesis of Cr_2O_3/ZnO composite photocatalyst for the degradation of organic dye. Heliyon 7(7)
Zhang H, Han X, Yu H, Zou Y, Dong X (2019) Enhanced photocatalytic performance of boron and phosphorous co-doped graphitic carbon nitride nanosheets for removal of organic pollutants. Sep Purif Technol 226
Zulkifili AN, Fujiki A, Kimijima S (2018) Flower-like $BiVO_4$ microspheres and their visible light-driven photocatalytic activity. Appl Sci 8(2)

Chapter 9
Biological Synthesis of Metallic Nanoparticles and Their Application in Photocatalysis

Soma Das

Abstract Removal of dyes from wastewater using the photocatalysis process gets more attention because of its suitability to completely remove of dyes under normal or moderate temperature and pressure, and thus, it overcomes the drawback of the conventional water treatment processes. Metal and metal oxide nanoparticles show valuable magnetic, electrical, optical and catalytic properties. Their unique properties allow them to be used in wastewater treatment as photocatalyst. Firstly, this chapter will focus on the biological synthesis of metallic nanoparticles. Mechanism for the bio-synthesis of metallic nanoparticles, advantages and disadvantages of biogenic route will also be discussed. Secondly, the mechanism of photocatalysis, factors affecting photodegradation, and role of photocatalysis against different water pollutants shall be explained. Different types of photocatalysts will be briefly presented in the next sections, and finally, the discussion will be highlighted with the future perspective and conclusions.

9.1 Introduction

Our drinkable water is under a serious threat. The reason for this threat is not only due to global climate change, but different industry activities and continuous population growth that limit the current natural water reserves. It is stated that about 1.2 billion people are unable to access drinkable water and millions died due to polluted water (Ahuja et al. 2021). Pointing out to industrial activities shows that the existing organic and inorganic pollutants from chemical accidents, or waste and illegal agriculture practices contaminant the available water source and arise the main hazardous for our ecosystem (Liang and Zhang 2019; Mane et al. 2018). Dyes are considered one of examples of organic pollutants that are toxic to our environment. On one side, they represent a major concern because their wide use in different industries such as food, textiles, plastic and pharmaceuticals industries. However, on the other side,

S. Das (✉)
GN Group of Institutes, Knowledge Park II, Greater Noida, India
e-mail: somachem17@gmail.com

S. Dave and J. Das (eds.), *Trends and Contemporary Technologies for Photocatalytic Degradation of Dyes*, Environmental Science and Engineering,
https://doi.org/10.1007/978-3-031-08991-6_9

it is observed that about 15% of the dyes used in textile manufacturing process are wasted and are discharged into the nature. Thus, it represents a serious damage to the environment due to their recalcitrance nature (Ratna 2012). The contaminated water from toxic chemicals, hazardous textile dyes and pesticides causes long-term adverse effects on aquatic and human life. For instance, the organic dyes used in textile industry are highly toxic, carcinogenic and non-degradable. They can cause several diseases such as skin diseases, allergies and cancer (Daniel and Shabudeen 2014; Vasantharaj et al. 2019). Therefore, there is a crucial need to treat this polluted water before disposal to environment.

Currently, there are various conventional wastewater treatment methods like chemical transformation, biological treatments, distillation, reverse osmosis, coagulation and flocculation, ultraviolet treatment and many others. These methods are efficient with various pollutants. However, they are costly, require specific equipment and high energy input (Sahu and Singh 2019; Saharan et al. 2014). Thus, a demand for an alternative water treatment process has risen to tackle these challenges. One of the alternative processes is called Advanced alternative processes (AOPs). In this process, a generation of the hydroxyl radical is needed to be used as an oxidant to destroy the polluted compounds and degrade to carbon dioxide and water (Saharan et al. 2014; Deng and Zhao 2015). In different contexts, AOP could also refer to a series of processes such as photocatalytic oxidation, electron-beam irradiation, ultrasonic cavitation and Fenton's reaction (Samuel et al. 2011). A concern about using AOP in water treatment process is that most of the dyes are resistant to AOPs degradation (Deng and Zhao 2015). The photocatalytic process is another alternative way to treat wasted water and it has more advantages over AOPs because the use of visible light or near-UV as irradiation is cost-effective and there is no sludge produced during the process (Singha et al. 2019). The definition of photocatalysis or photodegradation is process in which the chemical reaction rates change under the action of light and in the presence of substances called photocatalysts.

Photocatalysts absorb light and contribute to transform of the reactants. They must be stable, nontoxic, inexpensive and highly photoactive in order to attract more attention. Removal of dyes from wastewater using the photocatalysis process gets more attention because of its suitability to completely remove of dyes under normal or moderate temperature and pressure and thus, it overcomes the drawback of the conventional water treatment processes. Furthermore, it includes interaction between the catalyst and visible or UV light to generate reactive species such as hydroxyl radicals and oxygen radicals which could interact with organic pollutants and result in removal of organic pollutants (Fagier 2020). Involving catalysts in photocatalysis process opened the door to study the effect of different kinds of catalysts in water treatment processes. It is noticed that using nanoparticle catalysts showed superior photocatalytic effectiveness if they are compared to normal photocatalysts (Osuntokun et al. 2019).

Nanoparticles (NPs) refer to materials with sizes dimensions between 1 and 100 nm. Their sizes, distributions, crystal structures and surface-to-volume ratio give them superior properties (Rathnasamy et al. 2017). Metal and metal oxide nanoparticles (M/MO-NPs) show valuable magnetic, electrical, optical and catalytic properties. Their unique properties allow them to be used in several different applications such as solar cells, light-emitting devices, biomedicine, catalysis, soil stabilization, biosensors and water treatment (Afsharian and Khosravi-Daran 2019; Hardani et al. 2015). Next to Gold and silver metal oxide NPs, TiO_2, ZnO, SnO_2 and CeO_2 are used as photocatalysts in water treatment process.

Synthesis of nanoparticles can be described from structure point of view. Following structure approach, there are two main approaches; Top-down and Bottom-Up. Top-Down approach involves break of bulk or big materials down into nano-sized materials. However, with such approach, it is difficult to control the size of generated particles. It results in wide size distribution and variation of morphologies. The second approach, Bottom-Up approach is opposite to Top-Down approach as it involves growing of nanoparticles from single atom into bigger size material. Bottom-up approach is more common as it results in better-controlled size nanomaterials (Dhandapani et al. 2012). Regardless of the structured approach for synthesizing nanoparticles, Those two approaches may include physical and/or chemical methods. Synthesize nanoparticles by physical approach (it is also called mechanical approach) follow Top-down structure approach. Examples of physical methods are pyrolysis, using laser ablation and attrition. Contrary to physical methods, chemical methods usually follow Bottom-Up structure approach. Physical methods have an advantage of producing narrow particle size nanoparticles. However, it requires expensive instruments such as lasers, it also consumes high energy and the production of nanoparticles is lower if it is compared to chemical method. In chemical methods, we usually use wet chemical protocols. Common chemical methods may include sol–gel process, sonochemical methods, polyol process and reverse micelles method. The chemical methods are better than physical methods in terms of the nanoparticle production volume and they are relatively cheaper. The drawbacks of chemical methods are the use of toxic chemicals and the formation of by-products that might be hazardous and dangerous. Due to the limitations of physical and chemical methods, there is a need to find a green (or biological) method that can overcome the cost, production yield, energy consumption and eco-friendly. Not only because of its environmental benefits, but the green synthesis allows to produce nanoparticles in large quantities and with defined shape and size than the regular production methods offer (Isa et al. 2021a).

The green method of synthesizing nanoparticles means the use of natural materials like the extracts of plants and microorganisms (i.e. bacteria, fungus, yeast and algae) in the synthesis process. It is possible to classify the synthesize process into two categories; bioreduction and biosorption. Bioreduction means that chemical reduction is used to achieve stable metal ions using biological means and it is accomplished by dissimilatory metal reduction. Biosorption means that the metal ions are bonded to the cell wall or synthesized peptides by microorganisms or plants assembles into stable nanoparticle structures. A decision over which method to follow for biosynthesize nanoparticles depends on several variables (Dan et al. 2020).The green method of preparing nanoparticles is Bottom-Up approach. Various plants such as Aloe vera, lemon, Coriander and Oat contain heavy metals and biomolecules. They are utilized to synthesize silver and gold nanoparticles. Bacteria is easy to manipulate and so it is widely used in biotechnological applications. Bacteria such as Escherichia coli, Bacillus amyloliquefaciens, Bacillus cecembensis and Lactobacillus casei have been used to synthesize different nanoparticles silver, zinc and cobalt nanoparticles. Metal and metal oxide nanoparticles can be biosynthesized by fungi that play the role of biological agents. Biosynthesis of nanoparticles using fungi yields bigger amounts of nanoparticles in comparison to bacteria. eukaryotic cells have yeasts such as Saccharomyces cerevisiae and it is used to synthesize silver and gold nanoparticles. Like yeast, marine algae is used to prepare gold nanoparticles (Jagpreet et al. 2018).

Firstly, this chapter will focus on the biological synthesis of metallic nanoparticles from plant, microorganisms and animal-derived products. Mechanism for the biosynthesis of metallic nanoparticles, advantages and disadvantages of biogenic route will also be discussed. Secondly, the mechanism of photocatalysis, factors affecting photodegradation and role of photocatalysis against different water pollutants shall be explained. Different types of photocatalysts will be briefly presented in next sections and finally, the discussion will be highlighted with the future prospective and conclusions.

9.2 Preparation of Metallic Nanoparticles

Three methodologies such as; physical, chemical and biological are used to prepare metallic nanoparticles. Again in a broader way based on the structural point of view, the above-mentioned methods are classified as 'bottom-up' and 'top-down' approaches. In top-down approach slicing or a bulk material undergoes successive cutting or slicing to achieve nano-sized particles and this can be done by various mechanical or chemical methods. The major drawback of top-down approach is the produced metallic nanoparticles often have structural defects, which significantly alter the physical properties and surface chemistry behaviour.

Bottom-up approach means building up of nanomaterials from the bottom, i.e. by self-assembly of atoms, molecules or clusters. Here in this approach at first, the building units are produced and consequently, those building units are transformed into the nanostructured materials. The major advantages of bottom-up approach are nanomaterials are produced by this method have homogenous chemical compositions and no structural defects. Both approaches play very important role in modern industry and most likely in nano technology as well (Adam and Gabriela 2011).

9.2.1 Physical Methods

There exist four methods such as evaporation/condensation, pyrolysis, laser ablation and attrition that are mostly used as physical or mechanical methods. In evaporation/condensation technique the inert gas atoms are condensed and formed metallic nanoparticles. In the beginning, very high temperature is applied into reaction chamber to evaporate the metal atoms from the substrate. Evaporated metal atoms gain very high kinetic energy and collide with inert gas (carrier gas) atoms into the reaction chamber. These interatomic collisions resulted into loss of kinetic energy of metal atoms and condense in the form of small crystals. In pyrolysis method, the precursor is burnt into the furnace at very high temperature in presence of inert gases and with limited supply of oxygen under atmospheric pressure conditions. The precursor atomizes inside the furnace and then produced metallic nanoparticles by following either of the following methods; condensation, precipitation, thermal decomposition and intraparticle collisions. In laser ablation technique pulsed laser is used to produce metallic nanoparticles. The metallic precursor is irradiated by pulsed laser in liquid or gas environment, which causes formation of high mobility metal oxide atoms and finally, metallic nanoparticles are produced by condensation in presence of liquid nitrogen. All the above-mentioned methods are typical representatives of 'bottom-up' approach. The advantage associated with these methods is metallic nanoparticles with narrow particle size distribution can be produced while the limitation is these are costly methods as very expensive equipments like lasers, furnaces, inert atmosphere, etc. are necessary to proceed the reactions.

In attrition method, a size-reducing mechanism like ball milling is applied to ground the bigger particles into the nano-sized particles. The obtained metallic nanoparticles by this method are affected by various conditions, like the nature of starting material, time of drilling and reaction medium. This method typically represents 'top-down' approach (Adam and Gabriela 2011).

The limitations associated with these methods are low production rate as compared to chemical methods, higher energy consumption to maintain the pressure and temperature during the reactions.

9.2.2 Chemical Methods

The second approach is chemical method which is usually a 'bottom-up' method. Chemical method is actually the wet chemical procedure where the metal salt precursor is dissolved in a suitable solvent and undergoes controlled reduction procedure in defined reaction conditions such as at particular temperature and pH. This procedure allows the successive formation of clusters or aggregates of metal oxide nanoparticles in gel or precipitation form. In next step, the gel or precipitate is subjected to ageing and calcinations at very high temperature in muffle furnace, where it is transformed into a solid mass by vaporizing completely the water molecules and other volatile solvents. In this step, the gel network is contracted and solvent from gel pores gets ejected. At high temperatures ($T > 800$ °C) densification and decomposition of the gels occur and during these steps, the porous parts of the gel network get distorted and residual organic solvents are volatilized. Several methods such as the sonochemical method, the polyol process, the solvent-reduction method, the template method and the reverse micelles method have been developed for the preparation of nanoparticles. Seed-mediated growth is another chemical method to prepare metallic nanoparticles. In this method, small particles which are prepared by other methods are used as seeds and reducing agents are utilized to reduce metallic ions. Reduced metallic ions grow on the surface of the seed particles (Samanta et al. 2010). Based on the purposes of the metallic nanoparticles synthesis, required properties, types, sizes of n- MOs different reducing agents are used. Examples of few reducing agents are $NaBH_4$, $SnCl_2$, sodium citrate, methoxy polyethylene glycol, ascorbic acid, sodium hydroxide, etc. For instance, to prepare nano ZnO, NaOH can be used as reducing agent. The chemical methods are associated with few advantages such as; (i) these are relatively cost-effective methods for high volume, (ii) very simple to execute.

However, the disadvantages are; (i) Highly toxic chemicals are used (ii) Greater chance of contamination between precursor materials and reducing agents, (iii) There is a chance of development of hazardous by-products (Adam and Gabriela 2011).

9.2.3 Biological Methods

Both physical and chemical methods are associated with few limitations and therefore there is a burgeoning need to develop a new method, which will be eco-friendly, rapid, high percentage of yield, occur under normal air pressure and room temperature, energy-saving and nontoxic by-products. Biogenic route for metallic nanoparticle synthesis serves all these purposes. Recently, biological ways mediated by living organisms have become an easy way to synthesize metallic nanoparticles as we have variety of organisms available in nature. Plants and plant products, algae, yeast, fungi, virus and bacteria play important roles in this purpose. Many inorganic materials are derived by using intra or extracellular organisms and it has been a well-known method for almost 30 years (Wilbur and Simkiss 1979). Biomaterials are not only used for the easy, fast synthesis of nanoparticles but also they have a great application field in the removal of toxic materials. Many microorganisms such as bacteria and algae are used in this purpose (Pérez-de-Mora et al. 2006). In this way, a comparatively novel nature-derived process for the development of metallic nanoparticles by microorganisms and other living beings has been established. Biogenic synthesis of nanoparticles is known as an emerging route of synthesis which is designed by overlapping nanotechnology and biotechnology. In the last few decades, it has gained an ample of attention of the researchers because of its capability to develop environment-friendly methodologies in material science. Till now plethora of both unicellular and multicellular organisms have been utilized to yield metallic nanoparticles. However, this method also has many limitations and few of these are listed below:

(i) Biological approach using living cells are an extremely complicated method as cells contain thousands of molecules with variety of functional groups such as; hydroxyl, emine, amine, hydrides, etc. Each of these functional groups has separate role in the reduction of metal salts. Therefore, it is quite hard to explain a specific mechanism responsible for the growth of metallic nanoparticles.

(ii) The resulting solution contains a mixture of various biological components, metallic nanoparticles and other molecules. Therefore, the purification method of metallic nanoparticles is very much complicated.

(iii) Presence of large number of protein molecules in the resulting solution destabilized the nanoparticles and also can influence their properties.

(iv) Few salts are toxic for biological materials and therefore cannot be allowed to use at higher concentrations (Fig. 9.1).

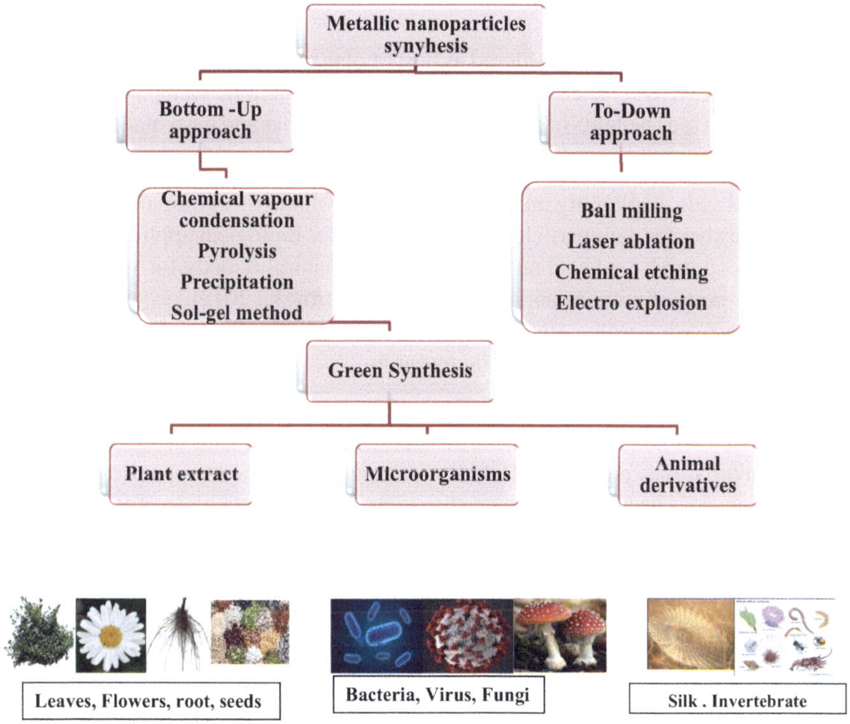

Fig. 9.1 Synthesis route of metal nanoparticles

9.3 Biological Synthesis

9.3.1 *Plant Components in Metallic Nanoparticles Synthesis*

It has been noticed that plants are the better option for the synthesis of metallic nanoparticles as plant molecules act as a natural capping agent and the particles present in plant sources are nontoxic. Extensive research on plant-medicated synthesis of metallic nanoparticles with different parts of plants belonging to different taxonomic groups has divulged the ability of green synthesis and capping activities as well (Vithiya and Sen 2011).

9.3.1.1 Angiosperms Mediated

Angiosperms are generally considered as best of plant evolutionary family and therefore they are extensively used in the green synthesis of metallic nanoparticles (Ratul et al. 2017). Various benefits of angiosperm joined together and have made this group more attractive in pursuing scientific studies. Few of those are listed below:-

(i) Worldwide availability and easy access have made this group of plants more persuasive for nanoparticle synthesis and other biological applications.
(ii) Angiosperm has played an essential role as natural remedy for various kinds of diseases of humans and other animals (Petrovska 2012).
(iii) Angiosperm species are edible in nature and that makes them even more important material for scientific research.

In terms of biogenic synthesis of metallic nanoparticles, angiosperms, being extremely rich in nontoxic plant bioresources, have better natural reducing capability of metallic ions and this pushed it more towards the rapid expansion of green field of science. It has been observed that high reducing potentials of plant phytochemicals react with metallic cations (M^{n+}) and reduced them to neutral atoms (M) for metal nanoparticle synthesis. Metallic cations with higher standard reduction potentials are reduced by the plant molecules with lower ionization potential values. This also proves nature's ability of reduction without the use of hazardous reducing materials.

Highly explored angiospermic plant species for the synthesis of metallic nanoparticles are Azadirachta indica, Camellia sinensis, Aloe vera, Centella asiatica, etc. as capping behaviour of the phytochemical molecules present in the above-mentioned plants can produce biocompatible metallic nanoparticles, have excellent medicinal values (Afaepour et al. 2009; D'Britto et al. 2012). However, overall biocompatibility of metallic nanoparticles depends on the size and shape of them (Alkilany and Murphy 2010). Metallic nanoparticles having size range less than 100 nm are highly desirable for biomedical applications and managing desired size and shape of metallic nanoparticles during the synthesis is a real challenge for the scientists working on green synthesis. Researchers suggested that morphology of nanoparticles is dependent on reaction parameters like plant extract concentrations, metallic ion concentrations, pH, temperature and, time taken for the reaction (Mishra et al. 2014).

Cost-effective synthesis of metallic nanoparticles with Asparagus racemosus under sunlight is an amazing attempt because external energies are not required to carry out the reaction and the whole synthesis can be done at room temperature. Other plant species Camellia sinensis (tea) and Coffea Arabica (coffee) are well-explored plants for green synthesis mechanism study because of isolation and purification of the tea and coffee biomolecules, viz. catechins, thea-flavins, other phenolic compounds, fibres and alkaloids are easy and less time-consuming. Moreover, effective capping and reduction can be done by teac and coffee biomolecules. Commercially available pure tea polyphenol (from Sigma) has been applied for the green synthesis of Pt nanoparticles (Alshatwi et al. 2015). These results support in designing large-scale production of metallic nanoparticles. Involvement of phenolic compounds in the synthesis of metallic nanoparticles such as; Fe_2O_3, ZnO, CuO, Ag and Au are reported in literature (Ratul et al. 2017; Wang et al. 2014). Other plant species like Jatropha curcas L, Azadirachta indica Kernel, Camellia sinensis leaves are also used to reduce metal cations using their biomolecules at room temperature. The reaction mechanisms using angiosperms claim that common phytochemical constituents, such as phenols, alkaloids, terpenoids and some pigments, are mainly responsible for the green synthesis of different metallic nanoparticles (Ratul et al. 2017). It has been reported that gold and silver nanoparticles can be derived from geranium extracts (Chandran et al. 2006). Literature has also reported that gold nanotriangles and silver nanoparticles can be synthesized using Aloe Vera plant extracts (Singaravelu et al. 2007). Huang et al. (2007) derived silver and gold nanoparticles in a simple way by using the sundried Cinnamomum camphora leaf extract (Huang et al. 2007). Most of the reports available on the derivation of silver or gold nanoparticles utilize broths obtained from the boiling fresh plant leaves. Well-defined silver nanowires were synthesized by reducing silver nitrate solution in presence of broth of sundried Cassia fistula leaf at room temperature without any additive (Lin et al. 2010). However, it should be noted that not all randomly selected plant species can be used for biological synthesis of metallic nanoparticles.

9.3.1.2 Bryophytes Mediated

Bryophytes are known as the most ancient and second-largest group of land plants in the world. Bryophytes are essential in understanding the plant origin and following the plant transition to land. These are the less explored plant groups in green synthesis of metal nanoparticles. Bryophytes can emit biologically active compounds which are used to protect themselves from other living organisms and because of the presence of these active biological components bryophytes are used in the synthesis of metallic nanoparticles. Studies showed that simple organization of makes the process facile. Literature showed that gold nanoparticles of various shapes viz.; spherical, triangular and hexagonal with the size in the range of 42–145 nm can be synthesized using the extract bryophyte gametophyte at 37 °C. Various metal oxide nanoparticles have been synthesized using plant bodies (thallus) of bryophytes (Acharya and Sarkar 2014).

9.3.1.3 Pteridophytes Mediated

Pteridophytes plants have excellent antibacterial activity and therefore over 350 million years, they have been investigated in research studies as an interesting topic. Extracts of these plants are used in the green synthesis of nanoparticles to get nanoparticles with antibacterial activity. Studies showed that (Britto et al. 2012) three different pteridophyte plants of pteris genus were used in the green synthesis of silver nanoparticles with excellent antibacterial activity. Antibacterial activities against bacterial pathogens have been shown by various nanomaterials derived from fern, Nephrolepis exaltata (Bhor, et al. 2014). Literature studies have shown that nanoparticles synthesized from pteridophytes have excellent antibacterial activity and this might be due to the different natural antioxidant and antibacterial properties of the plants themselves. However, the actual mechanism of biosynthesis of metallic nanoparticles from different plant species and their antibacterial properties are yet to be studied.

Presence of various phytochemicals, such as; flavonoids, alkaloids and terpenoids in pteriodophytes makes the plant extracts highly oxidant and helps in the production of metallic nanoparticles. Literature studies also revealed that flavonoids act as both reducing agents and stabilizers during the production of nanoparticles and have enhanced antioxidant effects (Singh 2013). Extract of Azolla pinnata can be used to derive silver-based nanoparticles with different size and shape with enhanced antibacterial effects (Korbekandi et al. 2014).

9.3.1.4 Gymnosperms Mediated

Gymnosperms can be found in everywhere on earth and a large diversity of them makes each plant family unique, well organized and different valued. The biomolecules present in plants can reduce metal ions to form metal nanoparticles and recent research suggested that, like prokaryotes, eukaryotes are also capable of producing metal nanoparticles by biosorption and bioreduction methods. Literature has shown many examples of the formation and bioaccumulation of metal nanoparticles by gymnosperms mediated methods using various reducing and stabilizing agents. Literature showed that these reduced and stabilized metal nanoparticles can be utilized as catalyst in detoxifying pollutants (Cirtiu et al. 2011). Researchers claimed that copper (Cu) is essential as micronutrient for plants, but high concentration of Cu is detoxified by plants by reduction of Cu ions into Cu(0) and ultimately into Cu nanoparticles (Ratul et al. 2017). Research revealed that, the mechanism of the formation, size, morphology and quantity of metal nanoparticles dependent on the plant type, phytochemicals, pH of the solution and bioavailability of metals (Ratul et al. 2017). The general mechanism of metal ion reduction, stabilization and formation of nanoparticles by phytochemicals of angiosperms has been studied thoroughly but not in case of gymnosperms but as the phytochemicals are almost similar among plant groups, therefore by studying the mechanisms of angiosperms the reaction mechanisms of gymnosperms can be explained. Jha and Prasad (2010) showed the formation of AgO nanoparticles from $AgNO_3$ solution using antioxidative system of Cycas plant. Researchers also claimed that leaves of Cycas plant contain phenolic compounds, carbonyl and thiols groups, which are responsible for the biosynthesis of metal nanoparticles. Green synthesis of spherical Au nanoparticles is reported by Noruzi et al. (2012) utilizing the aqueous extract of cypress leaves and they observed that the reaction got completed within 10 min at room temperature and the average size of the synthesized nanoparticles depended on the pH and concentration of leaves extract (Noruzi et al. 2012). An eco-friendly synthesis of Cu nanoparticles using Ginkgo biloba Linn leaves extract as a reducing and stabilizing agent at room temperature has been studied (Nasrollahzadeh and Sajadi 2015). Kalpana et al. (2014) derived Ag nanoparticles from Torreya nucifera leaves extract and they found that temperature and concentration of extract played a vital role in the determination of size and shape of nanoparticles (Kalpana et al. 2014). Velmurugan et al. (2013) reported that in pine family of gymnospermic plant leaves and bark contain high concentration of phenolic compounds and hydroxyl and carboxyl groups of phenolic compounds are involved in metal nanoparticles synthesis. Another study proved that polyphenolic functional groups and proteins present in plant biomolecules are responsible for the metal ion reduction and as stabilizing agents of the metal nanoparticles and high temperature, pH and total time is taken for the growth can positively influence the properties of the synthesized nanoparticles (Ratul et al. 2017). Hence, it can be said that the biomolecules of gymnospermic plant extract have an essential role in the formation and stabilization of metal nanoparticles and the route is eco-friendly and efficient alternative to conventional methods.

9.3.1.5 Algae Mediated

Algae-mediated method provides eco-friendly reducing agent, nanoparticle stabilizing agent and capping agent but still a very few algae-mediated synthesis of nanoparticles are reported in literature. In the synthesis of gold-based nanoparticles, marine algae were used and the process took relatively short time period compared to other biosynthesizing processes. Similarly, Palladium and platinum-based nanoparticles using meta chloride salts have been investigated (Jagpreet et al. 2018).

Bioreduction process of algae showed a great potential in the green synthesis of different metallic nanoparticles, such as copper oxide, zinc oxide, iron oxide, silver, gold, palladium, etc. (Jagpreet et al. 2018). Algae is capable of controlling size and shape of the synthesized nanoparticle and therefore gaining more interest in the biological synthesis of metallic nanoparticles. During the synthesis, metal ions are entrapped on the surface of the plant cells by an electrostatic force of attraction between the positive metal ions and negatively charged carboxylate groups present on the cell surface. In next step, metal ions get reduced by cellular enzymes and form new nuclei, which later on undergo self-assembly process to grow and converted to nanoparticles. Single-celled green algae have a strong binding ability toward gold and silver nitrate salts and afterwards reduced the metal ions into Au (0)/Ag (0).

Various functional groups, such as hydroxyl (–OH) from polysaccharides and different amino acids (such as tyrosine) and carboxyl anions (–COOH–) from various amino acids such as aspartic acid (Asp) present on the cell surface are the most active functional groups for the reduction of metal ions to form metal oxide nanoparticles. Literature reported that several metallic nanoparticles, such as; Fe_2O, MgO, NiO, Ag, Au, Pt and Pd can be synthesized by using algeas (Jagpreet et al. 2018).

9.3.2 *Microbial Components in Metal Oxide Nanoparticle Synthesis*

9.3.2.1 Fungi Mediated

Use of fungi extract for the green synthesis of metal/ metal oxide nanoparticles is an efficient method and in this route nanoparticles with mono-dispersed phase, well-defined morphology and excellent properties can be synthesized. Fungus are having a large variety of intracellular enzymes, which make them a good mediator for the green synthesis of nanoparticles as, those enzymes are used as reducing, capping and stabilizing agents for the synthesized products. The major advantages associated with fungi mediated route are;

(i) Total yield by fungi mediated route is much more higher than bacteria mediated route.
(ii) Fungus have variety of enzymes, proteins and reducing agents on their cell surfaces, which help in the synthesis.

The probable mechanism for the synthesis of metal/metal oxide nanoparticles using fungi extract is enzymatic reduction on the cell surface or inside the fungi cells.

To synthesize gold nanoparticles Fusarium oxysporum releases enzymes into the aqueous solution of $AuCl_4^-$ ions with NADH and these enzymes act as reducing agents. Binding of protein through cysteine and lysine linkage provides long-term stability to the synthesized nanoparticles (Mukherjee et al. 2002). Fusarium oxysporum can also synthesize crystalline zirconia nanoparticles by hydrolyzing metal salts in presence of K_2ZrF_6 aqueous solution. Synthesis of silica and titania nanoparticles from their aqueous anionic complexes by using Fusarium oxysporum are also reported in literature. When F. oxysporum gets exposed to equimolar solution of AuCl4 and $AgNO_3$, it can derive Au–Ag nanoparticles (Senapati et al. 2005). Platinum nanoparticles can be derived by inter and extracellular formation in the presence of H_2PtCl_6 (Riddin et al. 2006). The size and shape of all nanoparticles derived by using fungi can be manipulated by altering pH and temperature during growth conditions. Aspergillus flavus and C. versicolor, a white-rot fungus, are used in the large-scale production of high-stability Ag nanoparticles (Ratul et al. 2017; Vigneshwaran et al. 2007). The silver nanoparticles can be synthesized by fungus, where glucose is used as stabilizing agent. Researchers proved that these silver nanoparticles could be used as water-soluble metallic catalysts for living cells. Advantages associated with fungal mediated synthesis are,

(i) regenerative capability
(ii) eco-friendly and energy-conserving nature
(iii) large-scale production of metal nanoparticles
(iv) commercial feasibility of the synthesized nanoparticles (Ratul et al. 2017).

9.3.2.2 Bacteria Mediated

Bacteria has a wide range of applications in biotechnology field, such as; bioleaching, bioremediation, genetic engineering, etc. Because of the relative ease of manipulation of bacteria, a variety of it is widely used in the biogenic synthesis of metal/metal oxide and other novel nanoparticles.

Bacterial cell wall plays an essential role in the synthesis of metallic nanoparticles. During the reaction, metal ions penetrate through the cell into the cytoplasm and again transferred back to the wall for extracellular liberation. The cell wall contains a large number of metal-binding sites, such as; carboxylic acid, amines, etc. and they are responsible for stoichiometric interaction between metal and other biomolecules followed by deposition of metal. The crystallinity and non-crystallinity of nanoparticles are affected by the morphology and intra-extracellular environment of bacteria. The formation of metal nanoparticles depends on various parameters, such as; pH, temperature, composition of growth medium and growth in light or dark conditions. The biosynthesized metallic nanoparticles can be produced using metal-containing bacteria and they have excellent optical and electrical properties with potential future applications. Metals can be recovered from waste by biosorption process onto the bacteria which ultimately resulted in n bioreduction of metals into nanoparticles and therefore, this method is gaining more interest for the recovery and production of nanoparticles by industries.

Prokaryotic and actinomycetes bacterium are mostly used bacterial strains for the green synthesis of nanoparticles. *Escherichia coli, Pseudomonas proteolytica, Lactobacillus casei, Bacillus amyloliquefaciens, Arthrobacter gangotriensis, Bacillus indicus,* etc. are few examples of bacteria strains extensively used in the synthesis of silver oxide nanoparticles with distinct morphologies and properties. Similarly, to synthesize gold oxide nanoparticles *Bacillus subtilish168, Escherichia coli* DH5a, *Rhodopseudomonas capsulate,* etc. are extensively exploited (Jagpreet et al. 2018).

9.3.2.3 Virus Mediated

In bacteria and fungi mediated routes for the synthesis of nanoparticles protein cages, DNA recognizing linkers and surfactant assembled pathways are used but all these techniques have their own limitations. It has been observed that to produce metallic nanoparticles the above-mentioned routes are facing difficulties and hence virus-mediated method comes under consideration. This method can produce self-assembled semiconductor materials with exceedingly oriented quantum dots structures with mono-dispersed shapes and sizes in nanoscale dimensions. Literature recorded the use of genetic selection and molecular cloning techniques by the genetically engineered phage-based tobacco mosaic virus to yield 3D inorganic nanoparticles (Shenton et al. 1999). The storage stability of the fabricated viral films is upto 7 months without any bacterial infection and it can be utilized to store high-density engineered DNA and in other medicinal applications (Mao et al. 2003).

9.3.2.4 Yeast Mediated

Yeasts can be found in eukaryotic cells and they are unicellular microorganisms. Till date around 1500 varieties of yeasts is identified. Metallic nanoparticles have been synthesized biologically by numerous groups by using yeast as the mediator. Literature revealed that silver and gold-based nanoparticles can be synthesized biogenically by using silver-tolerant yeast strain and Saccharo-myces cerevisiae broth.

9.3.3 Animal Components in Metal Oxide Nanoparticle Synthesis

9.3.3.1 Animal-Derived Materials Mediated

Recently, many animal-derived materials such as; bones, shells, horns, etc. have gained attention of the researchers in the synthesis of metallic nanoparticles. In addition to that cellular organisms are also capable of producing inorganic nanoparticles in oxide form. For instance, various unicellular organisms are used to produce magnetite nanoparticles and inorganic composite materials can be yielded from many multicellular organisms.

9.3.3.2 Invertebrate Mediated

Various kinds of sponges and starfishes have mineralized biological materials which facilitate the green synthesis of nanoparticles and the synthesized materials have medicinal uses also. These biological materials are used to stabilize amorphous minerals and to yield metallic nanoparticles. The most used biomaterials are the hydroxyapatite and the bio-silica. The hydroxyapatite can be obtained from fish bones and it has natural minerals in the form of calcium apatite. This mineral is the main compound in nano-sized collagenous and noncollagenous proteins and has biomedical applications. Many aquatic organisms, such as sponges, diatoms, radiolarians and choanoflagellates can produce glassy amorphous biogenic silica and in sponges, it is formed by the enzyme silicatein (Stupp and Braun 1997). This method happens by appositional layering of lamellae consisting of silica nanoparticles (Müller et al. 2009). Gunduz (2014) synthesized nano-hydroxyapatite from corals by thermo-gravimetry method and used it in medical applications such as osteogenesis (Gunduz 2014). Many nano metaloxide particles have been synthesized from different kinds of worms. In this method, the extracts of worms are used as reducing agents as well as stabilizer in the production of the nanoparticles. Invertebrate chitin can release chitosan peptides which can be used to produce metallic nanoparticles/chitosan nanocomposites and can be used as photocatalysts.

9.3.3.3 Silk Protein-Mediated

Silk fibroins are the amino acids, such as glycine, alanine and serine, produced by a variety of insects and spiders. Different kinds of silks are produced by spiders, such as non-sticky, dry, strong and spiralling threads. Silkworms can also produce silk fibroin, which has commercial uses also. Other species like the larvae of Hymenoptera and Trichoptera can produce silk in their adult stage during metamorphosis. Silk also can be obtained from South American tree ants. All these kinds of silk fibroins have various applications in textile and clinical fields. As silk fibrons are nontoxic and bio-degradable in nature, therefore, in medicinal field they are utilized as material for tissue engineering for blood vessels, skin, bone, ligament and nerve tissue regeneration. Several nanocomposites are produced using fibroin as component, as fibroin-TiO_2 (Feng et al. 2007), nano-hydroxyapatite/ silk fibroin (Wang et al. 2008).with near about 100 nm length. Sericin, a waste material, which is discarded into the wastewater from silk industries, can be used for production of metallic nanoparticles. In this method, sericin is extracted from wastewater and reduced by rotation to yield a concentrated sericin solution. Nano sericin powder can be obtained by ultrasonication (Ratul et al. 2017). Metallic nanoparticles produced by these kinds of fibers show antibacterial activities, biocompatibility, oxidation resistance and UV resistance.

9.4 Mechanism of Biological Synthesis of Metallic Nanoparticles Using Plant Extract

For the synthesis of metallic nanoparticles heated plant extract (A) and metal salt solution (B) is taken and heated to 70–80 °C using a stirrer-heater for about 30 min. NaOH is added sometimes drop-wise to the mixture for adjusting the pH of the solution to 10–12 and then heated for more than one and half hours until it is reduced. The solution is then filtered using filter paper and the residue is kept overnight in an oven. Next day the material is carefully collected and mashed in a mortar-pestle to get a finer powder on metallic nanoparticles. Various parameters are responsible to control the rate of the reaction, size and stability of the nanoparticles. Those parameters are pH, temperature, concentration of plant extract, concentration of metal salt extract and phytochemicals. Bacteria, algae, fungi mediated methods require longer time of incubation to reduce metal ions, whereas phytochemicals present in plant extracts take lesser time to reduce metal ions and therefore metallic nanoparticle synthesis using plant extract is considered as an excellent method. In addition to that, plant extract has dual character in whole procedure. They act as a reducing agent as well as stabilizing agent during the reaction. Phytochemicals play an important role in this process. Various kinds of phytochemicals like sugars, ketones, aldehydes, carboxylic acids, terpenoids, flavonoids, etc. are responsible for the synthesis of metallic nanoparticles. Sugars such as glucose and fructose help in the synthesis of metallic nanoparticles of different shapes and sizes. Flavonoids contain large number of functional groups and during the synthesis because of tautomerism active hydrogen atoms get released from enol and converted it to keto form. This enol-keto transformation reaction is responsible for the synthesis of many plant extracts, like basil extract. Plant extract also contains protein biomolecules with functionalized amino acids, heterocyclic compounds which also help in the synthesis of metallic nanoparticles. The steps involved in the synthesis of metallic nanoparticles using plant extract are (i) the activation phase—bioreduction of metal salts followed by nucleation process of the reduced metal ions, (ii) the growth phase—self-assembly of small particles and (iii) termination phase—the final shape and size of the nanoparticles are defined (Jagpreet et al. 2018) (Fig. 9.2).

9.5 Advantages and Disadvantages of Biological Synthesis of Metallic Nanoparticles

Chemical methods for the synthesis of metallic nanoparticles are based on harsh chemicals, non-polar solvents and therefore chemically synthesized nanoparticles have some restrictions in clinical and biomedical applications. Toxic chemicals are used to reduce metal ions, as capping agents, to stabilize nanoparticles, etc. and those chemicals and their by-products contaminate soil and water. Another drawback of

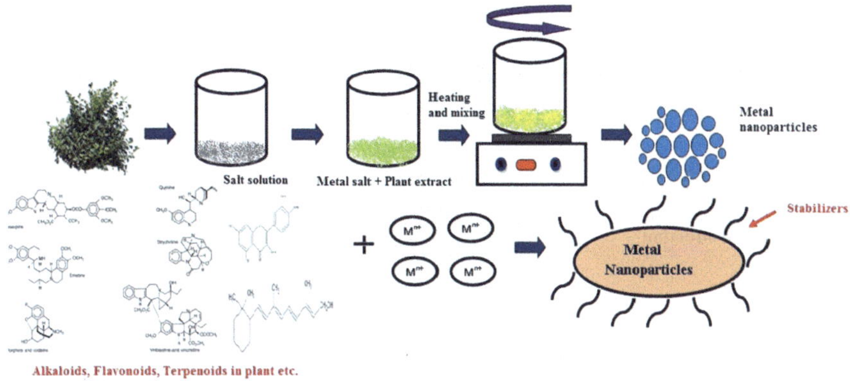

Fig. 9.2 Biological synthesis of metallic nanoparticles using plant extract

chemical methods are those are more capital and energy-consuming. Therefore, eco-friendly and green methods for the synthesis of metallic nanoparticles are in demand now and unlike chemical methods, lack of toxic chemicals in biological methods makes it a more environmentally friendly, consumes no energy and cost-effective route. The biogenic route involves synthesis of metallic nanoparticles from plant extract and the use of different microbes, such as fungi, bacteria and yeast. The whole process of synthesis depends on the enzyme and biomolecules activities when the microbes or the gigantic phytochemicals grab the target metallic ions and reduce them by controlling their sizes. Highly stable, well-characterized nanoparticles can be synthesized by green method. Biologically synthesized metallic nanoparticles are more polydispersed than chemically synthesized metallic nanoparticles. Controlling shape and size of the metallic nanoparticles are easier in biological method and as the fundamental properties of nanoparticles such as; optical, catalytical, magnetic, electronic, etc. are size-dependent properties, therefore, biogenic route is more convenient than chemical route for the researchers. Moreover, the bioproducts of these methods are non-hazardous and can be utilized for clinical and biomedical applications.

Nonetheless, the limitation associated with the biological synthesis is the total time required for the production as the microbes and other phytochemicals grow or work under natural conditions to synthesize metallic nanoparticles (Fig. 9.3).

Toxicity issue during biological synthesis

Compared to chemically synthesized metallic nanoparticles, green synthesized metallic nanoparticles have a broad range of applications such as; biosensors, water purification, drug delivery, cancer treatment, DNA analysis, gene therapy, magnetic resonance imaging, antibacterial agents, slow vaccine release, tissue engineering, etc. As chemical methods use harsh chemicals for the synthesis of metallic nanoparticles, therefore, this kind of vast field of applications cannot be accomplished by chemical synthesis. In biological method for the synthesis of metallic nanoparticles hazardous chemicals like, organic solvents and inorganic metal salts are used in very

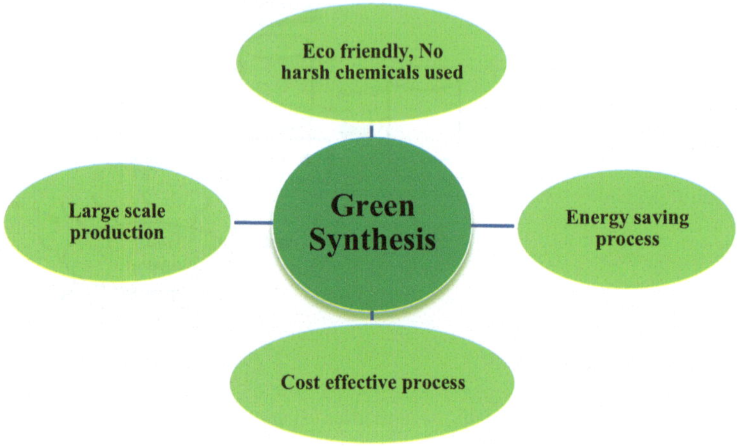

Fig. 9.3 Benefits of green synthesis

low concentrations, therefore, this kind of synthesis is known as green synthesis. However, it is quite impossible to prepare extracts of microorganisms and few plants without any organic solvents. Use of water as only one solvent during the extraction of biological substances and during the synthesis of metallic nanoparticles is still under investigation.

9.6 Photocatalysis

Release of industrial wastewater containing organic dyes into the main water bodies is highly objectionable because the coloured dye materials and their dissociated products are extremely toxic and carcinogenic in nature. Various methods have been evolved to remove dye particles from water but they all are non-destructive and as a result of that toxic dye particles are removed temporarily and in due course of time those get transformed into other toxic substances. Advanced oxidation processes (AOPs) have gained interest in this regard because of the formation of nontoxic by-products after oxidation of the pollutants. Degradation of organic pollutant through AOPs occur by using reactive oxygen species such as hydroxyl (·OH) and superoxide ($\cdot O_2^-$) radicals which are generated in several steps.

AOP follows four different pathways and those are, photolysis, ozonation, Fenton process and photocatalysis. To generate active radical species photolysis and ozonation methods use hydrogen peroxide in presence of ultraviolet and ozone, respectively. Hydrogen peroxide along with ferrous ions as catalyst are used in Fenton process to get radicals. In photocatalysis, radical species are generated in presence of light and semiconducting materials, where semiconductors absorb photons from light and produce active radicals. Among these four methods, recently photocatalysis process using semiconducting nanoparticles has attracted much more attention of the researchers as nanoparticles can degrade very low concentration organic pollutants in presence of photons and produces nontoxic degraded products, utilizes renewable solar energy and do not use harsh chemicals. Therefore, photocatalysis is considered as safe, green and sustainable process (Eleen et al. 2021).

9.6.1 *Mechanism of Photocatalysis*

Photocatalytic reaction rate depends on the catalyst structure that is used in reaction and the photons energy of visible or UV light. Depending on the electronic structure of the catalyst, it plays a role as a sensitizer for the irradiation of light-stimulated redox processes. Reaction mechanism of pollutant degradation photocatalysis process follows two ways; direct and indirect ways but indirect way is the most commonly used to explain the reactions. Indirect reaction process starts with photoexcitation. The general mechanism is that the electrons of valance band will excite to conduction band upon absorbing photons only if the band gap of the catalyst is equivalent or less than the incident light energy. When electrons move from valence band to conduction band, they leave holes (h^+) which play the key role to oxidize the donor molecules. An additional example is a strong oxidizer hydroxyl radical (·OH) is produced when H_2O reacts with these holes. After the electrons reach the conduction band, they are absorbed by the water molecules and produce superoxide ions ($.O_2^-$) which are reducing agents. Consequently, it is acceptable to say that the free electrons and holes are responsible for the redox reaction with any kind of pollutants which come into the contact with the catalyst and convert it into the desired products (Muhammad et al. 2020). The schematic representation of pollutant degradation by photocatalysis is shown in Fig. 9.4

The photocatalytic reactions of pollutants (P) in the presence of photocatalysts (metallic nanoparticles n−M) and light source are given below:

$$P + h\upsilon \rightarrow P*$$

$$P* + n - M \rightarrow (n - M)* + P^+ \quad P* + n - M \rightarrow P + h^+ + e^-$$

$$(n - M)*\left(e^-\right) + O_2 \rightarrow .O_2^- + n - MOs \quad h^+ + H_2O \rightarrow OH^- + H^+$$

$$O_2 + H^+ \rightarrow HO_2 \quad 2HO_2 \rightarrow H_2O_2 + O_2$$

$$H_2O_2 + O_2^- \rightarrow HO\cdot + OH^- + O_2 \quad H_2O_2 + e^- \rightarrow HO\cdot + OH^-$$

Organic and inorganic pollutants + reactive oxygen species → Degradated into harmless products.

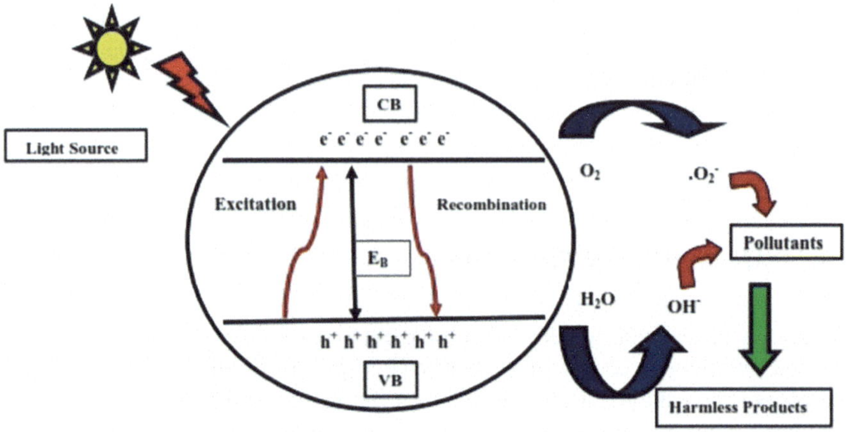

Fig. 9.4 Mechanism of photocatalysis process

9.6.1.1 Oxidation Mechanism

Light generates electron–hole pairs in the catalyst and it captures water molecules. When water molecule is oxidized and results into hydroxyl and oxygen radicals. Hydroxyl radicals have oxidizing powers that can decompose all organic pollutants. If oxygen radical is involved in the process, it creates chain reactions with organic molecules' intermediate radicals and results into carbon dioxide and water (Muhammad et al. 2020).

9.6.1.2 Reduction Mechanism

Oxygen is easily reduced by pairing reaction. When it is reduced, it produced hydrogen. The reduction reaction mechanism of oxygen is that electrons in the conduction band form superoxide ion by reacting with oxygen it is also possible that the anion attaches with intermediate to produce peroxides in an oxidation reaction and then produce water. Therefore, when the concentration of organic matter increases, the probability that number of holes which reduces the recombination rate of the carriers increases as well (Muhammad et al. 2020).

9.6.2 Factors Affecting Photocatalysis Process

There are different factors that can affect on the photocatalytic activity process for treating wastewater. Electron–hole pair separation rate, the structure of the catalyst, pH, amount of catalyst, intensity of the incident light and temperature are the factor that affects the performance of the photocatalyst. In case of electron–hole pair separation rate, a good catalyst should utilize in secondary reaction before recombination because the life of the electron–hole pair in semiconductor catalyst is small. In order to have a deal photocatalyst, it must have a wide band gap and electron recombination as low as possible. For instance, in TiO_2 holes are considered as good oxidizing agents while electrons are acted as reducing agents in valance band (Muhammad et al. 2020).

9.6.2.1 Structure of the Catalyst

The structure of the catalyst plays an important role in the photocatalytic reactivity. TiO_2 has three different structures, they are rutile, brookite and anatase. Each of them has different impact on the photocatalytic activity. It is observed that anatase is more effective than others. This is attributed to its structure stability, conduction band position and its high adsorption. Another example is zinc oxide which has two different shapes; spherical and rod shapes. The degradation efficiency is higher in spherical form than in rod form and this is because of its high surface area. In general, materials are used in their nano form instead of bulk state because their size dimension gives them an advantage in terms of reactivity rate. The enhancement of reaction rate in case of nanomaterials is because they have higher surface area or surface-to-volume ratio is high which permits reactants and catalysts to react together (Muhammad et al. 2020).

9.6.2.2 pH

Photocatalyst efficiency is pH-dependent. pH affects the surface charges of the photocatalyst and that can be observed in the electrostatic phenomena that are between pollutant and charged particles. The impact of different pHs on the photocatalysts has been studied. It is reported that in acidic medium, pH less than 5, photocatalyst does not have high-efficiency level l and the reason is that the high concentration of protons H which limits OH radicals that are used to degrade the dyes. If pH is in range of 5–10, the efficacy of the photocatalyst increase and it can reach its maximum if pH is at 10 which is alkaline medium because the amount of OH radicals is bigger and that help in degrading the dyes pollutant. However, if pH is above 10, such in range of 11–13, it is observed that there is a drop in the efficiency of the photocatalyst and it is attributed to the increased level of OH radicals in the process and that avoids them to react with the dyes in the medium (Muhammad et al. 2020).

9.6.2.3 Amount of Catalyst

There is direct relationship between the amount of catalyst and photocatalytic activity which means when the amount of catalyst increases, the degradation rate will increase as result of the increase of generated radicals. There is an optimal amount of catalyst after which the increase of amount of catalyst will not change the degradation rate because it will avoid the light to penetrate (Muhammad et al. 2020).

9.6.2.4 Intensity of Incident Light

The intensity of incident light on the catalyst surface has an impact on the photocatalytic degradation rate as well. When light incident increases, it leads to an increase in the quantum yield, i.e. the ratio of reaction rate and absorption rate. To illustrate, an example is TiO_2which has 3.2 eV bandgap and absorbs UV light. Comparing light intensity, when it is 0–20 mW/cm, the degradation rate increases. However, when high-intensity light helps in the recombination of electron–hole pairs then the rate of the reaction decreases (Muhammad et al. 2020).

9.6.2.5 Temperature

Temperature has an influence on the photocatalytic reaction rate. Each nanoparticle has an optimum temperature at which the reaction rate reaches its maximum. reported that 20–80 °C is the optimum temperature range for TiO_2. They observed that the rate of reaction of TiO_2 is decreased at 80 °C as a result of an increase in the recombination rate of electrons and holes increase. In addition, the absorption rate of the catalyst is reduced at this reaction rate. In addition to that, the rate of the reaction declines due to the reduction of the absorption rate of the catalyst at this particular temperature (Muhammad et al. 2020).

9.6.3 Role of Photocatalytic Process Against Different Wastewater Pollutants

Wastewater contains different kinds of pollutants such as heavy metals, microbes, organic and inorganic pollutants. Research has shown that major organic pollutants in wastewater are present in the form of carboxylic acid, alcohol, chlorinated aromatic compounds and phenolic derivatives. In addition to that, leather industry and textile industry are the main sources of dyes pollutants that have major role in polluting water. Dye industries contribute with roughly 20% of the annual dye consumption into the water bodies (Dequigiovanni et al. 2018).

TiO_2, ZnO and CuO semiconducting nanoparticles have a great potential to purify wasted water in presence of dyes (Danwittayakul et al. 2013; Chin et al. 2006). Researchers confirmed that such nanoparticles have an influence in treating wastewater during the photocatalytic reaction process. Liu et al. (2008) reported that TiO_2 reduced humic acid or natural organic matters by 80% and 65%, respectively (Liu et al. 2008). Humic acid substance is a yellow–brown material that has high molecular weight. Not only humic acid and other organic substances are removed by photocatalytic process but also various inorganic substances such as halides, ammonia, cyanide, nitrates and thiocyanate were also can be decomposed photolytically by metal oxide nanoparticles. Hong et al. (2009) reported that TiO_2 was used against $AgNO_3$ (Hong et al. 2009), Joshi and Shrivastava (2011), Bagabas et al. (2013) studied the effect of ZnO against Cr(VI) and potassium cyanide (Joshi and Shrivastava 2011; Bagabas et al. 2013) and Lee et al. (2002) showed that photo-oxidation of ammonia in water by nanotubes (Lee et al. 2002).

Heavy metals in water are a big concern due to its toxicity. The level of heavy metals in water varies and it can threaten the life of human and aquatic creatures. Thus, it is one of elements that is needed to be considered when treating wastewater. But as most of these metals are rare, expensive and valuable than others therefore very often it is preferred to recover those by photocatalysis from polluted water.

Microbes and bacteria are other kinds of pollutants that exist in polluted water. Photocatalysts have an anti-microbial effect. During the photocatalytic process, the catalysts generate radicals that destruct the cell wall of bacteria and thus prevent the growth of microbes and bacteria. Researchers studied the removal of microbes such as S. mutans, S. natuss, S. cricetus, E. coli, S. cerevisisas and L. acidophilus in wastewater by heterogenous photocatalyst. For instance, the growth of Chlorella vulgaris can be controlled by TiO_2 and ZnO is used to inhibit the growth of E. coli and S. aureus (Muhammad, et al. 2020).

9.7 Photocatalysts

9.7.1 Selection of Nanomaterials as Photocatalysts

Like typical redox reactions, photocatalytic reactions include oxidation and reduction steps. There is a need for a catalyst in photocatalytic reaction to support those steps. At electronic level, materials are classified as conductors, semiconductors and insulators. In conductors' class, the electronic valence band is overlapping with conduction band and the best conductors are alkali, alkaline earth metals and transition metals. They are the best conductors because there is no suitable band gap and thus, they are not appropriate for catalytic activity. In case of semiconductors, there is a moderate energy gap between valence band and conduction band which allows for redox reaction to occur. Similar to oxidation mechanism, free electron–hole pairs are generated upon exposure to light.

Semiconductors can be photocatalysts under certain conditions: there must be low recombination rate and the absorption wavelength must lie in range of 350–700 nm in visible region or band gap in 1.5–3.5 eV. The band gap 1.5–3.5 eV is not as large as for typical semiconductors, but this range is what is required to act as photocatalyst in UV–visible region. Examples of semiconductors are metal oxides. They offer properties such as light absorption, structure stability, superior morphology, high surface area, reusability, carrier transportation and band gap that make them good photocatalysts. It is crucial that band gap lies in UV–visible range in order to be an attractive photocatalyst. Examples of metal oxides that have these properties are chromium, vanadium, cerium, zinc and titanium and thus they are widely used as photocatalysts (Muhammad, et al. 2020).

Gold, silver, iron, zinc, nickel and cobalt are examples of metals that exist in metal nanoparticles and when they are combined with hydroxide, chloride, sulphides, phosphate, oxide and fluoride, they form compounds. The synthesis of nanoparticles by photochemical, chemical and electrochemical methods is prepared by using precursors of metals. The band gap of most metal nanoparticles lies in an infrared region and in ultraviolet region. However, this is not suitable to be a photocatalyst. Roy et al. (2015a; b) reported that the band gap of silver nanoparticles shows photocatalytic properties in visible region (Roy et al. 2015a; b). Metal nanoparticles of group IV for instance Si, Ge, group VI, for instance, Se, Te elements and group III–V, II–VI, I–VII, IV–VI, V–VI, II–V compounds such as; GaN, GaO, ZnS, CdSe, ZnO, TiO, Mgo, AgO are semiconducting in nature. They have different properties because of their wide band gap and thus they behave either as metal or non-metal. In addition to metal and non-metal nanoparticles, polymeric nanoparticles are other categories of nanoparticles. They are naturally organic compounds, have structural shapes such as nano-spheres or nano-capsular, the shape of polymeric nanoparticles will depend upon the synthesis method. The difference between nano-capsular and nanosphere is that the morphology of nano-capsular is like core–shell while nanosphere is like matrix structure. Polymeric nanoparticles are mainly used in medical field such in drug delivery and diagnosis (Muhammad et al. 2020). Generally, when metal oxide photocatalysts are irradiated with visible light they absorb photons and electrons from valance band get excited to conduction band and thus create electron–hole pairs. These electron–hole pairs decompose pollutants by redox reactions at the surface of the catalysts and because of this reason metal oxide photocatalysts are gaining more attention in the photodegradation of pollutants.

In case of insulators, they are electrons deficient and there is a band gap between valence band and the conduction band. Consequently, it is difficult for oxidation reaction to carry out in presence of insulators and there is a need for energy to perform redox reactions which make insulators are not good choice as catalyst. In the example of splitting water molecules using insulators, it cannot be achieved unless high energy inputs are utilized. Halogens and noble gases are examples of insulators.

9.7.2 Metal Oxide Nanoparticles Photocatalysts

Recently researchers are more focused on the green synthesis of nanomaterials as this route of synthesis is economical, easy, fast-paced and environment friendly. To produce metallic nanoparticles photocatalysts this methodology makes use of several biological species such as plants, animal products and microorganisms. Furthermore, scientists showed that through biogenic route of synthesis metallic nanoparticles photocatalysts of better size and morphology can be produced.

In this following section, we will be discussing on the green synthesized metal oxide nanoparticles such as zinc oxide, iron oxide and titanium dioxide nanoparticles for the application of photocatalytic degradation of pollutants.

9.7.2.1 ZnO Photocatalysts

A lot of literature suggests adsorption and advanced oxidation processes (AOP) are better methods in the treatment of textile industry disposed wastewaters and in the removal of dye particles from water. Several semiconductor materials have been used as adsorbents and photocatalysts in the removal of dye particles. Recently, an increasing attention has been directed to ZnO nanostructures in waste removal methods from polluted water where nano ZnO are found to be highly effective as it has a wide band gap of 3.37 eV and due to its low cost, low toxicity, antibacterial properties and high surface activity. ZnO can be found in nature within the earth's crust in the form of mineral zincite but it can be obtained through synthesis (Eleen et al. 2021). ZnO has three crystalline structures; rocksalt, wurtzite and cubic (zinc blend) and it can absorb larger fraction of the UV spectrum. AOPs and adsorption methods are used by ZnO in the waste removal method because the high surface area to mass ratios of ZnO nanoparticles can greatly enhance the adsorption capacities of sorbent materials. Various researchers have shown the removal efficiency of ZnO nanomaterials with different types of microorganisms and heavy metals, including Cu, Pb, Cd, Ni, Co, Pb, Hg and As. Literature also shows effective removal of dye particles and toxic organic contaminants from industrial and pharmaceutical wastewater by various groups (Eleen et al. 2021).

Sangeetha et al. (2012) produced ZnO nanoparticles by green route and used it against bacterial and fungal pathogens (Sangeetha et al. 2012). Varadavenkatesan et al. (2019) reported that *Cyanometra ramiflora* leaves extract can be used in the green synthesis of ZnO and the results showed nanoflowers morphology of the synthesized n-MO. This type of synthesized nanoparticles was used to degrade Rhodamine B dye up to 98% in 200 min (Varadavenkatesan et al. 2019). Eleen et al. (2021) have used biopolymer, pullulan to synthesize ZnO nanoparticles to degrade Rhodamine B and Methyl orange dyes completely within 60 min under UV irradiation (Isa et al. 2021b).

9.7.2.2 Fe(II/III) Oxide Photocatalysts

Iron oxide is known as a transition metal oxide, which can exist in three crystalline forms such as hematite (a-Fe_2O_3), magnetite (Fe_3O_4) and maghemite (γ-Fe_2O_3) but among all these, hematite is the most stable form and most commonly used as photocatalyst (Eleen et al. 2021). According to literature magnetic Fe_2O_3 nanoparticles being a semiconductor has a potential application in photocatalysis and under

UV radiation, electrons excite from valence to conduction band which ultimately resulting into the formation of electron–hole pairs, comparable to the electron–hole pair formation in ZnO nanoparticles. Fe_2O_3 nanoparticles having high specific surface areas, also have unique adsorption properties due to disordered surface regions as well as for different distributions of reactive surface sites. Shiying et al. (2017) reported that iron oxide /biochar nanocomposites loaded with photosynthetic bacteria are effectively used in wastewater treatment (Shiying et al. 2017).

The widespread applications of Fe_2O_3 in the removal of organic and inorganic pollutants by photocatalysis and adsorption methods are studied in detail by Santosh et al. Karunakaran et al. (2018) have shown that magnetism of Fe_2O_3 is a unique physical property that can influence the physical properties of contaminants in water and helps in water purification. Adsorption procedure combined with magnetic separation of Fe_2O_3 nanostructures has therefore been used widely in wastewater treatment (Karunakaran et al. 2018). Researchers reported that synthesized nanorods of iron oxide nanoparticles using Wedelia urticifolia DC. Leaf extract and the synthesized materials were used to degrade methylene blue dye up to 98% in 360 min under visible light irradiation. Literature showed that to photodegrade naphthalene iron oxide nanoparticles were synthesized using the leaves extract of spinny amaranth and 97% degradation efficiency within 150 min under UV irradiation was recorded (Eleen et al. 2021).

9.7.2.3 TiO_2 Photocatalysts

Titanium dioxide nanoparticles is another widely used photocatalyst in recent years and this is all because of their non-toxicity, cost-effectiveness, excellent photosensitivity, photocatalytic stability and plentiful availability. TiO_2 has two phases; anatase and rutile and the band gap value of those phases are 3.2 eV and 3.03 eV, respectively (Eleen et al. 2021). Sonker et al. (2020) reported that nanosheets of TiO_2 nanoparticles can be synthesized using Aloe vera extract. The synthesized materials showed degradation on Rhodamine B dye in 50 min under visible light irradiation and the degradation efficiency was 58% of Sonker et al. (2020). Literature showed that spherical TiO_2 NPs with particle sizes ranging from 50 to 120 nm were synthesized by using *Salvia officinalis* leaves extract. Produced n-MOs were utilized to degrade Reactive Black 5, Reactive Blue 19 and Brilliant Blue R dyes with degradation percentages of 69, 74 and 79%, respectively (Eleen et al. 2021).

9.7.3 Nanocomposites and Other Photocatalysts

Chitosan peptide can be derived from invertebrate chitin. Nanochitosan can be utilized for environmental applications such as removal of pollutants from water. For instance, Fe_2O_3/ graphene/chitosan is an useful material to remove organic dyes from water and this is due to the presence of the large number of hydroxyl and amino groups of chitosan and magnetic property of Fe_2O_3, which in together showed good adsorption capability for certain organic dyes (Sheshmani et al. 2014). Recently, TiO_2/chitosan composite was found to be an efficient photocatalyst in the removal of organic pollutants from wastewaters and also the photocatalytic activity of this catalyst remains almost same after 10 cycles (Xiao et al. 2015). A nanocomposite of Bentonite–chitosan can adsorb and remove synthetic dyes with great efficiency. Chitosan–metal oxide nanocomposites have several industrial and pharmaceutical applications like; in textile industry they can be used for colourizing textiles, in medicine, they can be used as a nano- capsule for slow release of vaccines and cancer treatment (Ratul et al. 2017).

The nanocrystalline semiconductors such as; HgSe, PbSe and CdSe colloids were studied thoroughly for their photocatalytic activity and it has been found that, for less than 50 Å diameters(d), optical absorption edge of HgSe and PbSe was blue-shifted by several volts. The enhanced photocatalytic effects of these above-mentioned quantum size semiconductors are due to the evolution of H_2. The nanocrystalline CdSe particles ($d < 50$ Å) have more stability against photo corrosion compared to HgSe and PbSe and hence were exploited for the reduction of CO_2 in formic acid. It has been observed that large particle-sized CdSe colloids did not show the same results under the same experimental conditions (Beydoun et al. 1999).

One research article showed that, when Zn is mixed with nanocrystalline ZnS, it can enhance the photoreduction of CO_2 by switching from formate to CO, without any loss in its efficiency. Another study on CO_2 photoreduction by the same group explained that CdS had enhanced photocatalytic capability due to the formation of sulphur vacancy on the surface and they also explained that, CO formation process occurred via adsorption of CO_2 to a Cd atom in the vicinity of a sulphur vacancy (Beydoun et al. 1999).

Sato et al. (1996a) showed the use of $H_2Ti_4O_9$/CdS nanocomposites for the photochemical reduction of nitrate to ammonia with and without methanol and it has been observed that addition of methanol increased the reduction rate. Photocatalytic activity of the nanocomposite could also be increased by doping Pt particles into the interlayer. This group also studied on the photocatalytic properties of layered hydrous titanium oxide/CdS–ZnS nanocomposites, where CdS–ZnS was incorporated into the interlayer and it was observed in presence of visible light the photocatalytic activity with the liberation of hydrogen on the surface of nanocomposites was more following compared to unsupported CdS–ZnS (Sato et al. 1996b).

9.8 Future Scope and Conclusion

Green synthesis of metal nanoparticles for their photocatalytic activity has become an interesting topic of research over the last decade. Various kinds of plants, microorganisms, fungi, algae and animal-derived products have been employed for the synthesis but it has been found that plant extract is more efficient in this regard as phytochemicals present in plants can be used as reducing as well as stabilizing agents. This chapter discussed over biosynthesis of metal nanoparticles by using various parts of plants, microorganisms and animal derivatives and also elucidated the mechanism of biosynthesis. Photocatalysts and their mechanism in photodegradation are explained thoroughly in this chapter.

Photocatalysis is initiated by light and completed with the help of catalysts, where nanomaterials are considered as most suitable photocatalyst for photodegradation. Semiconducting metal-based nanomaterials are considered as most efficient photocatalyst due to their excellent photocatalytic activity and wide range of band gaps in visible regions. As recently our focus is on cost-effective and environmentally friendly photocatalyst therefore biosynthesized nanoparticles play a vital role in this field of research. In water pollution treatment photocatalytic technique is the most desirable because of its cost-effectiveness, environmental friendliness and degradation efficiency towards organic, inorganic, heavy metal and microbe's pollutants. However, detail study on photocatalysis by biosynthesized metal nanoparticles has encountered with few existing problems, which can be resolved effectively in future studies.

(1) Few hours are required for complete degradation of pollutants by metal nanoparticles therefore in future more studies should be done to reduce the total time taken for complete photodegradation.
(2) More emphasis should be put on the study based on the concentration of pollutants, reproducibility and stability of the catalysts after several runs.

Biosynthesis of metal nanoparticles for the use in photodegradation studies, is a quite new field of research and it is in the developing stage and because of that, this methodology is encountering with several problems, which can be solved by taking some measures. The initial start problems faced by these methods are related to size, shape, self-assembly, stability and crystal growth of nanoparticles. These kinds of problems are quite common in metal nanoparticle synthesis but however, more and more attention should be drawn towards biological synthesis to wave out those issues in future. Some salient points of green synthesis of nanoparticles comprise the fact that:

(1) The mechanism of biogenic synthesis of metal nanoparticles is still not clear. Therefore future studies must put more concentration on reaction of biomolecules and enzymes during the growth phase. In addition to that details study on the properties of synthesized nanoparticles is much more needed.
(2) Another relatively unexplored area of research is purification of metal nanoparticles. Removal of unreacted metal salts, plant extract and microbes is very essential. For purification, chemical treatment must be avoided to keep the nanoparticles nontoxic. Therefore in future purification processes by physical methods, such as centrifugation, heating, ultrasound and osmotic shock can be investigated.
(3) Large-scale production is another important point to be considered in future as till date metal nanoparticles are synthesized only in laboratory scale via green method.

(4) Future studies must emphasize on cost-effectiveness of the large-scale production of metal nanoparticles. In biogenic synthesis, major expenses are coming from consumable metal salts and microbial growth. In this case, cost can be cut down by using recyclable waste materials and this can be a sustainable approach too.
(5) Future studies must put more stress in the rapid production of nanoparticles.

However, recent studies on biogenic route have suggested that principles of this method can be efficiently applied for the biological synthesis of metal nanoparticles and at the end, we can conclude that photocatalytic technique using green synthesized metal nanoparticles is a sustainable, emerging, efficient and green approach.

References

Acharya K, Sarkar J (2014) Bryo-synthesis of gold nanoparticles. Int J Pharm Sci Rev Res 29(1):82–86

Adam S, Gabriela K (2011) Biosynthesis of metallic nanoparticles and their applications. In: Prokop A (ed) Intracellular delivery: fundamentals and applications, fundamental biomedical technologies, vol 5. Springer, Ostrava, pp 1–38. https://doi.org/10.1007/978-94-007-1248-5_14

Afaepour M et al (2009) Green synthesis of small silver nanoparticles using geraniol and its cytotoxicity against fibrosarcoma-wehi 164. Avicenna J Med Biotechnol 1(2):111–115

Afsharian Z, Khosravi-Daran K (2019) Application of nanoclays in food packaging. Biointerface Res Appl Chem 10(1):4790–4802

Ahuja P, Ujjain SK, Kanojia R, Attri P (2021) Transition metal oxides and their composites for photocatalytic dye degradation. J Compos Sci 5(3):82–86. https://doi.org/10.3390/jcs5030082

Alkilany AM, Murphy CJ (2010) Toxicity and cellular uptake of gold nanoparticles: what we have learned so far? J Nanopart Res 12(7):2313–2333

Alshatwi AA, Athinarayanan J, Subbarayan PV (2015) Green synthesis of platinum nanoparticles that induce cell death and G2/M-phase cell cycle arrest in human cervical cancer cells. J Mater Sci Mater Med 26(1):1–9

Bagabas A et al (2013) Room-temperature synthesis of zinc oxide nanoparticles in different media and their application in cyanide photodegradation. Nanoscale Res Lett 8(1):516

Beydoun D et al (1999) Role of nanoparticles in photocatalysis. J Nanoparticles Res 1:439–458

Bhor G et al (2014) Synthesis of silver nanoparticles using leaflet extract of Nephrolepi sexaltata L. and evaluation antibacterial activity against human and plant pathogenic bacteria. Asian J Pharm Technol Innov 2(7)

Chandran SP, Chaudhary M, Pasricha R, Ahmad A, Sastry M (2006) Synthesis of gold nanotriangles and silver nanoparticles using *Aloevera* plant extract. Biotechnol Prog 22:577–583

Chin SS, Chiang K, Fane AG (2006) The stability of polymeric membranes in a TiO_2 photocatalysis process. J Membr Sci 275(1–2):202–211

Cirtiu CM, Dunlop-Briere AF, Moores A (2011) Cellulose nanocrystallites as an efficient support for nanoparticles of palladium: application for catalytic hydrogenation and Heck coupling under mild conditions. Green Chem 13(2):288–291

D'Britto V et al (2012) Medicinal plant extracts used for blood sugar and obesity therapy shows excellent inhibition of invertase activity: synthesis of nanoparticles using this extract and its cytotoxic and genotoxic effects. Int J Life Sci Pharma Res 2:61–74

Dan Z, Xin-lei M, Yan G, He H, Guang-wei Z (2020) Green synthesis of metallic nanoparticles and their potential applications to treat cancer. Front Chem 8:799. https://doi.org/10.3389/fchem.2020.00799

Daniel S, Shabudeen PSS (2014) Sequestration of carcinogenic dye in wastewater by utilizing an encapsulated activated carbon with nano MgO. Int J Chemtech Res 7:2235–2243

Danwittayakul S et al (2013) Enhancement of photocatalytic degradation of methyl orange by supported zinc oxide nanorods/zinc stannate (ZnO/ZTO) on porous substrates. Ind Eng Chem Res 52(38):13629–13636

De Britto AJ, Gracelin DHS, Kumar PBJR (2012) Biogenic silver nanoparticles by Adiantum caudatum and their antibacterial activity. Int J Univ Pharm Life Sci 2(4):92–98

Deng Y, Zhao R (2015) Advanced oxidation processes (AOPs) in wastewater treatment. Curr Pollut Rep 1(3):167–176. https://doi.org/10.1007/s40726-015-0015-z

Dequigiovanni G et al (2018) New microsatellite loci for annatto (Bixa orellana), a source of natural dyes from Brazilian Amazonia. Crop Bree Appl Biotechnol 18(1):116–122

Dhandapani P, Maruthamuthu S, Rajagopal G (2012) Biomediated synthesis of TiO_2 nanoparticles and its photocatalytic effect on aquatic biofilm. Jr Photochem Photobio B: Bio 110:43–49

Eleen DMI et al (2021) Photocatalytic degradation with green synthesized metal oxide nanoparticles—a mini review. J Nanosci Nanotechnol 2(1):70–81

Fagier MA (2020) Plant-mediated biosynthesis and photocatalysis activities of zinc oxide nanoparticles: a prospect towards dyes mineralization. Jr Nanotech 2021:1–12

Feng X-X et al (2007) Preparation and characterization of novel nanocomposite films formed from silk fibroin and nano-TiO_2. Int J Biol Macromol 40(2):105–111

Gunduz O (2014) A simple method of producing hydroxyapatite and tri calcium phosphate from coral (Pocillopora verrucosa). J Aust Ceram Soc 50(2):52–58

Hardani K, Buazar F, Ghanemi K et al (2015) Removal of toxicmercury (II) from water via Fe_3O_4/hydroxyapatite nanoadsorbent:an efficient, economic and rapid approach. AASCIT Jr Nanosci 1(1):11–18

Hong ZC et al (2009) Surface enhanced Raman scattering of nano diamond using visible-light-activated TiO_2 as a catalyst to photo-reduce nano-structured silver from $AgNO_3$ as SERS-active substrate. J Raman Spectrosc Int J Original Work All Asp Raman Spectrosc including Higher Order Processes Also Brillouin Rayleigh Scattering 40(8):1016–1022

Huang J, Li Q, Sun D, Lu Y, Su Y, Yang X, Wang H, Wang Y, Shao W, He N, Hong J, Chen C (2007) Biosynthesis of silver and gold nanoparticles by novel sundried *Cinnamomum camphora* leaf. Nanotechnology 18:105104–105114

Isa EDM, Shameli K, Jusoh NWC, Sukri SNAM, Ismail NA (2021a) Photocatalytic degradation with green synthesized metal oxide nanoparticles-mini-review. Jr Res Nanosci Nanotech 2(1):70–81

Isa EDM, Shameli K, Jusoh NWC, Hazan R (2021b) Rapid photodecolorization of methyl orange and rhodamine B using zinc oxide nanoparticles mediated by pullulan at different calcinations conditions. J Nanostructure Chem 11(1):187–202. https://doi.org/10.1007/s40097-020-00358-6

Jagpreet S, Tanushree D, KiHyun K, Mohit R, Pallabi S, Pawan K (2018) 'Green' synthesis of metals and their oxide nanoparticles: applications for environmental remediation. J Nanobiotechnol 16:84. https://doi.org/10.1186/s12951-018-0408-4

Jha AK, Prasad K (2010) Green synthesis of silver nanoparticles using Cycas leaf. Int J Green Nanotechnol Phys Chem 1(2):P110–P117

Joshi K, Shrivastava V (2011) Photocatalytic degradation of Chromium (VI) from wastewater using nanomaterials like TiO_2, ZnO, and CdS. App Nanosci 1(3):147–155

Kalpana D et al (2014) Green biosynthesis of silver nanoparticles using Torreya nucifera and their antibacterial activity. Arab J Chem 2014:1–11

Karunakaran C, Pazhamalai V (2018) CdO-intercalated TiO_2 nanosphere-clusters: synthesis and electrical Optical and Photocatalytic Properties. Catal Today 284:114. https://doi.org/10.1007/s12633-018-9832-1

Korbekandi H et al (2014) Green biosynthesis of silver nanoparticles using Azolla pinnata whole plant hydroalcoholic extract. Green Process Synth 3(5):365–373

Lee J, Park H, Choi W (2002) Selective photocatalytic oxidation of NH_3 to N_2 on platinized TiO_2 in water. Environ Sci Technol 36(24):5462–5468

Liang B, Zhang W (2019) $Sn_{21}Cl_{16}(OH)_{14}O_6$: a promising novel photocatalyst for methyl orange degradation. Mater Res Express 6:115066. https://doi.org/10.1088/2053-1591/ab4930

Lin L, Wang W, Huang J, Li Q, Sun D, Yang X, Wang H, He N, Wang Y (2010) Nature factory of silver nanowires: Plant-mediated synthesis using broth of *Cassia fistula* leaf. Chem Eng J 162:852–858

Liu S et al (2008) Removal of humic acid using TiO_2 photocatalytic process–fractionation and molecular weight characterisation studies. Chemosphere 72(2):263–271

Mane PV, Shinde NB, Mulla IM, Koli RR, Shelke AR, Karanjkar MM, Gosavi SR, Deshpande NG (2018) Bismuth ferrite thin film as an efficient electrode for photocatalytic degradation of Methylene blue dye. Mater Res Express 6:026426

Mao C et al (2003) Viral assembly of oriented quantum dot nanowires. Proc Natl Acad Sci 100(12):6946–6951

Mishra PM et al (2014) Biomimetic synthesis of silver nanoparticles by aqueous extract of Cinnamomum tamala leaves: optimization of process variables. Nanosci Nanotechnol Lett 6(5):409–414

Muhammad BT et al (2020) Role of nanotechnology in photocatalysis. In: Reference module in materials science and materials engineering. Elsevier. https://doi.org/10.1016/B978-0-12-815 732-9.00006-1. ISBN 9780128035818

Mukherjee P et al (2002) Extracellular synthesis of gold nanoparticles by the fungus Fusarium oxysporum. ChemBioChem 3(5):461–463

Müller WE et al (2009) Bio-sintering processes in hexactinellid sponges: fusion of bio-silica in giant basal spicules from Monorhaphis chuni. J Struct Biol 168(3):548–561

Nasrollahzadeh M, Sajadi SM (2015) Green synthesis of copper nanoparticles using Ginkgo biloba L. leaf extract and their catalytic activity for the Huisgen [3+2] cycloaddition of azides and alkynes at room temperature. J Colloid Interface Sci 457:141–147

Noruzi M, Zare D, Davoodi D (2012) A rapid biosynthesis route for the preparation of gold nanoparticles by aqueous extract of cypress leaves at room temperature. Spectrochim Acta A Mol Biomol Spectrosc 94:84–88

Osuntokun J, Onwudiwe DC, Ebenso EE (2019) Greensynthesis of ZnO nanoparticles using aqueous *Brassicaoleracea* L. var. *italica*and the photocatalytic activity. Green Chem Lett Rev 12(4): 444–457

Pérez-de-Mora A, Burgos P, Madejón E, Cabrera F, Jaeckel P, Schloter M (2006) Microbial community structure and function in a soil contaminated by heavy metals: effects of plant growth and different amendments. Soil Bio Biochem 38(2):327–341. https://doi.org/10.1016/j.soilbio.2005.05.010

Petrovska BB (2012) Historical review of medicinal plants'usage. Pharmacogn Rev 6(11):1

Rathnasamy R, Angasamy P, Angamuthu R et al (2017) Green synthesis of ZnO nanoparticles using Carica papaya leaf extracts for photocatalytic and photovoltaic applications. J Mater Sci Mater Electron 28:10374–10381

Ratna PBS(2012) Pollution due to synthetic dyes toxicity and carcinogenicity studies and remediation. Int J Environ Sci 3:940–55. https://doi.org/10.6088/ijes.2012030133002

Ratul KD, Mitra N, Vinayak LP et al (2017) Biological synthesis of metallic nanoparticles: plants, animals and microbial aspects (critical reviews). Nanotechnol Environ Eng 2:18

Riddin T, Gericke M, Whiteley C (2006) Analysis of the interand extracellular formation of platinum nanoparticles by Fusarium oxysporum f. sp. lycopersici using response surface methodology. Nanotechnology 17(14):3482

Roy K, Sarkar C, Ghosh C (2015a) Photocatalytic activity of biogenic silver nanoparticles synthesized using yeast (Saccharomyces cerevisiae) extract. Appl Nanosci 5(8):953–959

Roy K, Sarkar C, Ghosh C (2015b) Photocatalytic activity of biogenic silver nanoparticles synthesized using potato (Solanum Tuberosum) infusion. Spectrochim Acta A Mol Biomol Spectrosc 146:286–291

Saharan VK, Pinjari DV, Gogate PR, Pandit AB (2014) Advanced oxidation technologies for wastewater treatment. In : Vivek VR, Vinay MB (eds) Industrial wastewater treatment, recycling and reuse. Elsevier, UK, pp 141–191
Sahu O, Singh N (2019) Significance of bioadsorption process on textile industry wastewater. In: Shahid-ul-Islam, Butola BS (eds) The impact and prospects of green chemistry for textile technology. Woodhead Publishing, pp 367–416
Samanta S, Pyne S, Sarkar P, Sahoo GP, Bar H, Bhui DK, Misra A (2010) Synthesis of silver nanostructures of varying morphologies through seed mediated growth approach. J Mol Liq 153(2–3):170–173. https://doi.org/10.1016/j.molliq.2010.02.008
Samuel HSC, Ta YW, Joon CJ, Chee YT (2011) Recent developments of metal oxide semiconductors as photocatalysts in advanced oxidation processes (AOPs) for treatment of dye waste-water. J Chem Technol Biotechnol 86:1130–1158. https://doi.org/10.1002/jctb.2636
Sangeetha G, Rajeshwari S, Venckatesh R (2012) Green synthesized ZnO nanoparticles against bacterial and fungal pathogens. Prog Nat Sci 22(6):693–700. https://doi.org/10.1016/j.pnsc.2012.11.015
Sato T, Sato KI, Fujishiro Y (1996a) Photochemical reduction of nitrate to ammonia using layered hydrous titanate/cadmium sulphide nanocomposites. J Chem Technol Biotechnol 67:345–349
Sato T, Masaki SKI et al (1996b) Photocatalytic properties of layered hydrous titanium oxide/CdS–ZnS nanocomposites incorporating CdS–ZnS into the interlayer. J Chem Technol Biotechnol 67:339–344
Senapati S et al (2005) Extracellular biosynthesis of bimetallic Au–Ag alloy nanoparticles. Small 1(5):517–520
Shenton W et al (1999) Inorganic–organic nanotube composites from template mineralization of tobacco mosaic virus. Adv Mater 11(3):253–256
Sheshmani S, Ashori A, Hasanzadeh S (2014) Removal of acid orange 7 from aqueous solution using magnetic graphene/chitosan: a promising nano-adsorbent. Int J Biol Macromol 68:218–224
Shiying H, Zhong L, Duan J, Feng Y, Yang B, Yang L (2017) Bioremediation of wastewater by iron oxide-biochar nanocomposites loaded with photosynthetic bacteria. Front Microbiol 8:823. https://doi.org/10.3389/fmicb.2017.00823
Singaravelu G, Arockiamary J, Ganesh K, Govindaraju K (2007) A novel extracellular synthesis of monodisperse gold nanoparticles using marine alga, Sargassum wightii Greville. Colloids Surf B Biointerfaces 57:97–101
Singh P (2013) A simple, rapid, and green synthesis of capped gold nanospheres and nanorods using aqueous extract of azolla. Int J Green Nanotechnol 1:1–5
Singha J, Kaura S, Kaur G, Basu S, Rawat M (2019) Biogenic ZnO nanoparticles: a study of blue shift of optical band gap and photocatalytic degradation of reactive yellow 186 dye under direct sunlight. Green Process Synth 8(1):272–280
Sonker RK, Hitkari G, Sabhajeet SR, Sikarwar S, Rahul SS (2020) Green synthesis of TiO_2 nanosheet by chemical method for the removal of Rhodamin B from industrial waste. Mater Sci Eng B 258:114577. https://doi.org/10.1016/j.mseb.2020.114577
Stupp SI, Braun PV (1997) Molecular manipulation of microstructures: biomaterials, ceramics, and semiconductors. Science 277(5330):1242–1248
Varadavenkatesan T, Lyubchik E, Pai S, Pugazhendhi A, Vinayagam R, Selvaraj R (2019) Photocatalytic degradation of Rhodamine B by zinc oxide nanoparticles synthesized using the leaf extract of Cyanometra ramiflora. J Photochem Photobiol B 199:111621. https://doi.org/10.1016/j.jphotobiol.2019.111621
Vasantharaj S, Sathiyavimal S, Saravanan M, Senthilkumar P, Gnanasekaran K, Shanmugavel M, Pugazhendhi A (2019) Synthesis of ecofriendly copper oxide nanoparticles for fabrication over textile fabrics: characterization of antibacterial activity and dye degradation potential. J Photochem Photobiol B Biol 191:143–149. https://doi.org/10.1016/j.jphotobiol.2018.12.026
Velmurugan P et al (2013) Pine cone-mediated green synthesis of silver nanoparticles and their antibacterial activity against agricultural pathogens. Appl Microbiol Biotechnol 97(1):361–368

Vigneshwaran N et al (2007) Biological synthesis of silver nanoparticles using the fungus Aspergillus flavus. Mater Lett 61(6):1413–1418

Vithiya K, Sen S (2011) Biosynthesis of nanoparticles (review article). Int J Pharm Sci Res 2(11):2781–2785

Wang G et al (2008) Preparation of nano silk fibroin/hydroxyapatite biological composite by '"one-step"' method. Acta Materiae Compositae Sinica 6:027

Wang T et al (2014) Green synthesis of Fe nanoparticles using eucalyptus leaf extracts for treatment of eutrophic wastewater. Sci Total Environ 466:210–213

Wilbur KM, Simkiss K (1979) Carbonate Turnover and Deposition by Metazoa. In: Trudinger PA, Swaine DJ (eds) Studies in environmental science, vol 3. Elsevier, pp 69–106

Xiao G, Su H, Tan T (2015) Synthesis of core–shell bioaffinity chitosan–TiO_2 composite and its environmental applications. J Hazard Mater 283:888–896

Chapter 10
Mechanistic Aspect of the Dye Degradation Using Photocatalysts

Soumya Biswas, Saikat Ghosh, Suparna Maji, Soumyadipta Das, Subhrojeet Singha Roy, Rahul Bhattacharjee, Priya Mitra, Sumira Malik, and Abhijit Dey

Abstract It is estimated that there are approximately 1 lakh commercially accessible dyes, with an annual production of more than 750 thousand metric tonnes by a variety of industries such as textiles, carpets, paint, leather, and so on. Because of their widespread use, they have the undesirable side effect of causing up to 12% of their dyes to be wasted during the dyeing process. Additionally, around 20% of this waste reaches the environment, primarily through the water supply system. It should come as no surprise that a variety of ways have been developed to breakdown and eliminate these hazardous dyes from the environment. The dye removal from industrial and municipal water effluents using a cost-effective technique is thus becoming more important in light of increasing environmental problems and environmental consciousness. It has been discovered that these dyes are resistant to both biological and physical treatment procedures in the majority of cases. For dye degradation research, advanced oxidation processes (AOPs) seem to be the most used method. Over the last several years, a substantial number of works has been done in the area of AOPs, and as a consequence, a variety of different types of AOPs have been produced. An effective remedy for dye pollution is photocatalyzed degradation of dyes utilizing several types of combined or altered photocatalysts. As part of this chapter, we will cover the numerous kinds of dyes and their poisonous effects on diverse organisms. We will also explore the various mechanistic aspects of photocatalytic activity for the degradation of dyes, as well as the various types of synthesis procedures for photocatalysts. This chapter also includes techniques for analyzing

Soumya Biswas and Saikat Ghosh Both have the equal contribution as first author.

S. Biswas · S. Ghosh · S. Maji · S. Das · S. S. Roy · R. Bhattacharjee · P. Mitra
KIIT School of Biotechnology, Kalinga Institute of Industrial Technology (KIIT-DU), Bhubaneswar, Odisha, India

S. Malik
Amity Institute of Biotechnology, Amity University Jharkhand, Ranchi, Jharkhand 834001, India

A. Dey (✉)
Department of Life Sciences, Presidency University, 86/1 College Street, Kolkata, West Bengal 700073, India
e-mail: abhijit.dbs@presiuniv.ac.in

S. Dave and J. Das (eds.), *Trends and Contemporary Technologies for Photocatalytic Degradation of Dyes*, Environmental Science and Engineering,
https://doi.org/10.1007/978-3-031-08991-6_10

dye degradation and intermediate product identification in photocatalyst-mediated reactions, as well as variables impacting photocatalytic activity for degradation.

Keywords Phytoremediation · Nanoparticles · Phyto-degradadtion · Photolytic treatment

10.1 Introduction

Environmental contamination and degradation are today's most serious challenges impacting the globe. Continuous discharge of hazardous and poisonous industrial wastes into the air, water, and soil bodies, as well as the expansion of agricultural activities, have increased energy consumption, resulting in pollution-related ailments, global warming, and other difficulties. Owing to the vast amounts of water used in fabric processing, industries are the biggest polluters, with the textile industry generating the most liquid effluent pollutants. Water pollution has become a widespread problem that has harmed human health and exacerbated the scarcity of water supplies (Akpan and Hameed 2009). The presence of undesired industrial hazardous waste in water bodies, such as heavy metals, dyes, phenols, suspended solids, toxic organics, etc., is catastrophic as they pose a serious threat to microorganisms, living beings (Borker and Salker 2006; Dave et al. 2021a; Satapathy et al. 2021; Dave et al. 2021b). The textile sector being the leading consumer of dye is a major source of water contamination. Other industries that produce highly colored wastewaters include paper mills, the food industry, hair colorings, photoelectrochemical cells, distilleries, and tanneries.

Dye is a coloring agent that chemically associates with the substance to which it is applied to impart color (Nidheesh et al. 2013). It is colored because it absorbs light at a certain wavelength in the visible range of the spectrum (Shindy 2017). These are usually complex organic compounds that can withstand a variety of chemicals, including detergents. Dyes can be manufactured naturally from sources like vegetables, flowers, leaves, animals, and insects. However, they have a short shelf life, are easily affected by environmental aspects such as exposure to sunlight and rain as the dyes gets faded. Moreover, the quantity of production was low, and the variety of colors produced was limited. Thus, synthetic dyes eventually replaced natural dyes and were developed for coloring various materials in various industries such as leather, textiles, cosmetics, paper, and so on. Around 100,000 synthetic dyes are commercially available and are widely utilized in textile and other industries. The two primary sources of these dyes are petroleum intermediates and coal tar, with a combined annual output of more than 7×10^5 tonnes (Turhan and Turgut 2009; Mani and Bharagava 2018). Around 15% (1000 tonnes) of these non-biodegradable dyes are emitted into natural streams and aquatic bodies each year during manufacturing or processing activities by various industries (Javaid et al. 2021; Carmen and Daniela 2012; Yagub et al. 2014). The majority of dyes are synthetic and contain complex organic structures with two components, chromophores and auxochromes (Hasanpour and Hatami 2020). Chromophores are electron receivers containing nonbonding electrons on heteroatoms such as S, N, and O. These chromophores are primarily responsible for the dye's color. Most widely present chromophore groups are as follows N=N, =C=O, C=S, etc. Auxochromes, on the other hand, are electron donors and are responsible for color intensity. It elevates the dye's solubility, which improves its color. Auxochrome groups include NH_2, COOH, NR_2, OCH_3, and others. Due to the inclusion of these two components, synthetic dyes are stable and complicated. The color of these dyes do not fade away and are resistant to washing with soap, detergents, and other chemicals (Gupta and Mondal 2021). Synthetic dyes have a wide range of chemical properties due to their structural variety as well as physical properties. The most prevalent chemical groups of dyes used on a wide scale are azo, anthraquinone, indigoid, xanthene, aryl methane, and various phthalocyanine derivatives (Shindy 2017; Benkhaya et al. 2020a). These synthetic dyes are non-biodegradable and non-biocompatible, and they have a significant influence on aquatic ecology. Based on their chemical composition or the nature of the chromophore, dyes are classed as azo dyes, oxazine dyes, indigoid dyes, nitroso dyes, nitro dyes, phthalocyanine dyes, and so on. Besides, some dyes, such as basic, acidic, direct, mordant, and reactive dyes, are soluble in some solvents. Several techniques for both qualitative and quantitative dye degradation analysis have been evolved in years such as FTIR, Raman spectroscopy, GC/MS, HPLC. HPLC coupled with mass spectrometry has served greater results (Elert et al. 2017).

Among the primary dye groups, azo dyes are the most prevalent, accounting for more than half of all the dyes used in the industry. Azo dyes are distinguished by a double nitrogen bond, where one of the nitrogen atoms is linked to an aromatic family (rings of naphthalene or benzene). The azo bonds and their corresponding chromophores and auxochromes influence the pigment of azo dyes. The active bonds in azo dye compounds are those that may be oxidized by a positive hole or hydroxyl ions or reduced by an electron in the conduction band. The dissociation of N=N bonds results in the formation of dye decolorization (Ong et al. 2012). Azo dyes and other related dyes are carcinogenic owing to the emission of aromatic amines during the degradation process. Cloth dyes that release carcinogenic amines when in direct contact with the human body are leached/absorbed out of the cloth by perspiration. The absorbed dyes change the metabolic processes of the skin, which might cause allergic and carcinogenic reactions (Yusuf et al. 2015; de Wolf et al. 2013). When azo and aromatic amine-based dyes are digested by liver enzymes, they are shown to be hazardous. Water or sweat pores received via the skin and other exposed regions might mobilize aromatic amines. Ingestion of dye is also quicker and potentially more harmful since more dye may be absorbed in a shorter time (Sahoo and Patra 2020, 2018).

The usage of these chemicals also has a deleterious influence on the soil microbial populations as well as plant germination and growth (Imran et al. 2015; Rehman et al. 2018). These dyes in industrial effluents restrict the penetration of light in rivers, limiting the photosynthetic activities of aquatic vegetation and, as a result, the food supply of aquatic creatures is also restricted. Dye pollutants raise the biological oxygen demand (BOD) and chemical oxygen demand (COD) of the waterbody, which reduces the reoxygenation process and stymies the growth and development process of photoautotrophic organisms (Mani and Bharagava 2018; Tkaczyk et al. 2020).

Dyes containing carcinogens such as benzidine, naphthalene, and other aromatic intermediate compounds are reported to be poisonous, carcinogenic, genotoxic, teratogenic, and mutagenic, and may have long-term consequences on flora, fauna, and human health (Holkar et al. 2016). For example, Triphenylmethane dyes like crystal violet and acid violet are cytotoxic to mammalian cells, phytotoxic to plants, and may promote tumor development in some fish species (Suteu et al. 2011).

As a result, strategies for the degradation of dyes are progressively being researched and developed. Textile wastewater is treated using a variety of approaches, including biological, chemical, and physical processes, depending on the kind of contaminants present. Biodegradation of synthetic dyes by diverse bacteria is developing as a viable and promising option among existing pollution management solutions. Microbes and their extracellularly secreted enzymes such as laccases, azoreductases, and peroxidases work on synthetic dyes, these enzymes transform and even mineralize the fiber (Imran et al. 2015; Paździor et al. 2017; Lellis et al. 2019). Biological approaches have been shown to have bioremediation potential as future alternatives to traditional physicochemical techniques all over the world (Ali 2010). However, the biological treatments are time-consuming, generate a substantial volume of sludge, and need constant pH control as well as temperature. The chemical methods using activated carbon are very effective in removing the synthetic dye. However, it is suitable for treating only small effluent volumes (Fito et al. 2020). Besides, the operation requires a long time and high capital cost. However, it cannot remove all organic compounds as it is only appropriate for treating tiny volumes of wastewater. Furthermore, the operation involves a long time and a significant investment. Physical treatments such as coagulation, flocculation, and reverse osmosis were formerly used for dye degradation. The primary drawback of these approaches is the generation of additional waste products which causes further toxicity in the atmosphere (Madhav et al. 2018). Despite the presence of numerous successful systems utilizing a variety of physicochemical and biological methods, dye removal from effluents at a reasonable cost remains a major challenge. Because environmental concerns are becoming more widespread, it is critical to creating more cost-effective and ecologically friendly treatment solutions. To remove harmful contaminants from water and wastewater, several methods have been suggested and explored, including ion exchange, adsorption, membrane processes, advanced oxidation processes, and photocatalysis.

Among these methods, photocatalysis is regarded as a green water treatment technique since the mechanism depends on light irradiation to convert toxic pollutants to non-toxic pollutants. It is simply described as the acceleration of a photoreaction when a catalyst is present. Photocatalytic dye degradation has been studied since the early 1900s, and since the final part of the twentieth century, it is being used because it is a promising method for dye degradation (Sacco et al. 2012). Photocatalytic degradation of organic contaminants is a promising technique because it allows toxins to be degraded rather than transformed in the environment. The approach has the potential to remove a wide range of organic pollutants, including dyes, insecticides, herbicides, and micropollutants such as bacteria. During the dye degradation process, larger dye molecules are oxidized into smaller molecules such as water, carbon dioxide, and various mineral by-products. However, during photocatalysis some harmful and toxic chemicals are also produced, these by-products are considered more hazardous than the parent molecular compound. Several characterization techniques such as HPLC, GC/MS, LC/MS, UV-viz spectroscopy have been utilized to identify the physical and chemical properties of the by-products. Even at the lowest concentrations, this approach is effective in removing pollutants from industrial wastewater. As the dyeing process does not use all of the dye molecules, a significant number of dyes were found in the wastewater discharged by the industry. Dyes with a complex composition have poor photo degradability, and the presence of functional groups affects adsorption properties (Jorfi et al. 2018). Photocatalysis is an extremely advanced oxidation process (AOP) that is used to photodegrade hazardous chemicals. It is also utilized in water treatment (Deng et al. 2012). Photocatalysis is classified into two categories: (1) heterogeneous catalysis and (2) homogeneous catalysis. Heterogeneous photocatalysis is a contemporary technique that is commonly used for the degradation or bleaching of dyes (Pandit et al. 2015). In the presence of a semiconductor catalyst, it causes photoreactions to accelerate. When a semiconductor surface, principally oxides, and sulfides, is irradiated with a sufficient wavelength of light, electrons are transported from the valence band to the conduction band. When these excitons combine with oxygen or water, superoxide anions and hydroxide radicals are produced. These species have the ability to destroy a wide range of compounds, including commercial colors, due to their great oxidizing power (Elmorsi et al. 2010). Photocatalytic dye degradation is primarily concerned with factors such as dye concentration, photocatalyst quantity used, the influence of irradiated light intensity, frequency of irradiation, and effect of dissolved oxygen and other species. The photocatalytic degradation of dyes is thought to be a pseudo-first-order process, with kinetic data fitting to the equation $-\ln (C/C0) = kt$ (Yagub et al. 2014; Viswanathan 2018).

During the reaction, the photocatalysts are not consumed (Sreethawong 2012). TiO_2, ZnO, Fe_2O_3, SnO_2, CdS, and other powdered metal oxides have been used as photocatalysts for organic pollutant/dye degradation. TiO_2 and ZnO are the most often utilized bulk photocatalysts in photocatalysis. Photo-induced electrons and holes interacted with oxygen (O_2), water (H_2O), and hydroxyl groups to produce reactive oxygen species (ROS) with high oxidation capacities, such as hydroxyl radicals (OH) and superoxide radicals anions (O_2). These ROS are the primary species responsible for the breakdown of persistent organic contaminants in wastewater (Han et al. 2018). Photocatalysts are responsible for utilizing solar energy to degrade pollutants, making the photocatalysis treatment procedure economically feasible (Khalid et al. 2017). Photocatalytic semiconductors can be synthesized as powders, fibers, and films using a variety of synthetic methods such as the sol–gel process, hydrothermal and solvothermal techniques, direct oxidation reactions, sonochemical methods, microwave methods, chemical vapor deposition methods, and electrodeposition methods, among others (Medina-Ramírez et al. 2015). It has been observed that the preparation procedure utilized in the manufacturing of semiconductors has a significant impact on their physicochemical characteristics and photocatalytic activity. Conventional chemical, physical, and biological techniques have been widely used to remediate waste water containing color molecules.

These approaches have several drawbacks, including a high cost, a high energy need, and the creation of secondary contaminants during the treatment process. However, the most essential feature of photocatalysis is that it destroys pollutants at a lower cost (Viswanathan 2018). The deteriorated CO_2 and H_2O are the most common by-products, both of which are non-toxic (Wang et al. 2011a). It requires extremely active and affordable catalysts that are activated in specifically constructed reactors (Ibhadon and Fitzpatrick 2013).

10.2 Classification of Dyes

The dye is generally classified into different groups depending on the general dye structure, source, and fiber type shown in Fig. 10.1. Generally, dyes are classified

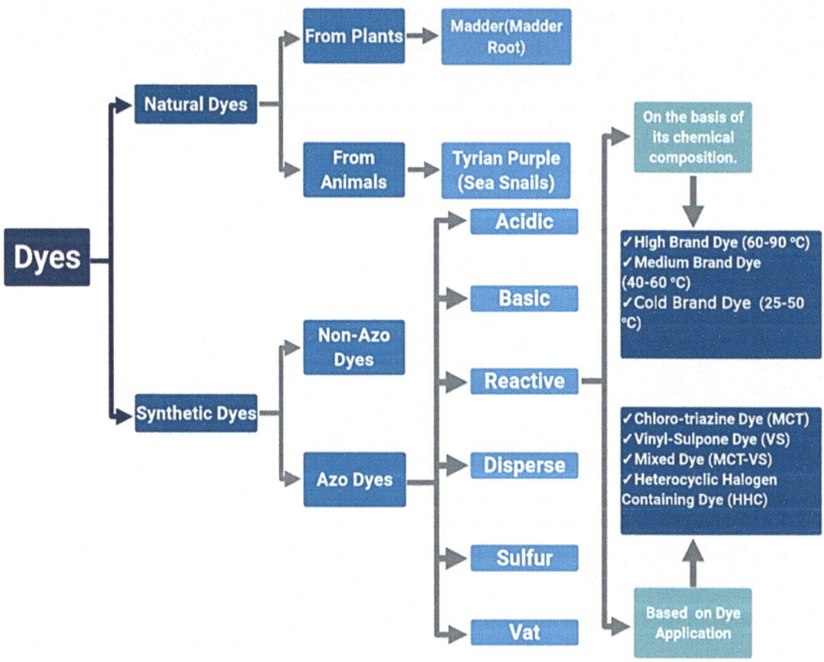

Fig. 10.1 Types of dyes adapted from Benkhaya et al. 2020b

based on their practical application in textile processing. Some of the common dyes used in industries are described below.

Azo dyes: The largest group of colorants are azo dyes, overall 50% of dyes used in industries are azo dyes (Ajmal et al. 2014). Azo dyes are the important colorants mostly used in textile, paper industries, manufacturing, etc. The largest production volume with a functional group (–N=N–) (Benkhaya et al. 2020b).

Reactive dyes: Water-soluble anionic dyes are also referred to as "Fiber reactive dyes". Generally applied to the cellulosic fiber, which from a solution with high pH or neutral solution, but today other fibers are also dyed using reactive dyes. Total 95% of reactive dyes are azo dyes.

Disperse dyes: Water-insoluble non-ionic dyes, with smaller size and planar structure. Usually solid with very fine particle size ($\leq 1\ \mu m$) and large surface area. It attached polar functional group.

Acid dyes: Water-soluble anionic dyes are used to dye natural protein, synthetic polyamide, and for small extent acrylic and blend fibers. Commercially available acid dyes are azo, triphenyl methane, and azine based.

Sulfur dyes: Water-insoluble non-ionic dye mostly used to produce bright shade on cellulosic fibers. Dye possesses sulfur linkage, applied in solubilized and reduced anionic state.

Basic dyes: Water-soluble cationic dye, which is salt of organic bases. Show very high affinity toward silk, wool, and cationic dyeable acrylic but basic dye has an affinity toward cellulosic.

10.3 Dye-Related Toxicity

Complex structural and chemical properties of dyes are demanding for biological decomposition. Dye is generally introduced into the surrounding environment through industrial effluents, textile, food, and cosmetic industries. Excessive use of dyes has become a concern for the environment. Textile dyes are highly toxic due to the presence of heavy metals (Sani et al. 2018). Textile industries mainly consume water and pollute water bodies by increasing chemical oxygen demand (COD) and biochemical oxygen demand (BOD). It also impairs photosynthesis, inhibits plant growth and bioaccumulation, and may promote carcinogenicity and toxicity (Lellis et al. 2019). Azo dyes are mostly used in textile and leather industries which contaminate soil and groundwater. To avoid azo dyes contamination researchers working on different groups of bacteria for bioremediation dye-polluted effluents (Mahmood et al. 2015). Studies have been conducted in different laboratories to reduce environmental toxicity and adverse effect on human health. Fugal laccase is such an example of a synthesis dye that is eco-friendly in nature (Polak et al. 2016).

10.4 Techniques for Investigating Dye Degradation

Dye degradation has a negative impact on the environment's equilibrium. Furthermore, these dyes are hazardous since they all contain aromatic groups (Rauf and Ashraf 2012). Many of the dyes such as azo dyes have shown several hazardous properties. Upon entering the water bodies these chemicals cause serious health concerns to all aquatic organisms as well as human beings (Khan and Malik 2014). Different experimental methodologies for both qualitative and quantitative examination of various dye degradation have been developed. These techniques often require instrumental methodology, such as HPLC (high-pressure liquid chromatography), ion chromatography, capillary electrophoresis LC/MS (Liquid chromatography-mass spectroscopy), GC–MS (gas chromatography-mass spectrometry), UV–vis spectrophotometry, radiometry, and other are used (Parmar and Sharma 2016). The molecular compounds are characterized by using commercial standards at a defined retention time. These methods may be linked to a mass spectrometer to give a high degree of chemical specificity (Beck et al. 2012). However, HPLC has been used in the investigation of natural dyes in ancient fabrics, it has permitted accurate identification of natural colors. The diode array UV–Vis detector is the most often employed detector in the HPLC study of dyes (Zasada-Kłodzińska et al. 2021). Because of spectral and chromatographic overlap, chromatographic detection and identification of dyes in a mixture takes a significantly longer time and also involves a prior separation. As a result, UV–vis spectrophotometric determination is favored, as compared to chromatographic technique, it is possible to get rapidly high accuracy and repeatability from complicated matrices using a very simple and affordable approach (Pirok et al. 2018). Though UV/vis spectrometry has been used to analyze dye degradation, it only provides a general classification of dye colors and is not specific enough for a detailed structural characterization (Carey et al. 2013; Jia et al. 2016). The Fourier transform infrared (FTIR) technique is a critical tool for determining the characteristics of current surface functional groups that are important in the adsorption and degradation of dye pollutants (Duy Nguyen et al. 2019). The "FT" approach, when integrated into an FTIR spectrometer, has the significant advantage of allowing small samples to be used due to the instrument's improved sensitivity and generating a nearly immediate spectrum for each sample. The identification of functional groups that correspond to phenolic compounds related to the $-CH_3$, –H, and $-OCH_3$ groups was made possible by FT-IR analysis. (Wrobel and Bhargava 2018). Raman spectroscopy techniques have been employed to characterize particles. The chemical nature of dye molecules has been studied using Raman spectroscopy. It has several benefits, including high sensitivity and chemical specificity for dye molecules in general, as well as a simple, rapid, and non-destructive analysis method. Raman spectrometric analyses, on the other hand, are frequently hampered by fluorescence and may be limited in their ability to distinguish between a variety of dyes (Massonnet et al. 2012). A high-performance liquid chromatography—iode array detector (HPLC–DAD) in combination with mass spectrometry is found to be excellent to study dye degradation. The process is delicate enough to examine single strands with a few millimeters in length millimeters or less, making it appropriate for forensic use. In a single analytical run, the HPLC–DAD-(high resolution) MS instrument can separate and detect a variety of dyes, including neutral, positively charged, and negatively charged dyes, dyes with multiple charges, dyes with single charges, hydrophilic and hydrophobic dyes, and dyes in a variety of sizes and molecular masses (Carey et al. 2013).When a heterogeneous catalyst or solid reactant is used, preparatory methods of extraction of the aqueous sample with an organic solvent or filtration are used. Because the majority of the dyes will eventually break down into smaller molecules, determining the total organic carbon of the system under study is also a crucial part (Rauf et al. 2011). A robustly churned batch photochemical cell, cylindrical flasks or jars, and glass plates are among the most common laboratory-scale experimental setups. On the other hand, photo-catalytic reactors for dye investigations have been described using up flow type, membrane based, and coated surfaces (Al-Dawery 2013).

10.5 Strategies for Dealing with Dye

Dyes are known to have their applications in a variety of fields. Categorically the azo dyes can be used for dyeing polyesters, while sulfur dyes are mostly used for dyeing cotton, and further dyes are even used in the pharmaceutical industries as a coloring agent for different drugs. Apart from these, dyes have their potential usages in different industries. Though dyes have many uses, the main disadvantage is that it is one of the most significant contributors to water pollution since it contains harmful poisonous compounds. The presence of large quantities of these toxic dye components in wastewater poses a significant environmental danger (Sarkar et al. 2017). While some dyes aren't that much hazardous, others, predominantly the azo dyes are carcinogenic. This is due to the fact dyes are composed of many compounds including heavy metals (lead, chromium), aromatics, amines, and other substances, promoting their toxic nature (Lazar 2005). Due to the toxic and carcinogenic effects of dyes, they must be handled carefully and degraded properly before being discharged as effluents into the water bodies by the industries. Thus, innovative physical, chemical and biological degradation technologies have been adopted to address this issue (Mallakpour and Rashidimoghadam 2019).

10.5.1 Physical Mechanistic Procedures

Physical treatments are simple procedures for physically transferring dyes from one phase to another rather than degrading them, with little or no change in their chemical or biological features. These technologies entail a mass transfer mechanism and are more efficient when the effluent volume is reduced (Samsami et al. 2020). Physical methods have several advantages, including low cost, straightforward operation, the ability to handle even very dangerous compounds, and increased efficacy in treating pollutants. In comparison to chemical and biological processes, this process is more predictable because of the lower use of chemicals and the absence of live creatures (Khan et al. 2018). Thus, since dyes do not decompose, this technique has the problem of causing secondary contamination. The contamination occurs when the pollutant gets transferred from one phase to another. Ion exchange, adsorption, and membrane filtration are among the most common physical dye removal procedures present (Katheresan et al. 2018).

10.5.1.1 Ion Exchange Dye Degradation

The type of physical method for dye removal from wastewater that has obtained traction is ion exchange. Ion exchange resins are polymer compounds that come in the form of beads or granules with diverse functional groups. It might be a cation exchange resin or an anion exchange resin that can bind oppositely to charged ions. The dye is removed by forming a connection between resins and the solute (Greluk and Hubicki 2013). Because of the ease of handling and regeneration of resins, this is a useful color removal approach. The pollutant level of the water received after the treatment process is almost non-existent. Commercial resins, on the other hand, are quite expensive, and synthetic ion-exchange resins are not biodegradable, so disposal is always an issue. In addition, the ion exchange degradation procedure is kind of ineffective toward different varieties of dyes (Samsami et al. 2020).

10.5.1.2 Dye Degradation Via Adsorption

Adsorption is a surface phenomenon that occurs when adsorbate molecules are drawn to the adsorbent's surface. By the ways of physisorption and chemisorption, the dye/effluent gets absorbed on the adsorbent surface (Gupta et al. 2018). In physisorption, creating a bond between adsorbate and adsorbent, physical forces such as Vander Waals force, hydrogen bond, static interaction, dipole–dipole interaction, and others are involved, but in chemisorption, electrons are exchanged (Kumar et al. 2014). Silica gel, zeolites, carbon nanotubes, activated carbon, and other adsorbents are frequently employed for dye removal (Gupta 2009). Any good adsorbent, on the other hand, should be porous, have a large surface area, must be easy to obtain, be inexpensive, and be regenerable (Tan et al. 2008). This technique can handle any dyestuff and leaves no hazardous waste behind. Additionally, the elements impacting adsorption (starting dye concentration, pH, temperature, contact time, and adsorbent quantity) should be correctly tuned for a better rate of adsorption. This increases the process' efficiency while also assisting in its scaling up.

10.5.1.3 Degradation by Membrane Separation

The separation of dyes from wastewater using a membrane or membrane assembly is known as membrane separation. A membrane is a porous substance that retains bigger solutes in wastewater while releasing clean water. The membrane's nature is determined by the substance used to make it, such as polymer, silica, zeolites, and so on. Pore size, surface charge, surface morphology, hydrophobicity, and hydrophilicity are some of the other features of a membrane. Microfiltration, ultrafiltration, nanofiltration, and reverse osmosis are some of the available membrane modules, depending on the pore size and surface features of the membrane. Microfiltration, ultrafiltration, and nanofiltration have pore sizes ranging from 0.1 to 10 μm, 0.1 to 0.001 μm, and 0.5 to 2.0 nm, respectively (Samsami et al. 2020). The membrane's rejection and penetration characteristics are used to evaluate its effectiveness (Ahmad et al. 2015). Highly efficient handling of dyes can be done using membrane filtration, but one of the key issues is membrane fouling, which produces a layer of cake and decreases membrane performance over time. This layer of cake must be removed at the same time as the other layers for the procedure to run smoothly (Pavithra and Jaikumar 2019). Other downsides include expensive installation costs and the need for high pressure.

10.5.2 Chemical Approaches

Chemical approaches for dye degradation entail the use of chemicals and other reagents. The pollutant's functional groups react with the reagent, causing the pollutant to be removed. These processes are often expensive, necessitating specialized equipment and a lot of electricity, further demanding a lot of chemicals, generating a lot of concentrated sludge, and so on (Santos et al. 2007). Though there are some cost benefits, such as near to complete dye removal, having a great efficiency, and a faster reaction time, to name a few, electrochemical destruction, coagulation-flocculation, ozonation, Fenton reagent, and photochemical and photolytic degradation are only a few of the chemical techniques that can be used to remove dyes.

10.5.2.1 Electrochemical Destruction

Electrochemical oxidation and electrocoagulation are the most common methods for destroying dyes electrochemically. Electrical energy is used in these systems to degrade contaminants through chemical reactions. It consists of two parts: anode and cathode, which perform oxidation and reduction, respectively. The dye's deterioration is mostly caused by the flow of electrons between them. It is extremely little or no chemical consumption in this process, hence no sludge is generated (Robinson et al. 2001). Thus, by simply altering the electric current, the process of the reaction can be easily controlled. Despite the great efficiency of the process for color removal and stubborn pollutant degradation, it still involves the use of expensive power, the generation of hazardous chemicals following degradation, and reduced dye removal as the flow rate increases.

10.5.2.2 Coagulation and Flocculation

Treatment of wastewater containing dyes components with coagulation-flocculation is a low-cost and reliable method. The addition of a coagulant changes the physical condition of both dissolved and suspended particles, allowing them to be removed through sedimentation. Moringa *oleifera*, Musa genus, Prosopis *juliflora*, Ipomoea *dasysperma* seed gum, Tamarindus *indica,* and Cactus species are some of the examples of plant-based natural coagulants while there can be coagulants containing chemical derivatives (Mathuram et al. 2018). For examples, aluminumsulfate solution ($Al_2(SO_4)_3.18H_2O$), MO-KCl (the seed extract of Moringa *oleifera Lam* in potassium chloride), and MONaCl (the seed extract of Moringa *oleifera Lam* in sodium chloride) are some of the chemical coagulants that can be utilized in this technique (Dotto et al. 2019). The addition of any kind of coagulants, is irrespective of their natural or chemical origin. The material is dependent on the dye properties present in the wastewater. This process has several drawbacks, which includes its futility against azo & acids, production of large amounts of sludge, being reactive, and presence of soluble dyes with a high reliance on pH, the addition of chemicals increases the number of dissolved constituents, and the process becoming expensive if the chemical costs are high.

10.5.2.3 Ozonation

Another effective chemical treatment method for removing colors from wastewater is ozonation. Because of its instability, ozone is a potent oxidant that may readily disintegrate into free radicals like O_2^-, OH, and others, which react with the dyes (Shriram and Kanmani 2014). Local production of ozone is accomplished by passing air or pure oxygen via a gap between two discharging electrodes. Though ozone has poor solubility, it is still a highly reactive gas. The process' response mechanism might be direct, indirect, or a mix of the two. When molecular ozone is used in the therapy, it is referred to as a direct reaction, however, when hydroxyl radicals are used in the treatment, it is referred to as an indirect reaction (Pavithra and Jaikumar 2019). The integrated process for deterioration, on the other hand, is dependent on the quantity of ozone and pH. Following this process, the molecules of ozone target the unsaturated bonds present in chromospheres, such as organic compounds containing double bonds, thus directly or indirectly degrading them to smaller intermediates (Soares et al. 2006). This procedure has several advantages, including a very minimal rise in the volume of the wastewater, a quick reaction time, zero production of sludge, and excellent clearance of the dye. While, this method also has several drawbacks, including the production of harmful by-products, a constant demand for electricity thus, giving up to an overall high operating cost in the end.

10.5.2.4 Fenton Reagent

Another non-irradiation catalytic approach for degrading dyes that are resistant to biological treatment is the Fenton reaction ($Fe^{2+} + H_2O_2$) (Khan and Yadav 2021). Hydroxyl radicals (OH^-) are formed when hydrogen peroxide and ferrous ions exchange electrons in this catalytic reaction. These radicals are incredibly reactive and function as the primary oxidant in the system (Oturan and Aaron 2014). Organic contaminants are attacked by these radicals, which breakdown them into intermediates that are smaller and have a lower molecular weight or water and carbon dioxide. This procedure is capable of removing both soluble and insoluble colors, as well as dye effluent with a solid component. This process has several drawbacks, including the formation of iron sludge, high efficiency at lower pH levels, such lengthy degradation time, as in the severely acidic zone, and ineffectiveness for vat dyes and dispersion (Katheresan et al. 2018).

10.5.2.5 Photolytic Degradation

Photolysis is an approach that causes breakdown of contaminants in the presence of a UV–visible light source (190–700 nm) thus, further irradiating within the system. Usage of chemicals does not take place in this system (Gmurek et al. 2017). In this technique, as a source of light, high-pressured xenon or mercury arc lamps with moderate emission in the near-ultraviolet spectrum could be utilized (Fasnacht and Blough 2002). Thus, photons are responsible for breaking the bonds of dye molecules. However, a molecule's ability to absorb light of a specific wavelength determines how quickly it degrades (Vela et al. 2012). Direct irradiation causes molecules to transition from their quiescent state to an agitated state with more potential energy, causing chemical bonds to break and the molecules to degrade. There are no harmful chemicals used, no sludge is produced, and bad odors are eliminated. However, it is an energy-intensive, expensive, and inefficient method that can only handle a restricted number of chemicals.

10.5.2.6 Photochemical Degradation

For the breakdown of dyes and other contaminants, photochemical degradation techniques need the inclusion of additional oxidants as well as light irradiation. Even though photolysis is an efficient method of degradation, some dye molecules are resistant to it. As a result, in addition to light, it requires other oxidants such as H_2O_2, O_3, Fenton reagent, and so on. These oxidants produce a lot of hydroxyl radicals (OH^-), that breaks down the dye components into H_2O, CO_2 and various other compounds (Robinson et al. 2001). The dye degradation rate is strongly fascinated by the structure of the dye, the intensity of radiation, dye bath constitution and pH. Thus, no sludge or foul odor is produced in this process, and in contrast to the direct photolysis method, it is a more efficient dye removal technique. But the disadvantage lies due to the generation of waste by-products and being a quite pricey technique.

10.5.3 Biological Remediation

Biological remediation techniques have acquired widespread recognition in recent decades due to several benefits over traditional treatments, including a safe and easy procedure, being relatively cost-effective, with almost zero sludge production, thus being environment friendly throughout the process (Ghosh et al. 2017; Ito et al. 2016). Usage of fungus, bacteria, different enzymes, and other microorganisms is taken place in this procedure to degrade dyes. Water & carbon dioxide, for example, are non-aggressive substances. According to the situations, dependable and risk-free procedures can be run in-situ or ex-situ as per need (Vikrant et al. 2018). The process can be performed using a single microbe, a mixed culture of microorganisms, or even a plant. Process conditions for microbial growth, such as pH, temperature & nutrient conditions must be tuned for greater dye breakdown efficiency. Because of substrate inhibition, the microorganisms' effectiveness drops at greater dye concentrations, resulting in less dye removal. In comparison to other physical–chemical processes, this technique has the benefit of using less energy and chemicals (Mallakpour and Rashidimoghadam 2019).

10.6 The Fundamentals of Photocatalysis

In the presence of a catalyst, photocatalysis accelerates a photoreaction. It occurs due to the interaction of light with the metallic nanoparticle. Inactivation depends upon nature, morphology, and concentration of catalysis. Advancement in photocatalysis research occurred when TiO_2 (Titanium dioxide) and Pt (Platinum) electrodes were introduced in the photolysis of water. TiO_2 nanoparticles show a crucial role in photoactivity by comparing the photocatalytic efficiency by deposition of TiO_2 nanoparticles on mesoporous material (Rahman 2019). The general type of photocatalysis is homogeneous and heterogeneous photocatalysis, transition metal oxide and semiconductor are the most common heterogeneous photocatalysis whereas ozone and Fenton are commonly used homogeneous photocatalysis.

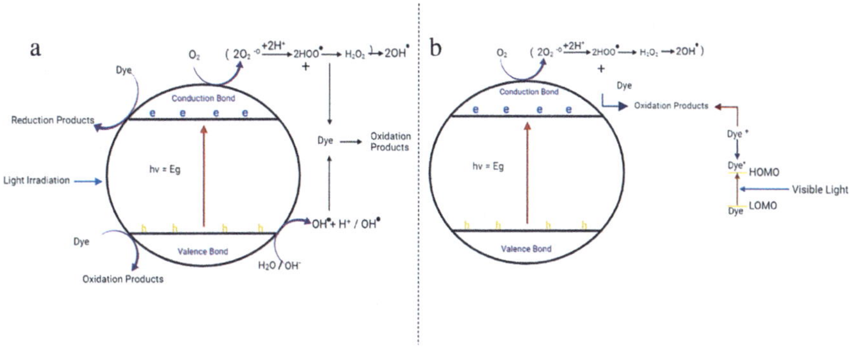

Fig. 10.2 **a** Indirect dye degradation, **b** Direct dye degradation, inspired from Ajmal et al. (2014)

10.6.1 Photocatalytic Routes and Their Mechanism

Photocatalyzed dye degradation: fundamental principles and processes are (i) Indirect dye degradation mechanism, which includes ionization of water, photoexcitation, protonation of superoxide, and oxygen ion sorption. (ii) Direct dye degradation mechanism, this happens when a visible light photon is stimulated from the ground state to the triplet excited state. The indirect mechanism was prevalent over direct mechanism (Ajmal et al. 2014) (Fig. 10.2).

10.6.2 Photocatalyst

A photocatalyst is a solid substance that can absorb UV–visible light and transport it to a higher energy level than the energy bandgap, allowing a chemical reaction to occur by creating electron–hole pairs. Photocatalysts, on the other hand, are not consumed throughout the reaction (Sreethawong 2012). Because of their increased photocatalytic efficiency, high surface area, capability of charge transfer, and other qualities, several powdered metal oxides such as ZnO, TiO_2, SnO_2, Fe_2O_3, CdS, and several others have been employed as photocatalysts for degradation of organic contaminants/dyes that are present in wastewater. However, it's tough to separate these photocatalysts from the reaction system, costly and time-consuming. Biocompatible, charge-generating when lit by light, stable, large surface area, easily accessible, non-toxic, and photocatalysts that can be reused are fantastic. The photocatalyst must be able to increase electron and hole transport while minimizing hole and electron recombination (Coronado, et al. 2013). Below are descriptions of the various photocatalyst preparation procedures.

10.6.2.1 Photocatalyst Preparation Techniques

Hydrothermal

Any heterogeneous reaction in the presence of aqueous solvents or mineralizers at high pressure and temperature is referred to as hydrothermal (Yoshimura and Byrappa 2008). It's a practical approach that uses a solution reaction using water as the solvent (Prasad et al. 2020). In most cases, the procedure is performed in an autoclave, which allows for precise control of pressure and temperature. The pressure within the autoclave is heavily influenced by the temperature and amount of solution supplied. This approach produces photocatalysts at a lower cost, smaller particle sizes and better grain development with a more equal distribution. When choosing an autoclave, the most important parameters to consider are the experimental temperature and pressure conditions, as well as the corrosion resistance in that pressure–temperature range in a given solvent or hydrothermal fluid. Corrosion resistance is a significant factor to consider when deciding an autoclave material when the reaction takes place directly in the vessel. The most effective corrosion resistant materials are high-strength alloys, such as 300 series (austenitic) stainless steel, nickel, iron, titanium and its alloys and cobalt-based superalloys (Yoshimura and Byrappa 2008). A titanium precursor's peptized precipitates can be hydrothermally treated with water to produce TiO_2 nanoparticles. Titanium butoxide was used as a precursor, and tetraalkylammonium hydroxide was used to create the precipitates (peptizer). The particle size decreased with increasing alkyl chain length at the same quantity of peptizer. Particle morphology was altered by the peptizes and their concentrations (Chen and Mao 2007). When exposed to visible light, TiO_2 is doped with nonmetals to boost its photocatalytic activity; in this case, the hydrothermal technique allows for the synthesis of NM-doped titania at low temperatures. For example, sodium borohydride can be used as a boron source to make boron-doped TiO_2. The hydrothermal B-doped sample had an increased adsorption ability, higher surface area, and enhanced photocatalytic activity for the degradation of reactive brilliant red and 4-chlorophenol when exposed to visible light when compared to the sol–gel technique (Xu et al. 2009). Hydrothermal treatment was recently used to produce P-TiO_2 (phosphorus-doped TiO_2) with a mesoporous structure. The titanium was obtained from tetrabutyl titanate, and the dopant was prepared from phosphoric acid. The hydrothermal treatment took place in a stainless-steel autoclave that is teflon-lined for 12 h at 200 °C. P-doped samples were generated in a similar manner using a high-temperature sol–gel technique. To measure the photocatalytic activity of the generated materials, the degradation of formaldehyde (HCHO) and the breakdown of methylene blue (MB) were utilized as test reactions. The mesoporous P-doped titania nanoparticles produced by both the sol–gel and hydrothermal techniques were well-crystallized. The hydrothermal samples, on the other hand, had a smaller average pore size, a greater surface area, a more uniform crystallite size, and higher photocatalytic activity than the sol–gel or Degussa-P25 catalyst samples (Guo et al. 2013). Additional lattice defects in the P-TiO_2 semiconductor are created by the hydrothermal treatment, which are active in trapping photogenerated electrons and enhancing catalytic activity. Additional lattice defects in the P-TiO_2 semiconductor are created by the hydrothermal treatment, which are active in trapping photogenerated electrons and enhancing catalytic activity. A hybrid of sol–gel and hydrothermal methods was used to make ternary Bi_2WO_6 photocatalyst microspheres. As precursors, we utilized $(NH_4)6W_7O_{24}.6H2O$ and $Bi(NO_3)_3.5H_2O$ solutions with EDTA. Before being transported to a 220 °C autoclave for 24 h, the pH 3.0 gel was matured for the whole night. Without the use of the sol–gel process, Bi_2WO_6 was generated directly by hydrothermal reaction. In the sol–gel-hydrothermal (SH) Bi_2WO_6, the monodispersed hierarchical microspheres were hollow in the middle, whereas the hydrothermal (H) Bi_2WO_6 exhibited an uneven plate-like structure. When methylene blue (MB) is degraded in the presence of other Bi_2WO_6 catalysts, SH-Bi_2WO_6 has better photocatalytic activity than H-Bi_2WO_6. SH increased the photocatalytic activity of Bi_2WO_6 was attributed to its unusual hierarchical structure, which was generated using a combination of sol–gel and hydrothermal techniques (Liu et al. 2013).

Sol–Gel

For many years, the sol–gel method has been the most well-known and promising approach. As a part of the production of semiconductor photocatalysts, the sol–gel approach has become the most used. It is a powerful method for tailoring metal oxides to fit specific applications because a large number of parameters can be controlled, including the nature of the precursors, reaction temperature, time, and pH, reagent concentrations, temperature and reaction time, the addition of organic additives, catalyst nature and concentration, and the amount of water added. This process ensures that the nanoparticles are evenly distributed throughout the supporting materials (Kim and Kan 2016). The fundamental benefit of the sol–gel process is that it allows for homogenous mixing of metal ions at the molecular level, which aids in the production of polycrystalline particles with unique characteristics. The sol–gel method also has the advantage of allowing different types of dopants to be incorporated at different stages of the process. When an active dopant is added to the sol during the gelation step, it permits doping components to interact directly with the support, improving the material's photocatalytic capabilities (Zaleska 2008). One of the most common methods for solidifying photocatalysts from their precursor solution is this method. The preparation of photocatalysts is divided into three processes, such as hydrolysis, polymerization, and drying of the gel (Liang et al. 2019). Metal oxides, metal salts and alkoxides are the precursors for this procedure. The precursor molecules' solution is hydrolyzed, resulting in colloidal sol. Following that, polymerization takes place, resulting in a gel-like form. Finally, depending on the use, the gel is calcined in ambient or hypercritical temperatures to generate xerogel or aerogel. Thermal treatment of the gel can provide photocatalysts in many forms, including films, fibers, monolits, and monosized powders (Medina-Ramírez et al. 2015). The sol–gel process starts with the production of a homogeneous solution, which is then converted into a sol using an appropriate reagent, ageing, shaping, and heat treatment/sintering (Niederberger and Pinna 2009).

Direct Oxidation

TiO_2 nanoparticles can also be made by oxidizing titanium metal with oxidants or by anodizing it. Using hydrogen peroxide to oxidize a titanium metal plate, crystalline TiO_2 nanorods were created. Anodic oxidation is a promising method for manufacturing immobilized TiO_2-based photocatalysts since it can result in the direct creation of immobilized self-organized TiO_2 on a variety of Ti substrates. The anodic oxidation technique is an electrochemical method for creating an oxide coating on a metallic substrate. It requires introducing a low-current electrical bias to a substrate that has been immersed in acid. Microstructural features of the films might vary, and they can be extremely thick and stable. The applied voltage and anodizing duration have a major impact on the phase development, thickness, and pore size of the films (Abdullah and Sorrell 2007). Originally, this approach was utilized to make highly organized nanoporous and nanotubular films (Wu et al. 2009). Anodic oxidation of titanium foil can produce TiO_2 nanotubes with a variety of morphologies (Su et al. 2013; Wang et al. 2009, 2011b). The physicochemical characteristics of zinc oxide (ZnO) and hence its use is dependent on its production technique. Despite the fact that ZnO has been produced in film form in a variety of ways, the electrochemical approach for the production of nanosized metal oxide films has attracted interest due to its ease of use, low working temperature, and economic viability (Wang et al. 2011c; Chandrappa and Venkatesha 2012). Using aqueous sodium bicarbonate electrolyte and sacrificial Zn anode and cathode in an undivided cell under galvanostatic mode, ZnO nanoparticles of different sizes (20, 44, and 73 nm) have been successfully produced by the electrochemical-thermal approach. The size and form of nanoparticles were unchanged by current density in all samples obtained at all current densities. When exposed to UV radiation, the materials were calcined at different temperatures (60, 300, and 600 °C) and tested for photocatalytic degradation of a methylene blue solution. The photocatalytic activity of the as-prepared and calcined (600 °C) ZnO photocatalysts was found to be higher, which is likely due to the reduced particle size and improved crystallinity of the latter (Chandrappa and Venkatesha 2012).

Solvothermal

The solvothermal technique is a nonaqueous solvent-based derivative of the hydrothermal technique. Due to the wide range of organic solvents with high boiling points available, the reaction temperature may be raised significantly higher than in the hydrothermal approach. The solvothermal approach permits excellent control of photocatalyst such particle size, shape, distributions, crystallinity, and the agglomeration pattern (Wu et al. 2014). This method has been utilized to make TiO_2 nanoparticles and nanorods with and without surfactants, and it has been shown to be a viable method for the synthesis of a wide range of nanoparticles with narrow size distribution and dispersity. Varying chemical and physical properties of solvents can affect the reactivity, solubility, and diffusion behavior of reactants; in particular, the polarity and coordinating ability of the solvent can affect the final product morphology and crystallization behavior. The reaction temperature can be set based on the boiling point of the solvent. This process of preparation is separated into four steps: dissolution, nucleation, growth in a specified direction, and lastly recrystallization (Liang et al. 2019). Ethanol with higher concentrations can alter the solvent's polarity, have a significant impact on the ς potential values of reactant particles, and increase solution viscosity. Instead of nanowires, short and broad flake-like structures of TiO_2 formed in the absence of ethanol. TiO_2 nanorods were formed when chloroform was used (Urbain et al. 2021). Non-aqueous solvothermal techniques frequently provide greater control over crystallinity, nanoparticle size and agglomeration behavior than aqueous solvothermal approaches. For example, rough-surfaced mesoporous TiO_2 microspheres have a large specific surface area, which is important for semiconductor activity because it provides more active sites for photocatalytic reactions. Solvothermal procedures were used to make mesoporous TiO_2 samples of various sizes. Tetrabutyl was employed to make hierarchical mesoporous TiO_2 microspheres with good crystallinity and a BET-specific surface area. A polyethyleneimine solution containing 100% ethanol was treated with titanate as a precursor. Under simulated solar light irradiation, the activity of the TiO_2 microspheres generated in the degradation of methyl orange (MO) and phenol aqueous solutions was studied. The findings revealed a high specific surface area (118.3 $m^2\ g^{-1}$) and a narrow pore-size distribution centered at 2.4 and 10.1 nm, respectively. On both MO and phenol, the TiO_2 microspheres display good photodegradation activities. A good example is the development of novel photocatalytic semiconductors with activity in the visible spectrum. Bismuth oxyiodide (BiOI) has sparked a lot of interest in heterogeneous photocatalysis due to its significant photocatalytic activity under visible light (Zhu et al. 2019; Huang et al. 2014; Wang et al. 2016). The solvothermal synthesis process has been used in a variety of applications due to its compatible methodology such as the ability to control the shape and size of the host material. Moreover, this technique is simple and cost-effective.

Sonochemical

Over the last few years, there has been a rise of interest in the potential of ultrasound for use in a broader range of chemistry and processing, which has been grouped under the umbrella term sonochemistry. The majority of these applications rely on sonic cavitation in a liquid medium being created (Yusof et al. 2016). Ultrasound has proven to be effective in the synthesis of a variety of nanostructured materials, including high-surface-area transition metals, alloys, carbides, oxides, and colloids. The use of ultrasound under appropriate conditions allows nanocomposites to be synthesized in a short amount of time, under moderate settings, in the air, and without calcination.

Ultrasound's chemical effects are not caused by a direct contact with molecular entities. Sonochemistry, on the other hand, is caused by acoustic cavitation, which is the development, growth, and implosive collapse of bubbles in a liquid. Localized hot spots with transitory temperatures of 10,000 K, pressures of 1000 atm or more, and cooling speeds of 109 Ks^{-1} are created when cavitation bubbles burst. At high circumstances various physical changes and chemical reactions occur, and a wide range of structured nanomaterials, including metals, biomaterials, alloys, and oxides, can be produced with the desired particle size distribution (Shirsath et al. 2013). Source that stimulates the chemical reaction is innovative in the sonochemical process. It gives a long-term solution for reducing energy use during nanoparticle production. Template-free approaches, such as hydrothermal and sonochemical procedures, can be used to synthesize nanostructured TiO_2. Sonochemical approaches for the production of metal- and nonmetal-doped TiO_2 have been studied extensively (Shirsath et al. 2013; Ángel-Sanchez et al. 2013; Yu et al. 2012). N-doped TiO_2 nanocrystalline was obtained by sonicating a solution of tetra isopropyl titanium and urea in water and isopropyl alcohol at 80 °C for 150 min. The effect of sonication time on the formation of N-doped TiO_2 nanoparticles was investigated. The sample crystallinity increased as the sonication duration increased (60, 120, 150, and 180 min), while the optimal sonication time for TiO_2 crystallization was 150 min (Wang et al. 2011d).Mesoporous TiO_2 nanorods were made from commercial Ti powder using a unique template-free sonochemical technique. The preparation was carried out under normal atmospheric conditions and at a low temperature of 70 °C. To assess the activity of the prepared samples, photocatalytic degradation of toluene was carried out under UV light (Guo et al. 2009). The sonochemical method was also used to manufacture silica-coated ZnO nanoparticles. Ultrasonography was utilized to irradiate a mixture of dispersion of ZnO, tetraethoxysilane (TEOS), and ammonia in an ethanol–water solution media. After 60 min of ultrasonic irradiation with a TEOS/ZnO loading of 0.8, the silica coating layer was formed. HRTEM scans revealed a homogeneous covering that extended up to 3 nm from the ZnO surface. The photocatalytic activity of silica-coated ZnO nanoparticles in an aqueous solution against photodegradation of methylene blue dye was considerably reduced (Siddiquey et al. 2012).

Chemical Vapor Deposition

Another major chemical approach for producing high-purity, the high-performance photocatalyst is chemical vapor deposition (CVD). The procedure uses either grafting or coating procedures. When conformal deposition of the material is necessary, CVD is regarded as one of the most accurate procedures. In this regard, CVD has several advantages when it comes to the deposition of ordered nanostructured thin films with a variety of compositions (Vasudev et al. 2013). Film growth in CVD is dependent on volatile molecular precursors. The precursor molecules are conveyed in a stream of carrier gas, which may be chemically inert or reactive. Chemical processes in the gas phase at the surface turn them into a thin solid layer of the desired substance. Hot carrier gases fuel the deposition reaction in CVD procedures (Carlsson and

Martin 2010). The technology of chemical vapor deposition (CVD) has been used to manufacture supported titanium dioxide compounds. CVD has been used to create thin films (supported in, metallic, semiconducting substrates or, glass substrates), composite materials (activated carbon/TiO_2, amine polymer/TiO_2, SiO_2/TiO_2) mesoporous materials, and titanium dioxide nanorods, to name a few. Chemical vapor deposition is a material-production process that straddles the chemical and physical worlds. Chemical techniques are the most widely used methods for producing titanium dioxide materials because of their relative ease of control over synthesis, narrow nanoparticle size distribution, cheap cost, and consistent nanoparticle stability in the system. Physical approaches, on the other hand, have received less attention, particularly for large-scale material manufacturing, because material creation necessitates highly specialized equipment and personnel. CVD allows for the production of high purity materials while obviating the need for post-heat treatment to boost the crystallinity of the final product. However, ultrahigh vacuum and high vapor pressure precursors are required for material deposition. A variety of gas deposition processes have recently been devised, with slight differences in the type of precursor employed, the substrate used, the desired degree of thin-film uniformity, and the reaction conditions (Shan et al. 2010). Some examples of these methods include atmospheric pressure chemical vapor deposition (APCVD), plasma-enhanced chemical vapor deposition (PECVD), metal–organic chemical vapor deposition (MOCVD) using metal–organic precursors, and hybrid physical–chemical vapor deposition (HPCVD)(Marchand et al. 2013). CVD has been used to fabricate nonmetal (C, B and N)-doped TiO_2 materials. In TiO_2 nanotubes doped with boron a visible light photocatalytic activity was created (Nasirian et al. 2018). The TiO_2 nanotubes were created using an electrochemical anodization method. The TiO_2 nanotube electrode was anodized and then annealed in boron-containing vapor through CVD to carry out the boron doping. Depending on the physicochemical properties of these compounds, MOCVD and AACVD are commonly used to deposit ZnO, ZnSe, and ZnS thin films (Medina-Ramírez et al. 2015). ZnO is the only one of these materials that have been used in environmental degradation; due to their low chemical stability, ZnS and ZnSe have received less attention. To summarize, CVD is a potential method for producing thin films of photocatalytic chemicals on various substrates. Air pressure deposition of the components is made possible by the employment of organometallic precursors. Although the cost of large-scale photocatalytic materials manufacturing by CVD may be higher than other fabrication procedures, the physicochemical properties of the deposited films by CVD may be adequate to outweigh the cost of production.

Electrochemical

Electrochemical deposition (EPD) is a process in which metallic ions are converted to solid metal which later deposit on the cathode surface in presence of electric current in the electrolyte. This technology is used for fabrication (electrocoating, painting, etc.). EPD processes provide fabrication for titanium dioxide (TiO_2) photocatalysts. EPD is widely used in textile industries for manufacturing fibers, dyeing and effluents discoloration. It is mostly used in Vat and sulfur dyeing (Sala and Gutiérrez-Bouzán 2012).

10.6.2.2 Dye-Degrading Photocatalysts of Various Kinds

See Table 10.1.

10.6.3 Parameters Affecting Photocatalytic Degradation

Photocatalysis is a crucial step in the decomposition of various contaminants in wastewater. The photocatalyst is made using several methods such as sol–gel, hydrothermal, solvothermal, sonochemical, and several others that we have already discussed. However, several factors affect the activity of the photocatalyst. The major purpose of investigating these factors is to find the most efficient and optimal conditions for the photodegradation of organic contaminants. The various parameters are as follows.

10.6.3.1 Effect of pH and Temperature

During the degradation process, pH performs a variety of roles. The photodegradation effectiveness of dyes is greatly influenced by pH. The surface charge characteristics of the photocatalyst are heavily influenced by the pH of the solution. The type of dye, such as whether it is cationic or anionic, has an impact on dye degradation (Chiu et al. 2019). As a result, it's critical to figure out what the dyes are made of before they break down. The catalyst surface is positively charged at pH less than pHzpc and negatively charged at pH more than pHzpc, indicating that the surface charge density is neutral at pHzpc. Negatively charged dyes should be degraded in acidic medium, while positively charged dyes should be degraded in alkaline medium (Singh Rathore et al. 2017). The point zero charges of various photocatalysts vary (pHzpc). The point zero charges are defined as the limiting pH where the net catalyst surface charge is zero (Zawawi et al. 2017). According to a study, the degradation of orange G dye (negatively charged) demonstrated a strong color loss in acidic conditions, with 98.98% degradation at 2 pH and just 31.67% at pH 6.5. At lower pH, the photocatalyst's surface was positively charged, with pHzpc ranging from 6.25 to 6.9. High removal was accomplished thanks to an electrostatic interaction between the positively charged photocatalyst and the negatively charged dye. The color removal, however, diminished as the pH of the solution was dropped. This could be due to catalyst particle agglomeration, which reduces surface area, or the production of high amounts of H^+ ions, which binds the azo linkages and lowers the electron density, limiting dye degradation (Sun et al. 2008). The features of the surface charge on photocatalyst, the size development of aggregates, and the adsorption behavior of

Table 10.1 Impact of photocatlyst on dyes

S. No.	Photo catalysts	Involved with dyes	Observation	References
1	TiO_2	Methyl Orange, Methylene Blue	As compared to UV rays, solar light seems to have a stronger effect on photo-catalytic activity	Nguyen et al. (2018)
		Methylene Blue, Methyl Orange, Congo Red	Phase (Anatase) and dimension (size) are critical. Additionally, the dye's ability to adhere to the catalyst surface is critical (Freundlich isotherm)	Nasrollahzadeh et al. (2016)
		Indigo, Indigo Carmine	The colors have been completely mineralized. The elimination of color was the sole effect of irradiation with visible light	Zolgharnein et al. (2014)
		Acid Orange 7	In terms of quality, UV light is superior to visible light	Chen et al. (2012)
		Methylene blue Chromium (VI)	When compared to other adsorbents, the TiO_2/UV system is extremely effective	Diaz-Uribe et al. (2014)
		Reactive Orange 16 Dye	The impact of TiO_2 concentration was examined	Chatterjee et al. (2008)
2	ZnO	Methylene Blue	The trials were conducted using real industrial waste water	Ahmad et al. (2019)
		Remazol Brilliant Blue dye	First-order kinetics govern the decomposition	Zouhier et al. (2020)
		Acid Green 25	A medium that is acidic and basic at the same time	Salem et al. (2015)
		Remazol BlackB, Remazol Brilliant BlueR, Reactive Blue 222 and Reactive Blue 221	In a coupling setup, TiO_2 and ZnO had no synergistic effect	Han et al. (2012)

(continued)

Table 10.1 (continued)

S. No.	Photo catalysts	Involved with dyes	Observation	References
		Crystal Violet	ZnO nanonails' improved activity is due in part to their higher specific surface area, higher crystallinity, and improved optical quality	(Priya et al. 2020)
		Acid Red 27	The rate of reaction increases as the concentration of $K_2S_2O_8$, H_2O_2 and $KBrO_3$ rises	Gnanaprakasam et al. (2015)
3	Orthorhombic WO_3	AO7 dye	Oxalic acid improved the decolorization of the dye by inhibiting phenol, humic acid, and EDTA, but did not affect the dye itself	Mohagheghian et al. (2015)
4	Undoped and Fe doped CeO_2	Methyl Orange	Fe^{3+} doping at a level of 1.5% was ideal	Channei et al. (2014)
5	Ni/$MgFe_2O_4$	Malachite Green	Active light is visible to the naked eye	Nguyen-Phan et al. (2011)
6	Mesoporous CeO_2	Rhodamine B	Hydroxyl radicals are the primary species in action	Kaur et al. (2021)
7	$Bi_{24}O_{31}Cl_{10}$	Rhodamine B	High electronic mobility at the same time as suitable energy levels	(Li et al. 2014)
8	BiOI	Anionic reactive KN-Rblue Rhodamine B	Dye decomposition is dominated by h^+ ions	Zhang and Lu (2018)
9	Mn_3O_4 nano particles	Amido black 10B	Hydrogen peroxide, Peroxomonosulfate, and Peroxodisulfate were used to speed up the degrading process, as was hydrogen peroxide	Govindan et al. (2017)
10	Silver phosphate	Methylene Blue	Dye contaminant photodegradation triggered by visible light	Dhanabal et al. (2015)

(continued)

Table 10.1 (continued)

S. No.	Photo catalysts	Involved with dyes	Observation	References
11	Photo-Fenton system	Reactive orange M_2R dye	Acidic pH favors, a mechanism is proposed	Chaudhari et al. (2017)
12	MgO	Methylene Blue	There has been a deterioration of more than 90%	Bekena and Kuo (2020)
13	CaO	Indigo carmine dye	pH 9 was a good match	Yao et al. (2016)
14	BiVO4-rGO	Rhodamine B	More effective than pure $BiVO_4$ and P-25	Shad et al. (2021)
15	Anatase and rich brookite rich films	Acid orange 7 4-chlorophenol	Both showed the same level of interest	Viswanathan (2018)
16	$K_6Ta_{10.8}O_{30}$	ARG dye	Degradation of ARG dye with significant photocatalytic activity	Viswanathan (2018)
17	N-doped ZnO	Azure A	It has been shown to be an excellent catalyst for several chemical processes using N-doped zinc oxide	Pandian et al. (2018)
18	$CoFe_2O_4/C_3N_4$ hybrid	Rhodamine B	It is more effective than pure $BiVO_4$ and P-25	Yao et al. (2016)
19	CuS	Methylene blue, eosin Y, rhodamine B, and Congo red	Dye photodegradation generally follows pseudo-first-order kinetics	Ayodhya et al. (2016)

the compounds may alter when the pH of the solution changes (Emam and Aboul-Gheit 2014). Alternatively, changes in surface potential have an impact on the redox process. In general, increasing the amount of hydroxyl radicals in a solution increases the efficacy of pollutant removal, and vice versa. The degradation of cationic dye methylene blue was studied using a manganese-doped ZnO photocatalyst. pHzpc was pH-9 in the case of ZnO. The most severe deterioration was observed at pH 10 due to the electrostatic interaction between the negatively charged ZnO surface and the positively charged dye (Chanu et al. 2019). To summarize, choosing the best pH solution for the photocatalytic process should be done with caution. A variety of aspects must be examined, including the photocatalyst and pollutant's properties, as well as their reaction activity in different pH solutions.

There have also been several investigations on the effect of temperature on the rate of decomposition of organic molecules. The influence of photocatalytic activity on temperature has been studied by several researchers (Nolan et al. 2012). In general, increasing the temperature causes charge carrier recombination as well as the desorption of adsorbed reactant species, resulting in a reduction in photocatalytic activity. According to the Arrhenius equation, the apparent first-order rate constant K_{app} should rise linearly as $\exp(1/T)$ increases.

10.6.3.2 Effect of Light Intensity

The prefix (photo) in the phrase "photocatalysis" refers to light. The process can only be classified as a common catalytic process without the use of light. As a result, light is required for the photocatalytic degradation activity. While the light intensity can be low or high up to 25 mW/cm^2, the reaction rate is directly proportional or connected to the square root of the light intensity (half order) (Kanakaraju 2013). The impact of light intensity on the breakdown of photocatalytic dyes has been extensively studied. Photodegradation is linearly proportional to light intensity at low light levels. At decreasing light intensities, electron–hole production is projected to outnumber recombination. As a result, radical formation increases, hastening the decomposition process (Akbari et al. 2020). However, when the light intensity is strong, more radicals are created, speeding up the photodegradation process. The percentage degradation remains constant even when the light intensity is increased further, because the active sites for a given quantity of catalyst are constant. As a result, the reaction kinetics between the substrate molecules and the active sites became more active, increasing the probability of the substrate molecules being oxidized and, as a result, the rate of photodegradation. However, as light intensity beyond its ideal level, it was no longer a limiting component in photodegradation rate (Lam et al. 2010). Due to the increase in temperature caused by the light, the excessive light intensity would cause electron–hole pair recombination, which would eventually decrease photocatalytic activity (Kumar and Pandey 2017).

10.6.3.3 Effect of Inorganic Salts

Mineral ions are commonly found in wastewater dye contents. Some cations, such as copper, phosphate, and iron, have been shown to reduce photodegradation efficiency when present in high concentrations. One of the main reasons is that these chemicals may compete with dyes for active sites on the TiO_2 photocatalyst surface, causing the photocatalyst to deactivate, resulting in reduced dye degradation (Chong et al. 2010). Calcium, magnesium, and zinc, on the other hand, have been shown to not influence the photodegradation of organic molecules, which is since these cations are present at their maximum oxidation states, which means they have no inhibitory effect. Inorganic anions such as carbonates, chlorides, nitrate, and sulfates are present in significant amounts in dye factory effluent. Colloidal instability increased mass transfer, and decreased surface contact between the target dye molecule and the photocatalyst are all caused by the presence of these salts (Chong et al. 2010). The hole scavenging capabilities of chloride ions cause a reduction in degradation efficiency in the presence of chloride ions. The chloride radical anions produced on the catalyst surface might potentially obstruct the active sites. The photocatalytic degradation inhibitory impact of chloride and phosphate ions has also been investigated (Chi et al. 2020). Furthermore, the rate of degradation is affected by the type of salt employed. Several researchers have investigated this consequence (Chen et al. 2012).

10.6.3.4 Effect of Pollutant Adsorption Strength

Competitive adsorption between water molecules and the degraded target molecules is critical in heterogeneous photocatalysis. This is because photogenerated oxidizing species may not travel far from their creation centers, resulting in minimal or extremely slow deterioration within a few nanometer layers surrounding the catalyst particles' surface (Malato et al. 2016). However, a direct relationship between adsorption and photodegradation has not been found by several authors in diverse investigations. Adsorption of chemicals or intermediates on the catalyst surface might behave as a toxin in specific situations (Porter et al. 2013). According to research, efficient dye adsorption always necessitates electrostatic interactions between the dye molecule and the photocatalyst's hydroxyl groups. Strong adsorption has been linked to the formation of a multilayer of dye molecules around the surface of catalyst particles. In the case of a direct degradation process, this may result in restricted engagement between the excited dye molecule and the catalyst; in the case of an indirect degradation mechanism, this may result in limited interaction between the incoming light and the catalyst (Emam and Aboul-Gheit 2014). The photo-oxidation process is likely to slow down in each of these scenarios. This might explain why the dye's initial decolorization rate in acidic solutions is claimed to be lower. In alkaline solutions, on the other hand, a reduction in the yearly rate of degradation reflects the anionic dye molecules' difficulties approaching the catalyst surface.

10.6.3.5 Initial Concentration of Dye

An aspect to consider is the dye concentration at the beginning of a photocatalytic process. When the quantity of dye concentration is increased while the amount of catalyst remains constant, the percentage of degradation is often reduced (Macedo et al. 2007). The photocatalytic breakdown of dyes is also influenced by their initial concentration. Many researchers have shown that raising the initial dye concentration while maintaining the pH and catalyst loading constant results in a decrease in the percentage of dye degradation. The pace of degradation is mostly determined by the photocatalyst's production of radicals (hydroxyl radicals and superoxides) and their interaction with dye molecules. More organic pollutants are available to be degraded at higher dye concentrations, but some of the radiation received by the photocatalyst is absorbed by the dye molecules, resulting in a reduction in the radiation received by the photocatalyst, reducing the generation of oxidants and, as a result, degradation efficiency (Rauf et al. 2007). As more dye molecules are bound to the photocatalyst surface at greater dye concentrations, the proportion of dye degradation is inhibited, and the formation of oxidants that cause dye degradation is reduced.

10.6.3.6 Effect of Photocatalytic Loading

To promote photodegradation, the amount of photocatalyst injected throughout the photocatalytic process is critical. As the photocatalyst dose is increased, the overall catalyst surface area increases as well. The presence of more active sites on the semiconductor surface is indicated by a bigger surface area (Ahmad et al. 2021). To carry out the mineralization, additional reactive radicals (.OH and $.O^{2-}$) will be created. In other words, there is a link between photocatalyst dose and degradation rate. When the photocatalyst dose is exceeded, however, this connection becomes invalid. Thus, if there is an overabundance of photocatalyst, the process may slow down. The aggregation of the photocatalyst prevents light from entering deeper into the solution, which explains this condition (Saravanan et al. 2017). As a result, the photocatalyst clusters obstruct and confine light from reaching a broad catalytic surface, scattering additional light. The optimal dosage of photocatalyst for 100 mL of dye solution was 200 mg in the research of napthol blue black dye, and beyond that, the dye breakdown rate decreased. Increases in photocatalyst loading beyond the optimal value increase its suspension in the solution, which intercepts light irradiation, resulting in fewer oxidants and hence a reduction in degradation efficiency. Also, when the number of photocatalysts increases, it agglomerates and coagulates, reducing the photocatalyst's surface area, resulting in reduced photon absorption and so affecting the degradation rate and efficiency (Akpan and Hameed 2009; Borker and Salker 2006). Since it is clear that as the quantity of photocatalyst grows, the rate of photodegradation increases at first, then reduces as the catalyst concentration rises. Some of the possibilities can explain this tendency. This includes when all of the dye molecules have been adsorbed on the photocatalyst surface, adding more photocatalyst has no further influence on the degradation efficiency. Another reason might be because an overabundance of photocatalyst particles increases the opacity of the solution, slowing the rate of degradation (Reza et al. 2017). Furthermore, when the number of particles in solution grows, particle–particle interaction becomes more significant. This might lead to a faster deactivation of activated molecules due to collisions with ground state titanium dioxide particles, and therefore lower dye concentrations (Ajmal et al. 2014).

10.6.3.7 Effects Related to Dopant Content

Many researchers have experimented by adding dopants to photocatalysts to improve their photocatalytic performance during contaminant degradation. Dopants are compounds, used for modification of conductance of a photocatalyst, which essentially changes the light absorption spectrum from UV to Vis as the bandgap energy diminishes, allowing the process to be conducted with solar light, comprises approximately 5% and 43% of UV and visible light respectively (Liu et al. 2019; Davis et al. 2013). Dopants can create crystal defects by causing lattice deformation. The electron–hole pair recombination is inhibited by these imperfections, which increases the rate of deterioration (Akpan and Hameed 2009). Furthermore, the proportion of dopants administered to photocatalysts should be kept under control, since any more addition than the optimal limit have a potential to reduce efficiency. Excess dopant reduces the specific surface area of the photocatalysts, slowing the adsorption process and lowering activity, by hindering light from absorption on the surface of photocatalyst, inhibiting the transfer of holes and electrons to reduce photocatalytic degradation (Liu et al. 2005). Metals (Cr, Fe, SnO_2, Cu, etc.) and nonmetals (e.g., carbon) can be employed as dopants (neodymium, nitrogen, carbon, sulfur).

10.6.3.8 Effects of Additives

Photocatalytic dye decomposition is affected by the existence of additives in the solution system. These additives are frequently added as ions to the dye solution for the improvement of the industrial pigmentation process. When wastewater is released, these ions become a component of the discharge. The most common ions detected in dye effluent are Fe^{+2}, Ag^{+}, Zn^{+2}, Na^{+}, Cl^{-}, PO_4^{3-}, BrO_3^{-},SO_4^{-2}, HCO_3^{-}, persulfate, and CO_3^{-2}. Each of these additional ions reduces the rate of dye solution degradation by a certain amount (Yoon et al. 2001). ·HO radicals are quickly converted to OH^{-} in the presence of Fe^{2+}, lowering their concentration and resulting in less dye solution deterioration. Similarly, CO_3^{2-} and HCO_3^{-} are also commonly used during dye baths to modulate pH, which basically slowed down the breakdown mechanism. Adding ethanol to a dye solution also prevents it from photodegrading, due to the neutralization of hydroxyl radicals, which are major contributors of degradation reaction (Behnajady et al. 2004). The presence of oxygen or air has an impact on photocatalytic degradation. In the absence of oxygen, the deterioration slows down, which has been linked to the reaction of the photogenerated electron–hole pairs (Khan and Swati 2016). Because the rate of reaction is related to the proportion of adsorption sites occupied by dissolved oxygen, it is a determining factor in the photooxidative reaction (Chong et al. 2010).

10.6.4 *Intermediary Product Detection*

In photodegradation, not only does the dye color fade away but the dyes are broken down into smaller parts as well. Instead of CO_2 and water, it releases a lot of other intermediates, which is not optimal. There is a chance that these intermediates are even more dangerous than the parent molecule itself (Anwer et al. 2019).As a result, it is critical to preserve the identities of the intermediates formed during photocatalytic degradation of dyes. The following are the three primary forms of analysis:

High-Pressure Liquid Chromatography, Gas Chromatography-Mass Spectrometry, and Liquid Chromatography-Mass Spectrometry. Interpretation is done by looking at commercial standards for a certain amount of time to identify the chemicals. Chemical structure, molecular formula and mass spectra are all given by these findings, as well as their source. Several scientists have investigated the photocatalytic destruction of dyes in depth. Bansal et al. (2010) used TiO_2 as a photocatalyst to breakdown acid orange 7 and found that it made a lot of different intermediates, including 4-[(2,3-dihydroxy-1-naphthyl) diazenyl]benzenesulfonate, phenyl-diazene-carboxylic acid after 2 h of treatment. Zn-doped Cu_2O particles made 8 main intermediate products when Yu et al. broke down ciprofloxacin (Yu et al. 2019). A study by Anwer et al. (2019) says that the intermediates may be more threatening than the initial dye molecule. Catechol might be one of the by-products of the photooxidation of phenols. Unlike phenol, which can make animals hypertensive and have convulsions, catechol is more dangerous to their health in this way. Since the intermediate chemicals that come from photodegradation must be inspected before they can be thrown away in the right way this is essential.

10.7 Conclusion

This chapter introduces the fundamentals of photocatalysis, from the view of dyestuff degradation from different industrial sources. Dye effluents are one of the major environmental issues on the planet today. These effluents also represent a major hazard to all types of species as well as the environment if they are not adequately separated before being discharged into surface water or groundwater. Despite the fact that there are a variety of dye degradation procedures based on biological, chemical, and physical principles, photocatalysis has received considerable attention. This photocatalysis method is particularly effective because it results in the complete oxidation of dyes into smaller, less harmful or nontoxic by-products, demonstrating the responsivity of the degradation process to a variety of process variables, such as the concentration of dye and the solution pH. By doping the photocatalyst with a dopant, it is possible to get synergistic improvements in photocatalytic performance. Using dye samples to establish the nature of the dye to be destroyed, researchers may select the kind of photocatalyst to be utilized in the degradation process. For the production of photocatalysts, many techniques such as sol–gel, hydrothermal, and others have been reported. Additionally, several kinds of photocatalytic materials such as TiO_2, ZnO, MgO, CaO etc., as well as their interactions with various dyes have been examined in this chapter. The existence of intermediate molecules reflects the amount of toxicity of the by-product compounds that have been produced. For long-term reliability, it is necessary to fine-tune the operational process parameters at the lab and pilot scales. To be more effective at the industrial scale, it is necessary to do a product life impact and cost estimations. The human population as well as the ecosystem benefit from research in this field.

References

Abdullah H, Sorrell C (2007) TiO_2 thick films by anodic oxidation. J Australas Ceram Soc 43(2):125–130
Ahmad A et al (2015) Recent advances in new generation dye removal technologies: novel search for approaches to reprocess wastewater. RSC Adv 5(39):30801–30818
Ahmad F et al (2019) Biological synthesis of metallic nanoparticles (MNPs) by plants and microbes: their cellular uptake, biocompatibility, and biomedical applications. Appl Microbiol Biotechnol 103(7):2913–2935
Ahmad W et al (2021) Photocatalytic degradation of crystal violet dye under sunlight by chitosan-encapsulated ternary metal selenide microspheres. Environ Sci Pollut Res 28(7):8074–8087
Ajmal A et al (2014) Principles and mechanisms of photocatalytic dye degradation on TiO_2 based photocatalysts: a comparative overview. RSC Adv 4(70):37003–37026
Akbari A et al (2020) Effect of nickel oxide nanoparticles as a photocatalyst in dyes degradation and evaluation of effective parameters in their removal from aqueous environments. Inorg Chem Commun 115:107867
Akpan UG, Hameed BH (2009) Parameters affecting the photocatalytic degradation of dyes using TiO_2-based photocatalysts: a review. J Hazard Mater 170(2–3):520–529
Al-Dawery SK (2013) Photo-catalyst degradation of tartrazine compound in wastewater using TiO_2 and UV light. J Eng Sci Technol 8(6):683–691
Ali H (2010) Biodegradation of synthetic dyes—a review. Water Air Soil Pollut 213(1):251–273
Anwer H et al (2019) Photocatalysts for degradation of dyes in industrial effluents: opportunities and challenges. Nano Res 12(5):955–972
Ayodhya D et al (2016) Photocatalytic degradation of dye pollutants under solar, visible and UV lights using green synthesised CuS nanoparticles. J Exp Nanosci 11(6):418–432
Bansal P, Singh D, Sud D (2010) Photocatalytic degradation of azo dye in aqueous TiO_2 suspension: reaction pathway and identification of intermediates products by LC/MS. Sep Purif Technol 72(3):357–365
Beck KR et al (2012) Liquid chromatographic and mass spectrometric analysis of dyes for forensic purposes. AATCC Rev 12(1)
Behnajady M, Modirshahla N, Shokri M (2004) Photodestruction of acid orange 7 (AO7) in aqueous solutions by UV/H_2O_2: influence of operational parameters. Chemosphere 55(1):129–134
Bekena F, Kuo D-H (2020) 10 nm sized visible light TiO_2 photocatalyst in the presence of MgO for degradation of methylene blue. Mater Sci Semicond Process 116:105152
Benkhaya S, M'rabet S, El Harfi A (2020a) A review on classifications, recent synthesis and applications of textile dyes. Inorg Chem Commun 115:107891
Benkhaya S, M'rabet S, El Harfi A (2020b) Classifications, properties, recent synthesis and applications of azo dyes. Heliyon 6(1):e03271
Borker P, Salker A (2006) Photocatalytic degradation of textile azo dye over $Ce_{1-x}Sn_xO_2$ series. Mater Sci Eng, B 133(1–3):55–60
Carey A et al (2013) Identification of dyes on single textile fibers by HPLC-DAD-MS. Anal Chem 85(23):11335–11343
Carlsson J-O, Martin PM (2010) Chemical vapor deposition. Handbook of deposition technologies for films and coatings. Elsevier, pp 314–363
Carmen Z, Daniela S (2012) Textile organic dyes-characteristics, polluting effects and separation/elimination procedures from industrial effluents-a critical overview. IntechOpen Rijeka
Chandrappa KG, Venkatesha TV (2012) Electrochemical synthesis and photocatalytic property of zinc oxide nanoparticles. Nano-Micro Lett 4(1):14–24
Channei D et al (2014) Photocatalytic degradation of methyl orange by CeO_2 and Fe–doped CeO_2 films under visible light irradiation. Sci Rep 4(1):1–7
Chanu LA et al (2019) Effect of operational parameters on the photocatalytic degradation of Methylene blue dye solution using manganese doped ZnO nanoparticles. Results Phys 12:1230–1237

Chatterjee D et al (2008) Kinetics of the decoloration of reactive dyes over visible light-irradiated TiO_2 semiconductor photocatalyst. J Hazard Mater 156(1–3):435–441
Chaudhari AU et al (2017) Effective biotransformation and detoxification of anthraquinone dye reactive blue 4 by using aerobic bacterial granules. Water Res 122:603–613
Chen X, Mao SS (2007) Titanium dioxide nanomaterials: synthesis, properties, modifications, and applications. Chem Rev 107(7):2891–2959
Chen X et al (2012) Accelerated TiO_2 photocatalytic degradation of Acid Orange 7 under visible light mediated by peroxymonosulfate. Chem Eng J 193:290–295
Chi Y et al (2020) Effective blockage of chloride ion quenching and chlorinated by-product generation in photocatalytic wastewater treatment. J Hazard Mater 396:122670
Chiu Y-H et al (2019) Mechanistic insights into photodegradation of organic dyes using heterostructure photocatalysts. Catalysts 9(5):430
Chong MN et al (2010) Recent developments in photocatalytic water treatment technology: a review. Water Res 44(10):2997–3027
Coronado JM et al (2013) Design of advanced photocatalytic materials for energy and environmental applications. Springer
Dave S, Dave S, Das J (2021a) Photocatalytic degradation of dyes in textile effluent: a green approach to eradicate environmental pollution. The Future of Effluent Treatment Plants. Elsevier, pp 199–214
Dave S et al (2021b) Mathematical modeling and surface response curves for green synthesized nanomaterials and their application in dye degradation. In: Photocatalytic degradation of dyes, pp 571–591
Davis M et al (2013) Enhanced photocatalytic performance of Fe-doped SnO_2 nanoarchitectures under UV irradiation: synthesis and activity. J Mater Sci 48(18):6404–6409
de Wolf K, Devaere S, Albrecht E (2013) Study on the link between allergic reactions and chemicals in textile products
Del Ángel-Sanchez K et al (2013) Photocatalytic degradation of 2, 4-dichlorophenoxyacetic acid under visible light: effect of synthesis route. Mater Chem Phys 139(2–3):423–430
Deng F et al (2012) Preparation of conductive polypyrrole/TiO_2 nanocomposite via surface molecular imprinting technique and its photocatalytic activity under simulated solar light irradiation. Colloids Surf, A 395:183–189
Dhanabal R et al (2015) Visible light driven degradation of methylene blue dye using Ag_3PO_4. J Environ Chem Eng 3(3):1872–1881
Diaz-Uribe C, Vallejo W, Ramos W (2014) Methylene blue photocatalytic mineralization under visible irradiation on TiO_2 thin films doped with chromium. Appl Surf Sci 319:121–127
Dos Santos AB, Cervantes FJ, Van Lier JB (2007) Review paper on current technologies for decolourisation of textile wastewaters: perspectives for anaerobic biotechnology. Biores Technol 98(12):2369–2385
Dotto J et al (2019) Performance of different coagulants in the coagulation/flocculation process of textile wastewater. J Clean Prod 208:656–665
Duy Nguyen H et al (2019) Activated carbons derived from teak sawdust-hydrochars for efficient removal of methylene blue, copper, and cadmium from aqueous solution. Water 11(12):2581
Elert AM et al (2017) Comparison of different methods for MP detection: what can we learn from them, and why asking the right question before measurements matters? Environ Pollut 231:1256–1264
Elmorsi TM et al (2010) Decolorization of Mordant red 73 azo dye in water using H_2O_2/UV and photo-Fenton treatment. J Hazard Mater 174(1–3):352–358
Emam E, Aboul-Gheit N (2014) Photocatalytic degradation of oil-emulsion in water/seawater using titanium dioxide. Energy Sources Part A Recov Utiliz Environ Effects 36(10):1123–1133
Fasnacht MP, Blough NV (2002) Aqueous photodegradation of polycyclic aromatic hydrocarbons. Environ Sci Technol 36(20):4364–4369
Fito J, Abrham S, Angassa K (2020) Adsorption of methylene blue from textile industrial wastewater onto activated carbon of Parthenium hysterophorus. Int J Environ Res 14(5):501–511

Ghosh A, Dastidar MG, Sreekrishnan T (2017) Bioremediation of chromium complex dyes and treatment of sludge generated during the process. Int Biodeterior Biodegradation 119:448–460
Gmurek M, Olak-Kucharczyk M, Ledakowicz S (2017) Photochemical decomposition of endocrine disrupting compounds–A review. Chem Eng J 310:437–456
Gnanaprakasam A et al (2015) Characterization of TiO_2 and ZnO nanoparticles and their applications in photocatalytic degradation of azodyes. Ecotoxicol Environ Saf 121:121–125
Govindan K et al (2017) Electron scavenger-assisted photocatalytic degradation of amido black 10B dye with Mn_3O_4 nanotubes: a response surface methodology study with central composite design. J Photochem Photobiol, A 341:146–156
Greluk M, Hubicki Z (2013) Evaluation of polystyrene anion exchange resin for removal of reactive dyes from aqueous solutions. Chem Eng Res Des 91(7):1343–1351
Guo S et al (2009) Synthesis of mesoporous TiO_2 nanorods via a mild template-free sonochemical route and their photocatalytic performances. Catal Commun 10(13):1766–1770
Guo S et al (2013) Synthesis of phosphorus-doped titania with mesoporous structure and excellent photocatalytic activity. Mater Res Bull 48(9):3032–3036
Gupta V (2009) Application of low-cost adsorbents for dye removal–a review. J Environ Manage 90(8):2313–2342
Gupta GK, Mondal MK (2021) Fundamentals and mechanistic pathways of dye degradation using photocatalysts. Photocatalytic degradation of dyes. Elsevier, pp 527–545
Gupta GK et al (2018) Pyrolysis of chemically treated corncob for biochar production and its application in Cr (VI) removal. Environ Prog Sustainable Energy 37(5):1606–1617
Han J et al (2012) Comparative photocatalytic degradation of estrone in water by ZnO and TiO_2 under artificial UVA and solar irradiation. Chem Eng J 213:150–162
Han M et al (2018) Recent progress on the photocatalysis of carbon dots: classification, mechanism and applications. Nano Today 19:201–218
Hasanpour M, Hatami M (2020) Photocatalytic performance of aerogels for organic dyes removal from wastewaters: Review study. J Mol Liq 309:113094
Holkar CR et al (2016) A critical review on textile wastewater treatments: possible approaches. J Environ Manage 182:351–366
Huang Y et al (2014) Oxygen vacancy induced bismuth oxyiodide with remarkably increased visible-light absorption and superior photocatalytic performance. ACS Appl Mater Interfaces 6(24):22920–22927
Ibhadon AO, Fitzpatrick P (2013) Heterogeneous photocatalysis: recent advances and applications. Catalysts 3(1):189–218
Imran M et al (2015) Microbial biotechnology for decolorization of textile wastewaters. Rev Environ Sci Bio/technology 14(1):73–92
Ito T et al (2016) Long–term natural remediation process in textile dye–polluted river sediment driven by bacterial community changes. Water Res 100:458–465
Javaid R et al (2021) Subcritical and supercritical water oxidation for dye decomposition. J Environ Manage 290:112605
Jia Z et al (2016) Amorphous $Fe_{78}Si_9B_{13}$ alloy: An efficient and reusable photo-enhanced Fenton-like catalyst in degradation of cibacron brilliant red 3B-A dye under UV–vis light. Appl Catal B 192:46–56
Jorfi S et al (2018) Visible light photocatalytic degradation of azo dye and a real textile wastewater using Mn, Mo, La/TiO_2/AC nanocomposite. Chem Biochem Eng Q 32(2):215–227
Kanakaraju D (2013) Photochemical and solar degradation of pharmaceuticals in water. James Cook University
Katheresan V, Kansedo J, Lau SY (2018) Efficiency of various recent wastewater dye removal methods: a review. J Environ Chem Eng 6(4):4676–4697
Kaur G et al (2021) CeO_2 supported $Zn_{1-x}Mn_xS$ (x: 0.025, 0.05, 0.075 and 0.10) catalyst for photoremoval of rhodamine B dye. Solar Energy 225:773–783
Khalid N et al (2017) Carbonaceous-TiO_2 nanomaterials for photocatalytic degradation of pollutants: a review. Ceram Int 43(17):14552–14571

Khan S, Malik A (2014) Environmental and health effects of textile industry wastewater. Environmental deterioration and human health. Springer, pp 55–71

Khan H, Swati IK (2016) Fe^{3+}-doped anatase TiO_2 with d–d transition, oxygen vacancies and Ti^{3+} centers: synthesis, characterization, UV–vis photocatalytic and mechanistic studies. Ind Eng Chem Res 55(23):6619–6633

Khan NA, Bhadra BN, Jhung SH (2018) Heteropoly acid-loaded ionic liquid@ metal-organic frameworks: Effective and reusable adsorbents for the desulfurization of a liquid model fuel. Chem Eng J 334:2215–2221

Khan SH, Yadav VK (2021) Advanced oxidation processes for wastewater remediation: an overview. In: Removal of emerging contaminants through microbial processes, pp 71–93

Kim JR, Kan E (2016) Heterogeneous photocatalytic degradation of sulfamethoxazole in water using a biochar-supported TiO_2 photocatalyst. J Environ Manage 180:94–101

Kumar A, Pandey G (2017) A review on the factors affecting the photocatalytic degradation of hazardous materials. Mater Sci Eng Int J 1(3):1–10

Kumar R, Rashid J, Barakat M (2014) Synthesis and characterization of a starch–AlOOH–FeS_2 nanocomposite for the adsorption of Congo red dye from aqueous solution. RSC Adv 4(72):38334–38340

Lam S-M, Sin J-C, Mohamed AR (2010) Parameter effect on photocatalytic degradation of phenol using TiO_2–P_{25}/activated carbon (AC). Korean J Chem Eng 27(4):1109–1116

Lazar, T., *Color chemistry: synthesis, properties, and applications of organic dyes and pigments.* 2005, Wiley Online Library.

Lellis B et al (2019) Effects of textile dyes on health and the environment and bioremediation potential of living organisms. Biotechnol Res Innov 3(2):275–290

Li F-T et al (2014) In-situ one-step synthesis of novel $BiOCl/Bi_{24}O_{31}Cl_{10}$ heterojunctions via self-combustion of ionic liquid with enhanced visible-light photocatalytic activities. Appl Catal B 150:574–584

Liang Q et al (2019) Surfactant-assisted synthesis of photocatalysts: mechanism, synthesis, recent advances and environmental application. Chem Eng J 372:429–451

Liu G et al (2005) The preparation of Zn^{2+}-doped TiO_2 nanoparticles by sol–gel and solid phase reaction methods respectively and their photocatalytic activities. Chemosphere 59(9):1367–1371

Liu Y et al (2013) Synthesis of hierarchical Bi_2WO_6 microspheres with high visible-light-driven photocatalytic activities by sol–gel-hydrothermal route. Mater Lett 108:84–87

Liu J et al (2019) F/W co-doped TiO_2-SiO_2 composite aerogels with improved visible light-driven photocatalytic activity. J Solid State Chem 275:8–15

Macedo L et al (2007) Degradation of leather dye on TiO_2: a study of applied experimental parameters on photoelectrocatalysis. J Photochem Photobiol, A 185(1):86–93

Madhav S et al (2018) A review of textile industry: Wet processing, environmental impacts, and effluent treatment methods. Environ Qual Manage 27(3):31–41

Mahmood S et al (2015) Detoxification of azo dyes by bacterial oxidoreductase enzymes. Crit Rev Biotechnol 36

Malato S et al (2016) Decontamination and disinfection of water by solar photocatalysis: the pilot plants of the Plataforma Solar de Almeria. Mater Sci Semicond Process 42:15–23

Mallakpour S, Rashidimoghadam S (2019) Carbon nanotubes for dyes removal. Composite nanoadsorbents. Elsevier, pp 211–243

Mani S, Bharagava RN (2018) Textile industry wastewater: environmental and health hazards and treatment approaches. Recent advances in environmental management. CRC Press, pp 47–69

Marchand P et al (2013) Aerosol-assisted delivery of precursors for chemical vapour deposition: expanding the scope of CVD for materials fabrication. Dalton Trans 42(26):9406–9422

Massonnet G et al (2012) Raman spectroscopy and microspectrophotometry of reactive dyes on cotton fibres: analysis and detection limits. Forensic Sci Int 222(1–3):200–207

Mathuram M, Meera R, Vijayaraghavan G (2018) Application of locally sourced plants as natural coagulants for dye removal from wastewater: a review. J Mater Environ Sci 2508:2058–2070

Medina-Ramírez I, Hernández-Ramírez A, Maya-Trevino ML (2015) Synthesis methods for photocatalytic materials. Photocatalytic semiconductors. Springer, pp 69–102
Mohagheghian A et al (2015) Photocatalytic degradation of a textile dye by illuminated tungsten oxide nanopowder. J Adv Oxid Technol 18(1):61–68
Nasirian M et al (2018) Enhancement of photocatalytic activity of titanium dioxide using non-metal doping methods under visible light: a review. Int J Environ Sci Technol 15(9):2009–2032
Nasrollahzadeh M et al (2016) In situ green synthesis of Ag nanoparticles on graphene oxide/TiO_2 nanocomposite and their catalytic activity for the reduction of 4-nitrophenol, Congo red and methylene blue. Ceram Int 42(7):8587–8596
Nguyen CH, Fu C-C, Juang R-S (2018) Degradation of methylene blue and methyl orange by palladium-doped TiO_2 photocatalysis for water reuse: efficiency and degradation pathways. J Clean Prod 202:413–427
Nguyen-Phan T-D et al (2011) Synthesis of hierarchical rose bridal bouquet-and humming-top-like TiO_2 nanostructures and their shape-dependent degradation efficiency of dye. J Colloid Interface Sci 356(1):138–144
Nidheesh PV, Gandhimathi R, Ramesh ST (2013) Degradation of dyes from aqueous solution by Fenton processes: a review. Environ Sci Pollut Res 20(4):2099–2132
Niederberger M, Pinna N (2009) Metal oxide nanoparticles in organic solvents: synthesis, formation, assembly and application. Springer Science & Business Media
Nolan NT et al (2012) Effect of N-doping on the photocatalytic activity of sol–gel TiO_2. J Hazard Mater 211:88–94
Ong S-A et al (2012) Comparative study on photocatalytic degradation of mono azo dye acid orange 7 and methyl orange under solar light irradiation. Water Air Soil Pollut 223(8):5483–5493
Oturan MA, Aaron J-J (2014) Advanced oxidation processes in water/wastewater treatment: principles and applications. A review. Crit Rev Environ Sci Technol 44(23):2577–2641
Pandian L, Rajasekaran R, Govindan P (2018) Synthesis, characterization and application of Cu doped ZnO nanocatalyst for photocatalytic ozonation of textile dye and study of its reusability. Mater Res Express 5(11):115505
Pandit VU et al (2015) Solar light driven dye degradation using novel organo–inorganic (6, 13-pentacenequinone/TiO2) nanocomposite. RSC Adv 5(14):10326–10331
Parmar A, Sharma S (2016) Derivative UV-vis absorption spectra as an invigorated spectrophotometric method for spectral resolution and quantitative analysis: theoretical aspects and analytical applications: A review. TrAC, Trends Anal Chem 77:44–53
Pavithra KG, Jaikumar V (2019) Removal of colorants from wastewater: a review on sources and treatment strategies. J Ind Eng Chem 75:1–19
Paździor K et al (2017) Influence of ozonation and biodegradation on toxicity of industrial textile wastewater. J Environ Manage 195:166–173
Pirok BW, Stoll DR, Schoenmakers PJ (2018) Recent developments in two-dimensional liquid chromatography: fundamental improvements for practical applications. Anal Chem 91(1):240–263
Polak J et al (2016) Toxicity and dyeing properties of dyes obtained through laccase-mediated synthesis. J Clean Prod 112:4265–4272
Porter NS et al (2013) Shape-control and electrocatalytic activity-enhancement of Pt-based bimetallic nanocrystals. Acc Chem Res 46(8):1867–1877
Prasad C et al (2020) An overview of graphene oxide supported semiconductors based photocatalysts: properties, synthesis and photocatalytic applications. J Mol Liq 297:111826
Priya R et al (2020) Comparative studies of crystal violet dye removal between semiconductor nanoparticles and natural adsorbents. Optik 206:164281
Rahman MM (2019) Introductory chapter: fundamentals of semiconductor photocatalysis. In: Concepts of semiconductor photocatalysis, vol 4, p 1
Rauf MA, Ashraf SS (2012) Survey of recent trends in biochemically assisted degradation of dyes. Chem Eng J 209:520–530

Rauf M et al (2007) The effect of operational parameters on the photoinduced decoloration of dyes using a hybrid catalyst V_2O_5/TiO_2. Chem Eng J 129(1–3):167–172
Rauf M, Meetani M, Hisaindee S (2011) An overview on the photocatalytic degradation of azo dyes in the presence of TiO_2 doped with selective transition metals. Desalination 276(1–3):13–27
Rehman K et al (2018) Effect of reactive black 5 azo dye on soil processes related to C and N cycling. PeerJ 6:e4802
Reza KM, Kurny A, Gulshan F (2017) Parameters affecting the photocatalytic degradation of dyes using TiO_2: a review. Appl Water Sci 7(4):1569–1578
Robinson T et al (2001) Remediation of dyes in textile effluent: a critical review on current treatment technologies with a proposed alternative. Biores Technol 77(3):247–255
Sacco O et al (2012) Photocatalytic degradation of organic dyes under visible light on N-doped TiO_2 photocatalysts. Int J Photoenergy
Sahoo A, Patra S (2018) A combined process for the degradation of azo-dyes and efficient removal of aromatic amines using porous silicon supported porous ruthenium nanocatalyst. ACS Appl Nano Mat 1(9):5169–5178
Sahoo A, Patra S (2020) A magnetically separable and recyclable gC_3N_4/Fe_3O_4/porous ruthenium nanocatalyst for the photocatalytic degradation of water-soluble aromatic amines and azo dyes. RSC Adv 10(10):6043–6051
Sala M, Gutiérrez-Bouzán MC (2012) Electrochemical techniques in textile processes and wastewater treatment. Int J Photoenergy, vol 2012
Salem MA, Shaban SY, Ismail SM (2015) Photocatalytic degradation of acid green 25 using ZnO and natural sunlight. Int J Emerg Technol Adv Eng 5:439
Samsami S et al (2020) Recent advances in the treatment of dye-containing wastewater from textile industries: Overview and perspectives. Process Saf Environ Prot 143:138–163
Sani Z, Abdullahi I, Sani A (2018) Toxicity evaluation of selected dyes commonly used for clothing materials in urban Kano, Nigeria. Eur Exp Biol 8(4):26
Saravanan R, Gracia F, Stephen A (2017) Basic principles, mechanism, and challenges of photocatalysis. Nanocomposites for visible light-induced photocatalysis. Springer, pp 19–40
Sarkar S et al (2017) Degradation of synthetic azo dyes of textile industry: a sustainable approach using microbial enzymes. Water Conserv Sci Eng 2(4):121–131
Satapathy S et al (2021) Mechanistic aspects and rate-limiting steps in green synthesis of metal and metal oxide nanoparticles and their potential in photocatalytic degradation of textile dye. Photocatalytic Degradation of Dyes. Elsevier, pp 605–630
Shad NA et al (2021) Facile synthesis of Bi_2WO_6/rGO nanocomposites for photocatalytic and solar cell applications. Ceram Int 47(11):16101–16110
Shan AY, Ghazi TIM, Rashid SA (2010) Immobilisation of titanium dioxide onto supporting materials in heterogeneous photocatalysis: a review. Appl Catal A 389(1–2):1–8
Shindy A (2017) Problems and solutions in colors, dyes and pigments chemistry: a review. Chem Int 3(2):97–105
Shirsath S et al (2013) Ultrasound assisted synthesis of doped TiO_2 nano-particles: characterization and comparison of effectiveness for photocatalytic oxidation of dyestuff effluent. Ultrason Sonochem 20(1):277–286
Shriram B, Kanmani S (2014) Ozonation of textile dyeing wastewater—a review. J Inst Public Health Eng, p 2014
Siddiquey IA et al (2012) Sonochemical synthesis, photocatalytic activity and optical properties of silica coated ZnO nanoparticles. Ultrason Sonochem 19(4):750–755
Singh Rathore A et al (2017) Study on mass transfer characteristics for Cr (VI) removal by adsorption onto residual black toner ink. Environ Prog Sustainable Energy 36(4):1022–1029
Soares OSG et al (2006) Ozonation of textile effluents and dye solutions under continuous operation: influence of operating parameters. J Hazard Mater 137(3):1664–1673
Sreethawong T (2012) Mesoporous-assembled nanocrystal photocatalysts for degradation of azo dyes. Advances in water treatment and pollution prevention. Springer, pp 147–175

Su Z et al (2013) Formation of crystalline TiO_2 by anodic oxidation of titanium. Prog Natural Sci Mater Int 23(3):294–301
Sun J et al (2008) Photocatalytic degradation of Orange G on nitrogen-doped TiO_2 catalysts under visible light and sunlight irradiation. J Hazard Mater 155(1–2):312–319
Suteu D, Zaharia C, Malutan T (2011) Removal of orange 16 reactive dye from aqueous solutions by waste sunflower seed shells. J Serb Chem Soc 76(4):607–624
Tan I, Ahmad A, Hameed B (2008) Adsorption of basic dye on high-surface-area activated carbon prepared from coconut husk: equilibrium, kinetic and thermodynamic studies. J Hazard Mater 154(1–3):337–346
Tkaczyk A, Mitrowska K, Posyniak A (2020) Synthetic organic dyes as contaminants of the aquatic environment and their implications for ecosystems: a review. Sci Total Environ 717:137222
Turhan K, Turgut Z (2009) Decolorization of direct dye in textile wastewater by ozonization in a semi-batch bubble column reactor. Desalination 242(1–3):256–263
Urbain M et al (2021) On the reaction pathways and growth mechanisms of $LiNbO_3$ nanocrystals from the non-aqueous solvothermal alkoxide route. Nanomaterials 11(1):154
Vasudev MC et al (2013) Exploration of plasma-enhanced chemical vapor deposition as a method for thin-film fabrication with biological applications. ACS Appl Mater Interfaces 5(10):3983–3994
Vela N et al (2012) Removal of polycyclic aromatic hydrocarbons (PAHs) from groundwater by heterogeneous photocatalysis under natural sunlight. J Photochem Photobiol, A 232:32–40
Vikrant K et al (2018) Recent advancements in bioremediation of dye: current status and challenges. Biores Technol 253:355–367
Viswanathan B (2018) Photocatalytic degradation of dyes: an overview. Curr Catalysis 7(2):99–121
Wang N et al (2009) Evaluation of bias potential enhanced photocatalytic degradation of 4-chlorophenol with TiO_2 nanotube fabricated by anodic oxidation method. Chem Eng J 146(1):30–35
Wang C et al (2011a) Doping metal–organic frameworks for water oxidation, carbon dioxide reduction, and organic photocatalysis. J Am Chem Soc 133(34):13445–13454
Wang Y et al (2011b) Rapid anodic oxidation of highly ordered TiO_2 nanotube arrays. J Alloy Compd 509(14):L157–L160
Wang H-J et al (2011c) Porous zinc oxide films: controlled synthesis, cytotoxicity and photocatalytic activity. Chem Eng J 178:8–14
Wang X-K et al (2011d) A novel single-step synthesis of N-doped TiO_2 via a sonochemical method. Mater Res Bull 46(11):2041–2044
Wang J-C et al (2016) Indirect Z-scheme $BiOI/g\text{-}C_3N_4$ photocatalysts with enhanced photoreduction CO_2 activity under visible light irradiation. ACS Appl Mater Interfaces 8(6):3765–3775
Wrobel TP, Bhargava R (2018) Infrared spectroscopic imaging advances as an analytical technology for biomedical sciences. Anal Chem 90(3):1444–1463
Wu Z et al (2009) Synthesis of immobilized TiO_2 nanowires by anodic oxidation and their gas phase photocatalytic properties. Electrochem Commun 11(8):1692–1695
Wu Z et al (2014) Progress in the synthesis and applications of hierarchical flower-like TiO_2 nanostructures. Particuology 15:61–70
Xu J et al (2009) Low-temperature preparation of Boron-doped titania by hydrothermal method and its photocatalytic activity. J Alloy Compd 484(1–2):73–79
Yagub MT et al (2014) Dye and its removal from aqueous solution by adsorption: a review. Adv Coll Interface Sci 209:172–184
Yao Y et al (2016) Enhanced photo-Fenton-like process over Z-scheme $CoFe_2O_4/gC_3N_4$ heterostructures under natural indoor light. Environ Sci Pollut Res 23(21):21833–21845
Yoon J, Lee Y, Kim S (2001) Investigation of the reaction pathway of OH radicals produced by Fenton oxidation in the conditions of wastewater treatment. Water Sci Technol 44(5):15–15
Yoshimura M, Byrappa K (2008) Hydrothermal processing of materials: past, present and future. J Mater Sci 43(7):2085–2103
Yu C et al (2012) Sonochemical fabrication of novel square-shaped F doped TiO_2 nanocrystals with enhanced performance in photocatalytic degradation of phenol. J Hazard Mater 237:38–45

Yu X et al (2019) Photocatalytic degradation of ciprofloxacin using Zn-doped Cu_2O particles: analysis of degradation pathways and intermediates. Chem Eng J 374:316–327

Yusof NSM et al (2016) Physical and chemical effects of acoustic cavitation in selected ultrasonic cleaning applications. Ultrason Sonochem 29:568–576

Yusuf M et al (2015) Dyeing studies with henna and madder: a research on effect of tin (II) chloride mordant. J Saudi Chem Soc 19(1):64–72

Zaleska A (2008) Doped-TiO_2: a review. Recent Patents on Engineering 2(3):157–164

Zasada-Kłodzińska D et al (2021) Analysis of natural dyes from historical objects by high performance liquid chromatography and electromigration techniques. Crit Rev Anal Chem 51(5):411–444

Zawawi A, Ramli RM, Yub Harun N (2017) Photodegradation of 1-butyl-3-methylimidazolium chloride [Bmim] Cl via synergistic effect of adsorption–photodegradation of Fe-TiO_2/AC. Technologies 5(4):82

Zhang S, Lu X (2018) Treatment of wastewater containing reactive brilliant blue KN-R using TiO_2/BC composite as heterogeneous photocatalyst and adsorbent. Chemosphere 206:777–783

Zhu G et al (2019) Enhancing visible-light-induced photocatalytic activity of BiOI microspheres for NO removal by synchronous coupling with Bi metal and graphene. Appl Surf Sci 467:968–978

Zolgharnein J, Bagtash M, Asanjarani N (2014) Hybrid central composite design approach for simultaneous optimization of removal of alizarin red S and indigo carmine dyes using cetyltrimethylammonium bromide-modified TiO_2 nanoparticles. J Environ Chem Eng 2(2):988–1000

Zouhier M et al (2020) Preparation of $ZnFe_2O_4$/ZnO composite: Effect of operational parameters for photocatalytic degradation of dyes under UV and visible illumination. J Photochem Photobiol, A 390:112305

The manufacturer's authorised representative in the EU is Springer Nature Customer Service Centre GmbH, Europaplatz 3, 69115 Heidelberg, Germany. If you have any concerns regarding our products, please contact ProductSafety@springernature.com

Printed and bound by CPI Group (UK) Ltd, Croydon, CR0 4YY

30/09/2026

02225493-0007